2D and 3D Nanostructures

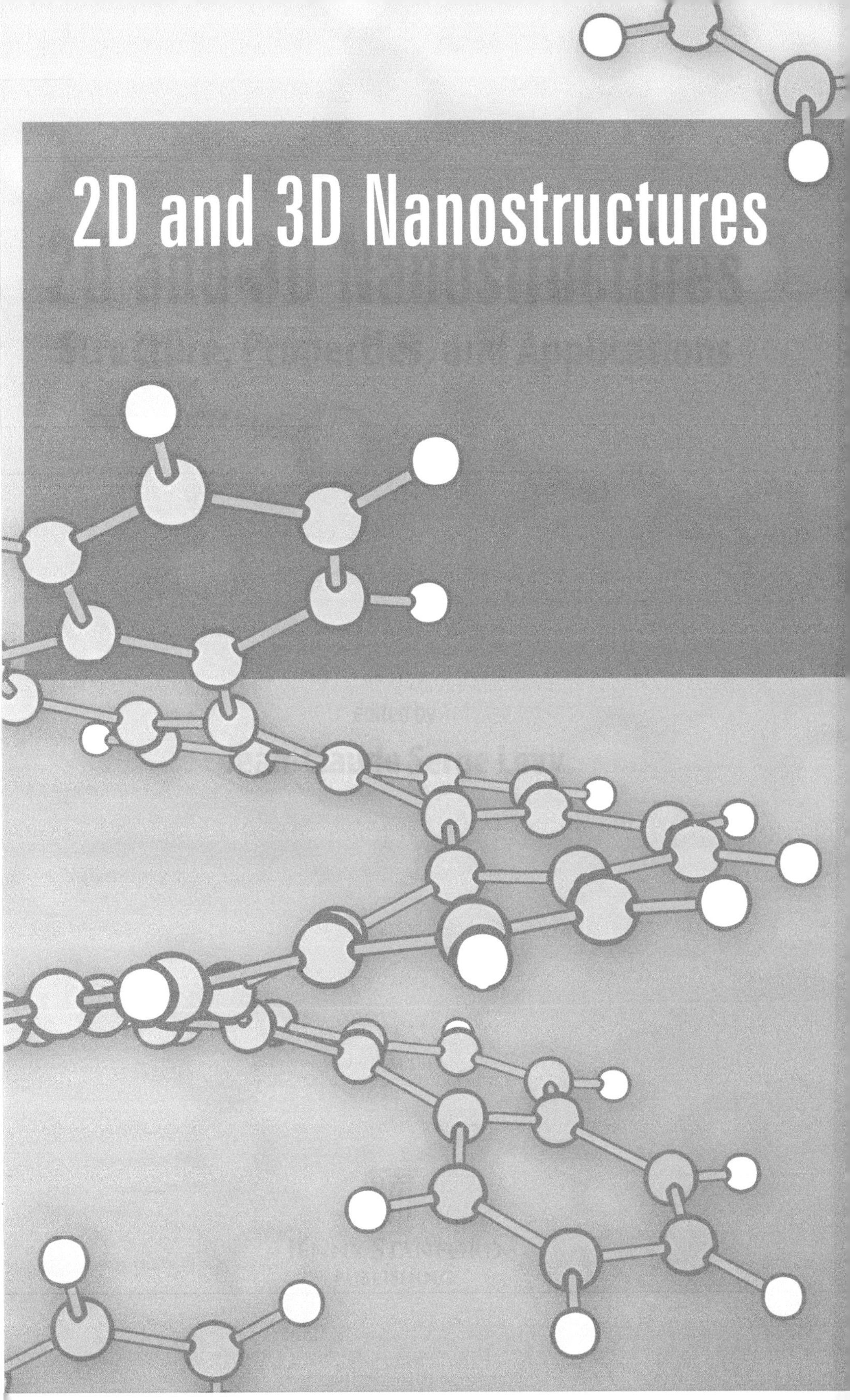

Preface

This book results from the intersection between the general history of scientific disciplines and the human history of scientific individuals. And this preface is devoted to introducing such intertwined histories, i.e., a complex task mixing science advancement and personal history of the authors.

For the case we are concerned about, i.e., nanostructures, the main rule is easy to understand: It consists of obtaining smaller and smaller components in order to achieve more and more complex scientific goals within less and less space, using less and less matter. This is the first reason for the present study of 2D and 3D nanostructures. There are also other reasons for such an interest. Another practical reason is the possibility of using significative computer simulations for deriving the behavior of such small structures. This point is largely developed here, even with a comparison between the results of different numerical methods. Such small amounts of matter are also interesting since they allow direct comparisons with biological cells which are of comparable sizes. That point opens the path for fruitful medical applications.

Of course, the advancement of the study of nanostructures requires a lot of time and a lot of successive innovations, as usual in the history of sciences. For instance, this volume follows the book *Magnetic Structures of 2D and 3D Nanoparticles*, published in 2016. It introduced new topics such as non-magnetic components, new materials of specific structure, and even biological views up to artificial intelligence.

In the introductory chapter, after a brief history of successive improvements, I introduce the possibility of conscious behaviors in artificial intelligence, its difficulties, and its advantages.

The next chapter is authored by Eric Bringuier, also from Paris City University, who takes advantage of his large experience in semiconductors to deal with the property of electric conduction within nanometric wires in light of recent experiments. Several different steps can occur in conduction, according to the density of electric carriers. This was observed and is analyzed here.

H. T. Diep from Cergy-Pontoise University has a very large experience in models of magnetism and computation. In his chapter, he begins with the starting point of his research: the magnetic surface structure of thin films, the case of a special phase transition. Then he reports on more recent works on criticality, which is a basic experimental and numerical feature of phase transitions, and later on skyrmions, new magnetic structures of practical interest as memory devices. His detailed introduction of frustration, a concept introduced in physics by Phil Anderson and developed by Gerard Toulouse and coworkers, including Hung Diep, himself is rich and powerful. And this concept is also useful for the next chapter.

Nicolas Rougemaille and Johann Coraux from Néel Institute in Grenoble, a famous place for magnetism, study a special case of frustration, the one of artificial spin ice, a new topic that these authors introduce very clearly with its long history. They give numerous practical, well-analyzed examples as well as many experimental and numerical referenced works on this fascinating topic, which is very well observed.

Nicolas Vukadinovic from Dassault Aviation and a team from IMPC in Strasbourg report on the static and dynamic magnetic properties of hollow spheres of ferrites. The interest of this special choice comes from its physical properties. Such samples are very light and exhibit a high surface-to-volume ratio. These characteristics induce special static and dynamic magnetic properties. These samples are also convenient for chemical properties such as catalysis, as well as for biological applications, as reported in the references.

Maciek Krawczyk and Mateus Gołębiewski from Adam Mickiewicz University in Poznan, Poland, study gyroid ferromagnetic nanostructures for magnonics. Maciej Krawczyk and his colleagues from Poznan already appeared in the 2016 book. Now, in this volume, their interest is focused on very special magnetic materials: gyroids. The structural complexity of these materials is expected to lead to fruitful applications in magnetism as well as in other disciplines. These authors found by means of several numerical computations quite interesting magnon spectra in such structures. These results pave the way for future applications in magnetism as well as in other disciplines.

Kamel Boukheddaden and his team from Université Paris-Saclay in Versailles (UVSQ) work on the spin cross transition in the case of a 2D active material surrounded by an inactive material. They continue their previous work reported in the previous volume and now focus on the magneto-elastic model of this thermally activated transition. The main point of this calculation is the change of atomic volume of the magnetic atom when changing from a low-spin to a high-spin state. The insertion of the whole magnetic sample within a non-magnetic shell induces a magneto-elastic effect which is controlled by temperature. This rich process is carefully analyzed within a 2D simple model. The results correspond to the observations and must induce careful interactions between experimentalists and theoreticians.

Finally, Denis Sabirov and co-workers from Ufa in Russia report on the structure of helicenes as obtained from a different DFT (density functional theory) method and compare their findings with experimental data. Here the reader deals with chemistry: helicenes (n) are structures of carbon and hydrogen involving n benzene rings and appear as helices. These helicenes are involved in petrochemistry, but not limited to this field. For instance, helicenes are well known in astrophysics. Their basic property used here to compare the different DFT models is the value of the string coefficient of these molecules which appear as parts of helix. The discussion between different numerical models demonstrates the numerical continuity between these different chapters.

Finally, it is a pleasure to acknowledge the continuous help of Professor Cristiano Ciuti, head of the MPQ lab, and Professor Atef Asnacios, head of the Physics Department at University Paris Cité.

Jean-Claude Serge Levy

Chapter 1

Introduction to "2D and 3D Nanostructures: Structure, Properties and Applications"

Jean-Claude Serge Levy
Laboratoire Matériaux et Phénomènes Quantiques,
Université Paris Cité and Unité mixte CNRS UMR 7162,
10 r. Alice Domon et Léonie Duquet 75205 Paris Cedex 13, France
jean-claude.levy@paris7.jussieu.fr

The first part of this chapter gives a short historical view of the evolution of sciences driven by the competition between general ideas and useful activities. This evolution explains the recent convergent interest towards nanostructures in physics after previous long studies of gas and solid states. The second part is devoted to the actual numerous different technical advantages of considering nanostructures. This abundance corresponds to the variety of potential nanostructures and explains their success. In the third part, we notice that numerous examples of minimal size such as cells already exist in nature. This explains why biological applications of nanostructures are so numerous. A biological example of time-dependent neuron behavior which can be useful for artificial intelligence (AI) is given. Finally, the different chapters are introduced.

2D and 3D Nanostructures: Structure, Properties, and Applications
Edited by Jean-Claude Serge Levy

ISBN 978-981-5129-97-7 (Hardcover), 978-1-003-74871-7 (eBook)
www.jennystanford.com

DOI: 10.1201/97810037487-1

1.1 Introduction

Sciences and, among them physics, are the result of a general human effort to understand our common environment, our world, to play with this environment and finally to optimize the results of our actions. So, from the very beginning, their extension has been limitless. But, at a given time, the growth of sciences is strongly connected with the possible practical use which can be made of their results. For instance, useful applied results generate an avalanche of quite numerous followers within a short delay. Science dynamics result both from the slow natural desire of a general knowledge and from this fast interaction with the real world which is accelerated by practical results. Of course, during this fast dynamic evolution, a lot of branching in quite different directions occurs with quite different intensities, according to the practical uses of such branches, i.e., to their success, to the need for complementary information they create. So before dealing with our specific topics, some global view of the physics development sounds useful to enjoy the complementarity between general knowledge and practical utility in our special case.

1.1.1 A Brief Historical Survey of Physics Advancement

A very short and basic historical review is useful to understand the basic lines of the actual research and its development. It happened that the first turning point of the increase of interest of science was the success of the search for a new mastery of the sea. This successful event resulted both from building optimal boats and from following the boat trajectories by means of efficient maps where the new discoveries were included step by step. The results of this campaign completely changed the human way of life.

1.1.1.1 The beginnings of modern science

This famous naval search was driven a long time ago by a Portuguese prince: Henri the Navigator (1394–1460), who created a naval school in Sagres [1]. His initial goal was to avoid the frequent invasions of his country by competing with the invaders and fighting them on their own ground, the sea, with an optimal boat and optimal navigation. This serious search was time consuming. At the end of this process, a new successful boat was created, the "Caravel." And a regular mapping of the sea and of the earth was also conceived. The long-term result of this exceptional campaign is famous: the discovery of the New World, of new travelling ways by boat, of new places, of new people, of new food, and of a very large number of new materials. This avalanche of newness completely changed everyday life of the layman in any place and even changed the view about sciences and techniques.

1.1.1.2 The development of modern science

This exceptional success induced a large change of mind in the whole world. Sciences and techniques appeared not only to be useful but also to be creative of such numerous and unexpected fruitful events for everyday life. The age of enlightenment was coming soon because of this exciting powerful success. This new era was surrounded with an enormous amount of experimentation everywhere.

For instance, Francis Bacon proposed some experiments to evidence or not the long-ranged character of gravity. This proposal was published a long time before Newton's intuition of gravity gave another example of observation. This was reported by Voltaire [2]. The naval research introduced by Henri the Navigator was the beginning of the modern era of mankind, in usual life as well as in sciences and philosophy. This new way was characterized by reinforced experimentation in all domains.

The next main step, the next main change both in mankind and in science and techniques occurred with the development of the steam engine as a new energy source. This decisive step occurred as the result of a collective invention as noted by Fernand Braudel [3]. The collective development of this basic technique giving common access to energy lasted nearly one century and resulted finally in a lot of applications of this new energy source to all domains of activity, transport as well as all kinds of industry. This new success introduced the "industrial era," with a complete change of common life everywhere in its local management [3], a deep social revolution.

This new industrial and scientific step came with the beginning of thermodynamics, its theoretical counterpart and especially with the useful concept of an "ideal gas," the driving actor of this efficient engine. This concept of a nearly free ideal gas was so productive that it introduced a basic way of thinking as well as a basis for numerous experiments.

Naturally different scientific developments occurred simultaneously in different branches of physics and technics as already introduced. Here we focus on the idea of an ideal gas with weak interactions with its environment. That enables to focus the interest on a single action, within a kind of vacuum. This basic idea of thermodynamics was developed during the eighteenth century [4], just before the discovery of electricity by Volta [5] following the experimental biological observations of Alexander von Humboldt in South America [6]. These two main discoveries introduced practically a new era of humanity and science.

1.1.1.3 The starting industrial era

Mixing the results of these two discoveries allowed for new early basic experiments. The introduction of an electric thread within a space filled by an inert gas led to the discovery of the electric light bulb [7], now a common source of light. On the other hand, electric discharges through a gas have also a lot of applications. Gas or vacuum diodes [8] mix the concepts of gas and electricity and practically give a clear example of a strong non-linear behavior. This discovery has led to early applications to logics gates, high-frequency detection, and signal amplification since the very beginning of the twentieth century with the successive improved versions of triode, tetrode and pentode, an innovative optimization of this initial device [9].

The success of these discoveries for telecommunication engineering as well as the success of its development for logic and scientific calculation resulted in a lot of applications. Among them, the first vacuum-tube computer by J. V. Atanasoff [10], featuring 300 vacuum tubes, introduced a decisive step in future development. These results happened progressively under also the influence of global human history, i.e., its political fluctuations and events at the world size which also play a decisive part in scientific evolution.

That starting point of computers brings us to the 1940s and was favored by the pressure of the Cold War, which quickly became the Second World War. This political climate introduced the search for new weapons with a never-reached intensity, and terrifying results. This significant technical advance led to new scientific perspectives for the peace time to come.

1.1.1.4 The "transistor" era and the solid-state physics

At the end of the Second World War, it was time to take advantage of the wonderful scientific and technical development achieved on weapons and to convert this progress towards the more common benefit. This transition was done according to different approaches. The first step to come in the improvement of this interesting computing technique was the change of size of the basic element, which was initially an electronic tube of a few centimeters. This progress led to a quite smaller element, to obtain more and more powerful computers within a reasonable space and a lower energy consumption. In this way, this first step consisted in a transition from the basic units, the vacuum tubes, towards new dense solid-state devices which took advantage of the already known semiconductors and the possibilities linked with their low conduction. In doing so, it was a complete change from gas applications with the limiting case of "vacuum," towards the high-density materials of solid-state physics with mow conduction.

The decisive discovery of this solid analog occurred early within the same forties: the transistor was found in 1947 [11]. This solid-state device made of variously doped semiconductors has the performance of a pentode, and simultaneously a very small size. Also, it works under low tension and has a low energy consumption. Thus, many transistors could be used within a very small space. Because of this enormous size reduction modern personal computers appeared.

This was a strong change for everybody. As noticed before, such a strong change induced also a change in human sciences and philosophy. The stabilization of electronic systems by feedback impressed largely. According to that principle, a new "systemic" philosophy was developed in many countries by many authors. A new approach was born.

The transistor revolutionary early discovery induced a quick size change in all electronic components and even with integrated printed circuits. This success opened the way for a lot of practical possibilities together with the scientific development of solid-state physics [12]. We will see within this book how much the initial idea of electron gas remained fruitful in the analysis of the behavior of these solid-state devices [13].

For instance, the largely used concept of electron mean free path refers to a gas-like model where this length is quite larger than the other lengths of the problem such as the atomic sizes within a semiconductor. In other words, in a weakly doped semiconductor, free electrons nearly behave as a gas!

After the decisive step of the creation of new semiconductor devices such as the transistor, the progress consisted in a progressive optimization and miniaturization of these components. This innovation occurred by means of a lot of new solid-state materials. This progression occurred at a regular rhythm, as proposed by Moore's law [14], an extrapolation of the beginnings of this development. This success was also due to a better and better mastery of sample preparation. The present research for nanocomponents occurs with a similar optimization of a large class of new materials in all physics domains and still follows approximately this rule.

As already said, in this technical progress different technical branches have been developed simultaneously, in parallel ways. In the principle of computers, the part of memory is essential. The simultaneous use of several memory units enables the computer to start different computations together and to use their results later, when needed. It became early obvious that magnetic devices can fulfill this memory part [15] and such magnetic memories were largely developed. Here too, the miniaturization of magnetic memories was a technical challenge. Again, the next steps of scientific development consisted in converting the results of these magnetic memories in electric signals in an optimal way [16]. The miniaturization of these magnetic components is a topic largely investigated in this book, with also interest on their excitations which cover a large frequency band of electromagnetic signals [17]. These magnetic nanostructures pave the way for future applications.

1.1.1.5 The era of nanomaterials

After this short historical frame, we will briefly consider the techniques which allow one to deal practically with nanomaterials, before looking at the new technical opportunities opened by nanomaterials.

A lot of technical progress was required to obtain well-defined nanomaterials with efficient properties. Different improvements occurred rather simultaneously. Scanning tunneling microscope [18] was decisive in controlling surfaces at the very atomic level. Improvements in electron microscopy allowed to reach the observation of atomic size in 3 D structures, giving thus an optimal control of the samples [19]. On the other side, sample preparation of nanomaterials was optimized by means of different techniques such as sputtering within a high vacuum [20] or chemical deposits within a specific chemical liquid phase for instance [21] with a control of the wanted properties by means of direct observation and numerical simulation. A lot of new techniques are involved in this approach as shown in the following chapters [22].

1.1.2 The Technical Advantages of Nanocomponents

The advantages of nanocomponents over bulk materials are quite numerous. Of course, the size advantage enables one to obtain more powerful units within a reduced space. A nanostructure contains less defects than an extended one. That can be a real advantage. The finite nature of nanomaterials is useful to define accurate numerical simulations. And the reference to biological models is more direct with nanocomponents.

1.1.2.1 The size advantage era

An obvious advantage of nanostructures is the final small size of the component. Reversely, it means that for a given volume of components one obtains a clear advantage in computation power, or in the chosen technical goal, in front of standard approaches.

The same effect of size reduction minimizes the required quantity of used matter for a given result. This reduction results in both an advantage in cost and in resulting pollution. But, of course, the realization of nanostructures requires an advanced technique mastering, i.e., an extra cost for high-level activity and a possible extra pollution for new components. The detailed economic balance must be observed in a nearly stationary regime if it happens, since improvements are coming very quickly in this high-level competition. Basically, nanocomponents must be cheap.

The main point to consider is the extreme need for AI, and thus for more and more components to reach a real high quality. Smaller these components are, better it is. Thus, the need for nanostructured components is essential for the future.

1.1.2.2 A single structure

A more technical advantage of nanostructures consists in the avoidance or the very restriction of the use of statistical average in a very small nanostructure. A first example of such an advantage of a single active element in a component over a statistical distribution of numerous elements within an active component is the use of a single color-center or of a very few color-centers in a detector [23] instead of a bulk color center sample. The advantage is clear. In this case, there are already many examples of nanodiamond color centers with a very high definition of spectral lines [24] as compared to the analysis for bulk materials where statistical average broadens the lines.

A similar case of avoidance of statistical average in a nanocomponent occurs in the case of the conduction electron scattering by magnetic impurities. With extended samples, this phenomenon is the Kondo effect [25] which is well known to be complex because of this unavoidable distribution of sites. Recently this classical Kondo effect has been revisited in the case of a single magnetic impurity within a nanostructure [26] with a clear advantage.

The example of giant magnetoresistance observed in multilayered nanostructures [27] is somewhat similar. Here the individual surface effect is reinforced by its repetition at numerous interfaces, thus the individual interface effect appears strongly also. In other words, this is a dual effect of the unicity of nanostructures: the repetition reinforces the effect of a single event, the interface scattering.

And there are numerous other advantages for nanocomponents as listed below.

1.1.2.3 The simulation help

Another main technical advantage of considering nanocomponents is the possibility of using realistic sophisticated numerical simulations with a realistic number of atomic sites which can be comparable to the actual site number. The results of such models are realistic and strongly help experimentalists. This is now a basic research tool which occurs between analytical theory and experiment. The examples of such applications are quite numerous. Numerical simulations sound now necessary to play with experiments. For instance, there are a lot of available recent micromagnetism programs as seen in a recent review [28]. These programs enable the searcher to derive both the magnetic structure of the sample and its magnetic excitations. In the following chapters [29] numerous examples of such programs are used such as MuMax3, tetmag and Comsol. Searchers also use homemade versions which enable to consider different sample shapes [30, 31] with both the magnetic structure and the magnetic excitations, the spin waves. And numerical results can be compared to experimental ones. More generally, all the details of nanomaterials are easily known and included in these simulations.

1.1.2.4 The nanostructure variety

The power of nanostructures lies in their extreme variety of possible chemical compositions, sizes, shapes and arrangements since set of independent nanostructures can be considered. There is an extraordinary creativity in such a use. Several concrete examples are shown in the following chapters [32–37]: from nano semiconductors [32] to helicenes [37] through spin ice models [33], hollow spherical nanoparticles [34], gyroid ferromagnetic structures [35] and core-shell spin transition nanoparticles [36]. As a matter of fact, the concrete nanostructures which are considered there for their magnetic properties sound able to have also many other properties. These nanostructures define a new world for physics and chemistry, which can be used as jigsaw pieces to obtain new properties.

1.1.2.5 The biological reference

Finally, it is interesting to compare the physics of nanostructures and biology. The basic size of the basic biological unit, the cell, is about a few tens of micrometers. This is rather large in front of samples of a few nanometers. So, nanomaterials can be easily introduced within cells. This is one basis for medical applications of nanomaterials [38]. In the case of magnetic nanomaterials, they can be driven within biological fluids from the outside by applying convenient external magnetic fields. This is another basis for medical applications [39].

Another use of the experimental knowledge of biology is the use of biomimetics [40] which consists in copying some features appearing in nature. Such a process can avoid many infructuous trials. The natural selection rules can be helpful by means of their previous experimental use. Here we will consider briefly one such approach. The use of circadian rhythm, i.e., the day and night alternance, is common to animals and leads to interesting properties [41], such as energy saving for instance, but it leads also to other properties which can be useful for AI and thus for the nanocomponents which are involved in AI.

1.2 Physical Applications of the Circadian Rhythm to AI

The circadian rhythm is common everywhere on the earth, even if this rhythm weakly changes a little bit with seasons and latitudes. The regular alternance of day and night has numerous consequences and particularly on living matters and animals who are forced to take advantage of this forced behavior in their common life [41]. As usual with adaptative systems, a well-integrated constraint becomes an advantage, a main component of common life tools.

1.2.1 The Two Basic Rhythmic Neural Activities

For diurnal animals, this regular alternance means that daytime is devoted to action while night is devoted to rest, even if some tasks such as breathing, or heartbeat, remain permanently to be achieved. In other words, during the day, diurnal animals are active, communicate with their environment and eat, while during the night they rest, have nearly no external contact, and digest. This clear alternance is imposed by the circadian rhythm.

From the viewpoint of intelligence and artificial intelligence (AI), which is here the goal of our mimetic task, this alternance also means two quite different neural types of activity. During daytime the diurnal animal's action and interaction with its environment is driven by awareness, a conscious behavior, while during night and rest, there is still a mind activity as evidenced by observed dreams, even for animals, but this neural activity is quite different from that of the daytime, since during night nearly no signal comes from outside. This neural night activity, an unconscious activity, mainly works with the already registered facts, the memory. This special feature of distinct night and day neural activities was proved by means of the analysis of the neural activity at least for humans [42]. In other words, the physical existence of Freud's unconsciousness [43] is clearly proved to occur because of the circadian rhythm.

For animals, their metabolism is quite like the human one, thus the existence of unconsciousness for animals as well as for humans sounds also to be well established [44] as it can be observed on dreaming animals. Thus, consciousness and unconsciousness are largely extended in nature as due to natural circadian rhythm.

The difference in neural activity between day and night deserves detailed observation before looking at AI applications, a strongly expected property for future [45] because of the hope for consciousness, which sounds to be a powerful tool, in a coming AI. This difference of activity is obviously linked with energy saving during night, a concept which is also relevant for AI and already economically interesting. The nearly exact lack of external contact at night is another difference with daytime behavior.

During rest and night, there is nearly no contact with the environment, no new information, thus the need for brain activity is weak. So, naturally during rest, the average level of activity of neurons is statistically weak. This feature has some strong consequences on physiology and on mind activity.

1.2.1.1 The night neural activity

The neural level of activity is driven by the neuron action thresholds which are statistically shifted during night in an internal process for minimizing the number of active neurons at a given time during night. Such statistical shifts enable us to save energy, i.e., here blood flows to the brain. This low neural activity results in an observed temperature lowering of the brain during night. These statistical shifts of the neural thresholds must continuously change during the rest time to share everywhere in the brain this low averaged level of activity, without specific localized peaks of activity, blood flow and temperature. In other words, it means that the active part of the brain is permanently changing during night, apart from permanent tasks.

Because of this repartition of energy saving, the unconscious mind, i.e., the restricted part of the brain, which is active at a given time, is regularly changing during the night or the rest time. Finally, this way of energy saving leads to the fact that there is not a single unconscious mind, but quite a lot of different unconscious minds which are completely changing during nighttime [46]. This strong instability and multiplicity of the unconscious mind explains a lot of well-observed features of our unconscious activity. And such features can be used for improving AI in a biomimetic view.

1.2.1.2 Consequences of the night neural activity on unconsciousness

The instability of the unconscious active part of the brain and thus its multiplicity and, at a given time, its restricted nature involving only a small part of the neural set, explains the metaphoric nature of the dream representations. Precisely, at a given time of sleep or rest, there is no access to the full brain memory, there is just access to a small part of it, the active part, so approximations and metaphors are required from this restricted memory when dealing with past details. The exception to this approximative treatment is the case of an exceptional event which has a very extended track. Such exceptional memory events were already noticed by behaviorists. And corresponding theories about the spatial extension of the memory tracks were already produced [47, 48]. The approximative treatment of the night neural activity gives place to some originality or poetry in its translations with the famous examples of the use of such an unconscious activity by surrealism as a creative process [49]. These forced approximations become often so complex that they require the long use of psychoanalysis to clearly understand their ways and the meanings of these dreams, if possible.

This metaphoric, approximate nature of the instantaneous unconsciousness has numerous consequences such as the universal taste for music where just a "smell," a taste of feelings appears, without full and clear understanding.

On another viewpoint, the approximations induced by this weak instantaneous memory strongly enrich the nature of the brain activity which, because of this lacunary behavior, becomes more open to other unexpected interpretations and involve many exploratory trials to try to come back to reality. In other words, the multiplicity of the unconscious mind breaks down the conscious inhibition and paves the way for more intuitive approximate translations with the potential risk for errors or discoveries in such trials. In that sense, this multiplicity and instability of the unconscious mind is an essential part of the intelligence, with its successes and its limits. The unconscious mind revisits the daytime experience in many ways during night.

Since nighttime lasts nearly the same time as daytime, this evidence means that nighttime provides a very long time of analysis of the day information. This length offers the opportunity for many different trials by means of this brain instability. There is time for numerous partial brain scans. Finally, this unconscious nighttime activity leads to the mastering of different partial views of the considered object. In other words, it sounds that the unconscious mind in its multiplicity achieves a main task of deep analysis, which results in the common sense [50] which cannot be reached during daytime because of the required need for a such long free reflection time. This circadian rhythm means a thorough duality between conscious and unconscious mind. And this double approach is essential for understanding facts with their own whole complexity.

That conclusion of the potential enrichment of thought by the internal unconscious numerous analyses of data and memory means the interest of introducing such a similar approach for AI, at the risk of errors, but with the power of a thorough analysis of complex data, together with the advantage of saving energy.

1.2.1.3 "Night neural activity" for AI

The need of a translation of this night activity for AI sounds clear and the words conscious intelligence, common sense are quite challenging for the future of AI [45, 50]. The basic translation of this circadian rhythm starts with a rest or sleep function with less activity, no entry and a randomly distributed neural activity which must continuously vary during this effective "sleep time." The result must be a rather free internal activity, modulated by the stochastic scan of activity thresholds. So, the unconscious behavior of AI must be controlled at any time and analyzed. Quite obviously, this introspection must be guided in a learning process, at least at the beginning.

In doing so, we must account for what happens nowadays for a baby who starts to interact with its surroundings. Some people [51] looked for that initial learning at the beginning of the whole human history. It is too much to learn at the beginning. For a young baby, the rest time is very large, and he is in contact with the present world. The child learning of its surroundings takes time and several layers, several sets of knowledge which correspond to the famous Freudian classes: id, self and later superego are involved [43]. This natural structure sounds to be unavoidable and must evolve according to the steps of human life and learning time, i.e., personal experience. In other words, the need for an early dialogue with other intelligences which can be human intelligence or other AI sounds to be unavoidable to induce both conscious and unconscious mind. Later a more extended dialogue will involve a more extended structure.

At the end of this learning process, there is the goal to obtain such an automatic unconscious behavior of AI. Of course, similarly to the observation of real life, this unconscious intelligence results from its experience, its learning during the interaction with its environment, i.e., its experience and its dialogue with humans. At the end, this is a new AI which is expected to occur. It must be added that this learning process is a difficult one and numerous troubles happen in reality and can happen in such AI projects [52].

1.2.1.4 AI learning of night neural activity

As we know there is a large variety of human minds and learning methods, and it has been true for thousands of years. So, the search for AI with both consciousness and unconsciousness requires many trials. Here we have just underlined some basic methods which can help us in copying the way used by nature.

1.3 Conclusions

This short introduction proved the present necessity of nanomaterials and their numerous advantages such as their enormous variety. So, it is time now to introduce the selected themes of the different following chapters. We know there is a large variety of human minds and learning methods. The next chapter, Chapter 2, is devoted to the study of the conductivity of nano semiconductors [32]. This choice follows the historical one of semiconductor transistor which made a scientific revolution. Then Chapter 3 [33] deals with the world of 2D magnetic memories which are strongly linked to the transistor revolution. And this chapter comes to the most recent progress in this field, the skyrmions. Chapter 4 [34] follows with the theme of frustration which appeared in magnetism and now leads to new applications with nanostructures. Chapter 5 [35] brings the idea of new structures such as hollow spheres for their interest in telecommunications, with both theoretical and experimental approaches. Such new structures, both light and very small, suggest many other applications. Chapter 6 [36] goes further in the research for optimal materials and deals with gyroidal systems and their magnetic properties, i.e., both their static and dynamic properties. The direct interest is the production of new materials for magnonics, with the hope of discovering metamaterials. The second goal is obviously the research of new materials. The next chapters follow this principle. Chapter 7 [37] is devoted to an extended theoretical approach of spin transition materials in the case of nanomaterials made with such a core with spin crossover, surrounded by a shell of another material. The spin transition induced by light in the core leads to an atomic expansion and thus induces a very strong magneto-elastic coupling which is carefully studied and can lead to many applications. Finally, chapter 8, [38] brings the new approach of theoretical chemists for building new nanomaterials, here helicenes. As considered before, one main advantage of nano-materials is the narrow collaboration between experimentalists and numerical computations. In this theoretical Chapter 8, the results of different density functional theory (DFT) approaches are compared and compared with experimental results. This discussion paves the way for new perspectives about nanomaterials and their conception.

References

1. Subrahmanyam, S., and Markovits, C. (2009). Navigation, exploration, colonisation. In: *Histoire du monde au XVe siècle*, ed. P. Boucheron, Part 4, Chapter 1, pp. 603–618.
2. Voltaire (1878). De François Bacon et de l'attraction. In: *Dictionnaire Philosophique*, Ed., Garnier Paris, France.
3. Braudel, F. (1979). In: *Le temps du Monde* Ed., Armand Colin Paris, France.
4. Carnot, S. (1824). *Réflexions sur la puissance motrice du feu et sur les machines* Ed., Bachelier Paris, France.
5. Volta, A. (1800). On the electricity excited by the mere contact of conducting substances of different kinds, *Phil. Trans of the Royal Soc.*, **90**, 403–431.
6. Von Humboldt, A. (1797). *Versuche über die gereizte Muskel und Nervenfaser nebst Vermutungen über den chemischen Prozess des Lebens in der Tier und Pflanzenweldt*, Berlin.

7. Edison, T. A. (1879). Electric light bulb patent.
8. Fleming, J. A. (1904). Fleming valve patent.
9. Holst, G., and Tellegen, B. D. H. (1934). Means for amplifying electrical oscillations, US patent.
10. Atanasoff, J. V., and Berry, C. (1940). ABC computer.
11. Shockley, W., Bardeen, J., and Brattain W. (1947). Transistor, US patent.
12. Seitz, F. (1940). *Modern Theory of Solids*, McGraw Hill.
13. Bringuier, E. (2024). Ohm's law of electrical conduction at the nanoscale, Chapter 2 of this book.
14. Moore, G. E. (1965). Cramming more components onto integrated circuits, *Electronics,* **38**.
15. Devol, G. C., and Hansell, E. B. (1956). Magnetic storage and sensing device, US patent.
16. Baibich, M.N., Broto, J. M., Fert, A., Nguyen Van Dau, F., Petroff, F., Etienne, P., Creuzet, G., Friederich, A., and Chazelas, J. (1988). Giant magnetoresistance of (001)Fe/(001)Cr magnetic superlattices, *Phys. Rev. Lett.*, **61**, 2472–2475.
17. Golebiewski, M., and Krawczyk, M., (2024). Gyroid ferromagnetic nanostructures in 3D magnonics, Chapter 6 of this book.
18. Binnig, G., Rohrer, H., Gerber, Ch., and Weibel, E. (1982). Surface studies by scanning tunneling microscopy, *Phys. Rev. Lett.*, **49**, 57–61.
19. Ricolleau, C., Nelayah, J., Oikawa, T., Kohno, Y., Braidy, N., Wang, G., Hue, F. Florea, L., Pierron-Bohnes, V., Alloyeau, D. (2013). Performances of a 200 kV microscope employing a cold_FEG and an aberration-corrected objective lens, *Microscopy*, **62**, 288–293.
20. Keywell, F., (1955). Measurement and collision-radiation damage theory of high-vacuum sputtering, *Phys. Rev.*, **97**, 1611–1615.
21. Fievet, F., Ammar-Merah, S., Brayner, R. Chau, F., Giraud, M., Mammeri, F., Peron, J., Piquemal, J.-Y., Sicard, L., and Viau, G. (2018). *Chem. Soc. Rev.*, **47**, 5187–5233.
22. Vazquez Bernardez, M. J., Stoeffler, Lefevre, C., and Vukadinovic, N., (2024). Magnetization dynamics of barium hexaferrite hollow spherical nanoparticles, chapter 5 of this book.
23. Thiering, G., Gali, A., (2020). Color centers in diamond for quantum applications. In: *Semiconductors and Semimetals,* eds. Nebel, C. E., Aharonovitch, I., Mizuochi, N., Hatano, M., Elsevier, Vol. 103, Chapter 1, pp. 1–36.
24. Liu, Y., Chen, G., Rong, Y., McGuiness, L. J., Jelezko, F., Tamura, S., Tanii, T., Teraji, T., Onoda, S., Ohshima, T., Isoya, J., Shinada, T., Wu, E., and Zeng,

Chapter 2

Ohm's Law of Electrical Conduction at the Nanoscale

Eric Bringuier
Matériaux et Phénomènes Quantiques,
Unité mixte 7162 CNRS & Université de Paris Cité, case 7021,
5 rue Thomas Mann, 75205 Paris Cedex 13, France
eric.bringuier@univ-paris-diderot.fr

This chapter is concerned with the electrical properties of Si- and GaAs-based nanostructures designed to investigate electronic conduction in wires of nanoscopic cross section. Two experimental studies are theoretically interpreted herein. In experiments on Si nanowires, Ohm's law was found to survive to the atomic scale in both its integral and local forms. In this chapter, the observed current–voltage relationship is shown to be quantitatively accounted for within the semiclassical theory of electron transport. In a GaAs nanowire exhibiting quantized conductance steps, Ohm's integral law was observed to be valid, whereas its local counterpart was not. Ballistic transport inside the wire is accounted for by means of a Vlasov-like kinetic equation. Joule heating occurs in the contacts outside the wire as evidenced by the flat profile of electro-chemical potential observed in the nanowire. Thermal conduction in insulating GaP nanowires is addressed, too. The reason why a fully quantum theoretical description is unnecessary is sketched.

2D and 3D Nanostructures: Structure, Properties, and Applications
Edited by Jean-Claude Serge Levy

ISBN 978-981-5129-97-7 (Hardcover), 978-1-003-74871-7 (eBook)
www.jennystanford.com

DOI: 10.1201/97810037487-2

2.1 Introduction

This chapter deals with the physics of Ohmic electrical conduction at the nanoscale, not in exhaustive fashion but in two experimental systems making use of well-controlled semiconductor structures [26, 83]; metallic systems will be mentioned, too. The nanostructures examined herein were designed to study electronic conduction at very short scales. In [26] and [83], carriers of electric charge have been confined in pipes of nanoscopic cross sections with the aim of investigating the response of the carriers to a voltage applied across the ends of the pipe. By Ohmic conduction, it is meant that the electric current flowing through these artificial wires is proportional to the voltage applied across the ends of the wires. Before broaching the topic, we cite Andrews' words of caution: 'very little is new in nanoscience' [3]. The challenge of nanoscale science is different. Interpreting phenomena occurring at the nanoscale compels us to reappraise the physics at larger scales which is presumed to be well understood. Implementing experimental procedures and/or theoretical concepts at the nanoscale can thus serve as a touchstone of presumably well-known physics. In view of this remark, Section 2.2 reviews fundamental macroscopic concepts of electronic conduction in a solid medium such as voltage and electrical resistance. The distinction is made between Ohm's integral law in a given sample of solid, involving its shape and size, and the local constitutive relation of the solid substance. Theoretical tools relevant to the constitutive relation are presented in Section 2.3; those tools are semiclassical in the sense that some elements of quantum theory are grafted upon a frame of thinking taken from the classical kinetic theory of gases as adapted to the assembly of movable electrons in a solid by Lorentz [51]. Quantum theory is invoked through the use of the energy-band structure, through the use of quantum-mechanical scattering cross sections, and eventually through the use of Fermi–Dirac statistics [5]. Section 2.4 interprets an experiment performed on Si wires of nanoscopic cross section claiming that 'Ohm's law survives to the atomic scale' [83]. Section 2.5 reviews Landauer's theoretical question, 'Can a length of perfect conductor have a resistance?' [41]. Section 2.6 interprets an experiment performed on a GaAs wire of nanoscopic cross section [26] designed to answer that question. The chapter is concluded in Section 2.7. Appendix 2.A gives an account of electron transport, including both energy and momentum balance. Appendix 2.B sketches the connection between a fully quantum-mechanical and the semiclassical description of transport.

The spirit of this chapter is rooted in Ockham's razor rule [1]. In Sagan's words, 'this convenient rule of thumb urges us when faced with two hypotheses that explain the data equally well to choose the simpler' [70]. Herein, when interpreting data obtained from a nanophysical system, we shall preferentially resort to knowledge gained from the study of microphysical systems, without invoking extraneous theoretical concepts specifically devised to explain nanoscopic phenomena. We stress that the interpretations of this chapter are based on the experimental study of three particular systems [26, 76, 83]. The chapter does not purport to provide a general frame of thinking for conduction in nanostructures.

2.2 Electrical Current, Voltage, and Resistance

The electrical resistance of a conducting wire is defined as the voltage across the ends of the wire needed to pass a unit electric current; this definition presupposes that the current linearly responds to the applied voltage. In SI units, the current (voltage) is reckoned in amperes (volts) and the resistance is reckoned in ohms. Throughout this chapter, we shall only consider steady, also called direct-current, regimes and disregard alternating currents. While it is easy to think of electric current as the flow of electric charge per unit time, the concept of voltage is less straightforward. In the current edition of the Oxford English Dictionary, voltage is defined as 'the electromotive force (e.m.f.) reckoned or expressed in volt' and the e.m.f. is 'the force [. . .] which tends to make an electric current flow; the tendency to cause an electric current'. Varney and Fisher [79] remark that the concept of electromotive force as

[1] *Pluralitas non ponenda est sine necessitate* ('Plurality is not to be posited without necessity') or *Frustra fit per plura quod potest fieri per pauciora* ('What can be explained by the assumption of fewer things is vainly explained by the assumption of more things) [58].

devised by Volta [80] was not fully understood by Maxwell. The entry 'electromotive' of the Oxford English Dictionary was revised—albeit slightly—in 2008, which confirms that the meaning of this word is not so easy to grasp. The operational definition of the voltage across two points is as the work done to displace a unit electric charge from one point to another [34]. In a vacuum, this work coincides with the drop in the electrostatic potential energy of a unit charge. In a solid medium whose temperature T is well defined and held constant and where charge is carried by electrons, the work is a variation in the Helmholtz energy A of the assembly of movable electrons, per unit charge. The Helmholtz energy function is relevant when the volume Ω is insensitive to the ambient pressure under a given temperature, which is an excellent approximation in solids. We are concerned with unbinding a charge carrier at a given location and binding it at another one in the presence of an external electric force. In thermodynamics, an external force field is not considered as part of the system, here the electron assembly. The contribution of the field to the energy per particle is separately taken into account as $\bar{e}V$, where $\bar{e} \equiv -e$ denotes the signed electron charge, the electric potential $V(\mathbf{r})$ is governed by Poisson's equation and $\mathbf{r}$ denotes the position. Thereby we are concerned with the isothermal and isochoric variation of the augmented Helmholtz energy $\tilde{A} \equiv A + N\bar{e}V$, where N is the number of electrons. The variation per electron is $(\partial \tilde{A}/\partial N)_{T,\Omega}$. Now $(\partial A/\partial N)_{T,\Omega}$ is the chemical potential, or Gibbs energy per electron g, so that

$$\left(\frac{\partial \tilde{A}}{\partial N}\right)_{T,\Omega} = g + \bar{e}V. \tag{2.1}$$

This quantity, henceforth denoted by $\tilde{g}$, is called the electrochemical potential in contradistinction to the chemical potential g determined by the carrier density $n \equiv N/\Omega$ along with temperature. Thus, the voltage across two points in a solid medium which thermalizes the charge carriers is the drop in the value of $\tilde{g}/\bar{e} = g/\bar{e} + V$ between the points; it usually differs from the drop in the electric potential V. For example, while the voltage across the terminals of a Si-based p-n junction diode in equilibrium is zero, the electric-potential difference is about one volt [7]. The difference can become important as the wire's length shrinks, for then a gradient in the carrier density is expected near the entrance and the exit of the wire. The above reasoning assumes one single temperature at the two points; otherwise another tendency is observed to cause an electric current, namely the temperature difference besides the electrochemical-potential drop. Temperature differences and attendant thermoelectric effects will not be considered in this chapter, except incidentally in Section 2.6.

Let eU, with the voltage U reckoned in volts, denote the energy required to carry one electron from the entrance to the exit of the wire, and let I be the electric current reckoned in amperes. In a steady regime, the battery causing an electron current I/e, in s^{-1}, releases a power $eU(I/e) = UI$, in watts. If the wire has a given electrical resistance R independent of the voltage or the current, a power RI^2 quadratic in I is released in the form of heat, as was first observed by Joule [36] and Lenz [46]. To comply with the First Law of thermodynamics, a theory of conduction must account for the heating effect. It was not so in Drude's simple theory assuming a single electron speed [28] or in Lorentz's improved theory based on the distribution of speeds of Maxwell and Boltzmann [51]. Quotation from Uehling and Uhlenbeck: 'A well-known defect of the Lorentz theory is that it gives no account of the Joulian heat, since only elastic collisions between the electrons and immovable ions are considered' [75]. To obviate this defect, transport will be handled herein within a Fokker–Planck framework explicitly accounting for the interchange of energy between the charge carriers and the lattice hosting them [10, 12, 13]. The framework was originally devised to deal with transport in arbitrarily large electric fields such that the energy distribution of electrons can significantly deviate from the equilibrium one [39].

Two statements are intertwined in Ohm's monograph [59, 64]. In modern language, (i) the voltage U needed to pass an electric current I is proportional to the latter, i.e.

$$U = RI, \tag{2.2}$$

where the coefficient R is called the electrical resistance of the sample of conductor; and (ii) if the conductor is a wire of length L and cross-sectional area $\mathcal{A}$, its resistance is proportional to L and inversely proportional to $\mathcal{A}$, i.e.

$$R = \rho L/\mathcal{A}, \tag{2.3}$$

where the coefficient ρ is called the electrical resistivity of the conductor. Alternatively, one defines the electrical conductance $G \equiv 1/R$ and the conductivity $\gamma \equiv 1/\rho$. The first statement involves the integral coefficient G or R relevant to the whole conductor. In the second statement, γ or ρ is a coefficient connecting local quantities, namely the electron current density $\mathbf{j}_\mathrm{N}$ whose flux through the wire's cross section is the electron current I/e, and the driving force $\nabla(-\tilde{g})$ whose line integral over the wire's length is eU, i.e.

$$\bar{e}\mathbf{j}_\mathrm{N} = \gamma\nabla(-\tilde{g}/\bar{e}). \tag{2.4}$$

The concept of conductivity holds as an intrinsic, bulk material property, independent of its size and shape, if ρ in (2.3) is not a function of $\mathcal{A}$ and/or L as those geometrical parameters are varied. Where the chemical potential $g(n, T)$ is independent of position, the connection (2.4) reduces to an electrodynamical relation between the electric current density and the electric field $\nabla(-V)$, called the constitutive relation of the medium. In good conductors, i.e. metals and strongly doped semiconductors, the high carrier density is hardly sensitive to external influences, and the conductivity γ is a material constant. In undoped and moderately doped semiconductors, the carrier density may vary widely in space, and the concept of a conductivity is of little relevance.

The distinction between the integral relationship (2.2) and the local relationship (2.4) is crucial. In some French textbooks [61], the integral relation (2.2) is named after Ohm, while the relation (2.3) involving the local quantity ρ is named after Pouillet although actually the relations are intertwined in Ohm's monograph [59].

2.3 From Macro- to Microphysics

Upon writing the electron current density as the product of the density of movable electrons and their average (drift) velocity $\mathbf{v}_\mathrm{d}$, the local connection (2.4) yields

$$\mathbf{v}_\mathrm{d} = \mu\nabla(-\tilde{g}), \tag{2.5}$$

where $\mu \equiv \gamma/ne^2$ is called the carrier mobility. Most experimental papers make use of the electrical mobility $e\mu$, reckoned in $\mathrm{m^2V^{-1}s^{-1}}$

in SI unit, connecting the average velocity to the electrochemical field $\nabla(-\tilde{g}/\bar{e})$. In (2.5) the carrier velocity is proportional to the force, which is Aristotle's law of motion. It asserts that a force is required to keep an object in motion, while Newton's second law asserts that an external force is required to disrupt the motion of an object. Here the force is electrochemical rather than electrical. Upon rewriting (2.5) as a force balance,

$$\nabla(-\tilde{g}) - \frac{1}{\mu}\mathbf{v}_{\mathrm{d}} = 0, \tag{2.6}$$

the inverse mobility appears as a coefficient of viscous friction in a hydraulic analogy[2].

A consequence of (2.5) is that the Joule heat is released locally at a time rate $\nabla(-\tilde{g}) \cdot \mathbf{v}_{\mathrm{d}}$ per carrier. Equivalently, the Joule power per unit volume, reckoned in watt/m^3, is

$$n\nabla(-\tilde{g}) \cdot \mathbf{v}_{\mathrm{d}} = \frac{(ej_{\mathrm{N}})^2}{\gamma}. \tag{2.7}$$

The energy balance is in terms of the Helmholtz energy as we are considering a system—the assembly of movable electrons—whose temperature is held constant by an agency—the thermostat—which performs its function by bringing or removing energy in a disordered form (thermal energy). The thermostat is the lattice which hosts the movable electrons. The latter are endowed with a temperature because energy is interchanged between them and the lattice whose internal energy is that of atomic oscillators in thermal equilibrium [5, 21]. The temperature of the electron gas is set by the vibrating lattice insofar as the interaction between the two systems allows for interchange of energy. In the absence of an electrochemical gradient, the interchange is balanced on the average; it consists of fluctuations averaging out to zero. If the

[2]The current edition of the Oxford English Dictionary defines a voltmeter as 'an instrument for measuring the pressure of electricity in volts'. To recover the thermodynamic pressure p of the electron gas, multiply (2.6) by the electron density n so as to obtain forces per unit volume. The Gibbs–Duhem relationship, namely $n\,\mathrm{d}g = \mathrm{d}p$ under no $\mathrm{d}T$, entails $n\nabla g = \nabla p$, wherefore $n\nabla\tilde{g} = \nabla p + n\nabla(\bar{e}V)$ appears as a pressure gradient augmented with the electric force per unit volume. Since (2.6) implies $\mathbf{j}_{\mathrm{N}} = \mu[\nabla(-p) + n\nabla(-\bar{e}V)]$, an ideal voltmeter such that no current flows ($\mathbf{j}_{\mathrm{N}} = 0$) relates a pressure difference *stricto sensu* to an electrostatic-potential difference in the electron gas.

equilibrium is disturbed by an electrochemical gradient, imbalance in the interchange is manifested in a net release of thermal energy. The release occurs locally according to (2.7) as the electrons get scattered by the lattice of atoms.

In Bloch's quantum theory of solid lattices based on Hartree's one-electron self-consistent picture, it was understood that [5]

(i) in the absence of collisions, an electron moves with a group velocity given by a quantum-mechanical dispersion relation between its energy and a pseudomomentum;
(ii) the electron's motion is disrupted by imperfections of the lattice which cause a finite variation of the group velocity.

Lattice imperfections can be localized—e.g. foreign impurities such as a phosphorus atom in a silicon lattice—or not—e.g. the disordered vibrations embodying the thermal agitation of the lattice. More importantly, since a voltage is an energy released per unit charge, in a microscopic description the distinction between the energy-conserving (elastic) and energy-interchanging (inelastic) events which scatter the charge carriers as they flow along a conducting wire is mandatory. Unlike interaction with a static imperfection, interaction with a vibrational mode of the lattice results in energy interchange. Therefore, the electrical resistance is due to the latter interaction. In a quantum-mechanical description, a normal mode of lattice oscillation has its energy quantized as a half-integer multiple of the phonon energy. In Levich's words, 'the interaction of the electrons with the vibrating lattice and their scattering on the atoms of the lattice give rise to electrical resistance. This interaction [. . .] can formally be considered as a result of the collisions between electrons and phonons which, as a gas, fill the volume of the solid body' [48].

While Joule heating tends to enhance thermal agitation, the latter is stabilized, statistically speaking, if the lattice—the phonon gas—is large enough to behave as a perfect thermostat. Under a steady applied field, however, the lattice must continually remove more energy from electrons than it brings to them. As the lattice is finite, stability cannot hold at infinite times. On touching a commercial resistor in an electronic circuit made up of discrete components under operation, one can feel that the resistor is warmer than the environment such as the atmosphere. The higher temperature in the resistor is due to the production of heat co-occurring with the flow of charge. Although the resistor temperature is higher than the environmental one, it is stable. The discrepancy of temperatures is observed to depend on the wattage rating of the resistor (1/4 W, 1/2 W. . .); the larger the rating, the smaller the difference. To make the resistor temperature closer to the environmental one, besides choosing a larger wattage rating one will mount the resistor on a heat sink, which acts as an outer thermostat in continually removing energy.

Lorentz's linear variant of Boltzmann's kinetic equation was improved by Sommerfeld [5]. It is reformulated in the three-dimensional space of pseudomomenta once Bloch's energy-band structure of the crystal is taken into account. That improved kinetic equation involves a complicated (integral) scattering term which often is replaced by a much simpler, non-integral relaxation-time ansatz. Unfortunately that ansatz lacks mathematical justification [5, 62]. Besides, the ansatz can violate the law of conservation of particles [71, 74] and the First and Second laws of thermodynamics [18] so that it does not fit our purpose in this chapter. We shall instead use a mathematically simpler transport description formulated in the one-dimensional space of electron energies. It is basically a continuous-time implementation of the central limit theorem of probability theory for small position and/or energy steps [69], named after Fokker and Planck. The Fokker–Planck description is epitomized in Appendix 2.A, based upon [10] for a dilute electron gas and [12, 13] when Pauli exclusion is significant. The simplifying features adopted herein are to ignore the crystal anisotropy and take a single parabolic conduction band. That is to say, an electron of energy E reckoned with respect to the bottom of the conduction band travels at a group velocity $v_g(E) = (2E/m^*)^{1/2}$ with a scalar effective mass m^*. Besides, an electron is scattered by the lattice elastically at a rate $1/\tau_{el}$ and inelastically at a rate $1/\tau_{in}$. All scatterings are taken to be isotropic in that the rates depend on energy E instead of pseudomomentum. The effect of an electric force field F is to change energy smoothly at a rate $Fv_d(E)$ per electron, where $v_d(E)$ is the drift velocity of an electron of energy E. Under a small field, that velocity is $\mu(E)F$, where $\mu(E)$ is an energy-dependent mobility. This Fokker–Planck framework quantitatively reflects the fact that conductivity can be chiefly governed by elastic scatterings while conduction is basically a dissipative phenomenon involving inelasticity. In plain language, for a given length directed along the applied force, scattering by static imperfections, which is elastic and does not allow for energy interchange with the host medium, increases the undirected total path length of one electron. Thereby, for a given wire length and applied force, the time spent by a free electron to cross the wire from entrance to exit is enhanced by elastic events, and so too is the number of inelastic events due to lattice vibrations. Thus it is that the electrical resistance is enhanced by elastic scatterings although they do not produce heat. It is found that the mobility μ in (2.5) is given by a Maxwell–Boltzmann average of $\mu(E)$ if the electron gas is dilute, and by the value $\mu(E_F)$ at the Fermi energy E_F if the gas is degenerate.

The forthcoming section describes an experiment performed on Si wires of nanoscopic dimensions where Ohm's law in the strong sense, i.e. both (2.2) and (2.3), is obeyed. The measured mobility will be interpreted theoretically in terms of the existing scattering mechanisms.

2.4 Ohm's Law Survives to the Atomic Scale

2.4.1 The Experiment

Phosphorus-doped Si strips were grown in single-crystalline Si by Weber et al. [83]. The strips were patterned on a nominally undoped Si surface by means of scanning probe lithography and a gaseous dopant source. The strips were subsequently embedded within undoped Si deposited by means of molecular beam epitaxy. Doped strips of lengths L = 20–312 nm and widths w = 1.5–11 nm, as listed in Table 2.1 and schematically shown in Fig. 2.1, were thus fabricated; the symbol Ω will henceforth denote the SI unit of resistance instead of the conductor's volume. The strip thickness as such is not sharply defined; what is well defined is the penetration or segregation length of phosphorus after the Si surface has been exposed to phosphine (PH_3) and subsequently

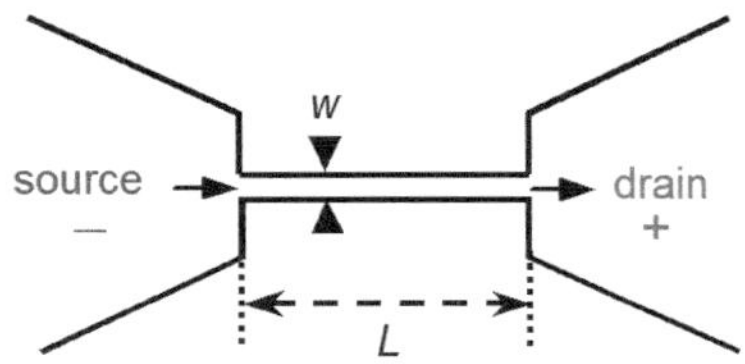

Figure 2.1 A phosphorus-doped Si strip of width $w = 1.5$–11 nm and length $L = 20$–312 nm (see list in Table 2.1 after [83]) is schematically shown. The strip thickness (perpendicular to the plane of the figure) is the characteristic penetration length, namely 2 nm, of the P dopant.

Table 2.1 Geometrical and electrical wire parameters, after Weber et al. [83]

L (nm)	w (nm)	R ($10^3\Omega$)	L/w	ρ ($10^{-6}\Omega$ m)	$e\mu$ (10^{-4} m^{2}V^{-1}s^{-1})
312	11	48.6	28.4	4.3	15
106	1.5	82.3	70.7	2.6	24
20	2.3	17.1	8.7	4.2	15

annealed. The phosphorus penetration decreases exponentially with a segregation length 2 nm. If the width is very small, the strip is a wire *stricto sensu*. A prominent feature is the atomically abrupt nature of the wires as shown by scanning tunnelling microscopy. The dopant density in the wire is $N_D \approx 10^{21}$ cm^{-3}, while in the nominally undoped bulk it is $\approx 10^{15}$ cm^{-3}. The typical distance between neighbouring P atoms in the wire is $N_D^{-1/3} \approx 1$ nm. The same estimate is obtained from the two-dimensional number density $\approx 2 \times 10^{14}$ cm^{-2} of P atoms.

The I–U relationship of each wire was obtained at $T = 4$ K. The I–U plot was linear for $|U| < 1$ mV. The wire resistance R reported in Table 2.1 was calculated by subtracting the series resistance of the two contacting regions whose sheet resistivity (530 $\Omega/\square$) and geometry were known. A scaling law $R \propto L/w$ is observed in Table 2.1. This suggests that resistivity ρ is well defined once a thickness (2 nm) of the P-doped layer is defined. It has been remarked [23] that the scaling law is rough; but complementary data taken from measurements on similarly grown P-doped Si wires [68], reported in Table 2.2, support the scaling law over the wider range

Table 2.2 Geometrical and electrical wire parameters, after [67] and [68]

L (nm)	w (nm)	R ($10^3\Omega$)	L/w	ρ ($10^{-6}\Omega$ m)	$e\mu$ (10^{-4} m^{2}V^{-1}s^{-1})
900	90	10	10	0.6	70
320	27	50	11.9	2.5	11
310	50	34	6.2	3.3	10

L = 20–900 nm. The resistance of the other wires was obtained in a different way by means of a four-point measurement where the current leads and the voltage probes are different as shown in Fig. 2.2. Two points are used to inject and eject current, while the two other points are connected to a high-impedance voltmeter. From

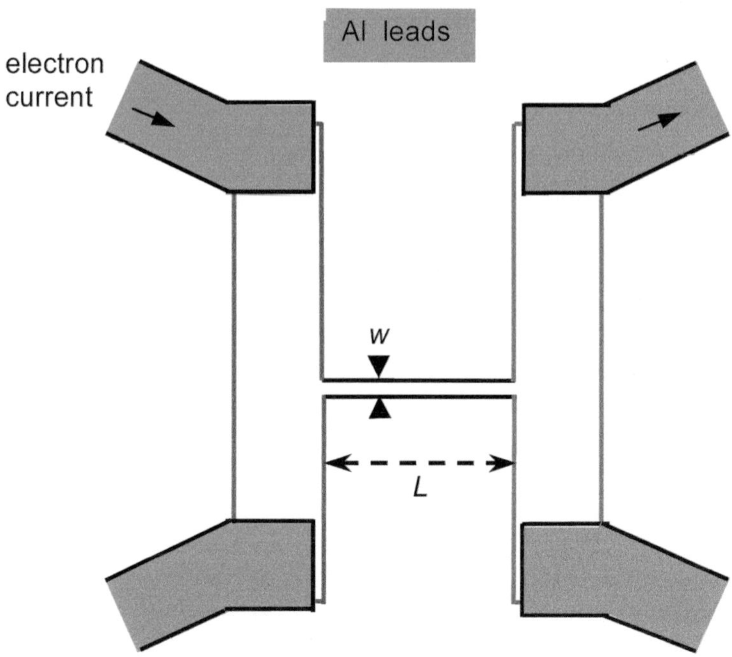

Figure 2.2 Schematic picture of the four-point measurement device of Rueß et al. [68] where current leads (above) and voltage probes (below) are distinct. The latter are connected to a high-impedance voltmeter. The wire of width w and length L and the contacting regions are made of highly-doped Si. The measured resistance is essentially due to the wire.

now on, we shall take the scaling law for granted despite the scatter in the experimental data. That is to say, we take the wire resistances to be accounted for by the relationship $R \propto \rho L/w$ involving a resistivity ρ whose value is calculated by setting thickness $\mathcal{A}/w$ equal to 2 nm in [83] and 0.6 nm in [67]. Weber et al. conclude that 'the average value of the wire resistivity is near that for the Si bulk resistivity at equivalent doping density'. The electrical mobility relating the drift velocity to the electrochemical force is inferred from $\rho = (e^2 n\mu)^{-1}$ and $n = N_D$ (full ionization $P \rightarrow P^+ + e^-$), and it is $e\mu \approx 15$ cm^{2}V^{-1}s^{-1}. Our goal in what follows is to provide a theoretical explanation of the mobility observed in those wires.

2.4.2 Interpretation of the Experiment

We begin with the thermostatics of the electron gas. Given a density-of-states effective mass m^* equal to 1.08 m_0 in Si, where m_0 is the free electron mass, the calculated thermal de Broglie wavelength of electrons at $T = 4$ K is 37 nm. As it largely exceeds the typical inter-electron distance $N_D^{-1/3} \approx 1$ nm, the gas is degenerate. In the conduction band ($E \geq E_c$), the low-energy states are filled by ionization electrons until a Fermi level lying at $E_c + E_F$ so that

$$n = \int_{E_C}^{E_C+E_F} N(E)\mathrm{d}E \quad \text{at} \quad T \ll \frac{E_F}{k}, \tag{2.8}$$

where

$$N(E) = \frac{(2m^*)^{3/2}(E - E_C)^{1/2}}{2\pi^2\hbar^3} \tag{2.9}$$

is the density of states per unit volume and per unit energy range, reckoned in m^{-3} J^{-1} or cm^{-3} eV^{-1}, including both spin projections. Letting $n = N_D$ gives $E_F = (2m^*)^{-1}h^2(3N_D/8\pi)^{2/3} = 0.338$ eV. The speed of Fermi-level electrons,

$$v_F = \left(\frac{2E_F}{m^*}\right)^{1/2} = 3.32 \times 10^5 \text{ m/s}, \tag{2.10}$$

largely exceeds the thermal speed $(3kT/m^*)^{1/2}$ of non-degenerate electrons at $T = 4$ K.

We next turn to the kinetics of the electron gas. Mobility $e\mu = 15\ \mathrm{cm^2V^{-1}s^{-1}}$ and the Drude formula $\mu = \tau_F/m^*$ provide a velocity-relaxation time $\tau_F = 0.92 \times 10^{-14}$ s of electrons lying at the Fermi level. The mean free path, denoted by λ_F, is $v_F\tau_F = 3$ nm. The usual formula of kinetic theory,

$$\lambda_F = \frac{1}{N_D\sigma_\mathbf{v}}, \tag{2.11}$$

provides the velocity-relaxation scattering cross section, $\sigma_\mathbf{v} = 0.33\ \mathrm{nm^2}$. Now electrons are scattered by (i) acoustic phonons via deformation-potential coupling; (ii) impurities (ionized donors P^+); and (iii) other conduction electrons. Scattering by conduction electrons is negligible because degeneracy entails a phase-space factor $(kT/E_F)^2 \approx 10^{-6}$ weakening the scattering rate [1, 66]. Given the large dopant density, scattering by ionized donors occurs at a much larger rate than by phonons [22, 66]. Therefore, the experimental cross section should be compared to the theoretical value expected from the screened Coulomb potential of an ionized impurity, namely

$$v(r) = v_0\left(\frac{\lambda_{TF}}{r}\right)\exp\left(-\frac{r}{\lambda_{TF}}\right), \tag{2.12}$$

whose range is given by the Thomas–Fermi formula for the screening length,

$$\lambda_{TF} = \left(\frac{2\varepsilon E_F}{3ne^2}\right)^{1/2} = 0.38\ \mathrm{nm}, \tag{2.13}$$

where $\varepsilon = 11.7\,\varepsilon_0$ is the static permittivity of Si, and the strength of the screened Coulomb potential is assessed by

$$v_0\frac{-e^2}{4\pi\varepsilon\lambda_{TF}} = -0.315\ \mathrm{eV}. \tag{2.14}$$

For scattering by this potential, at vanishing kinetic energies the differential cross section becomes independent of the scattering angle so that the velocity-persistence ratio, defined as the average direction-cosine of the velocities before and after the event, vanishes. At increasing energies, the differential cross section falls off increasingly rapidly as a function of the scattering angle. For $p \gg p_{TF} \equiv \hbar/\lambda_{TF}$, only a forward peak of angular width $\approx p_{TF}/p$ is left so that the persistence ratio approaches unity and the collision becomes ineffective. In the first Born approximation of quantum scattering theory [40], it is calculated that

$$\sigma_{\rm v}(p) = 2\sigma_{\rm TF}\left(\frac{p_{\rm TF}}{2p}\right)^4\left\{\ln\left[1+\left(\frac{2p}{p_{\rm TF}}\right)^2\right]-\left[1+\left(\frac{p_{\rm TF}}{2p}\right)^2\right]^{-1}\right\}, \tag{2.15}$$

where

$$\sigma_{\rm TF} \equiv 4\pi\lambda_{\rm TF}^2\left(\frac{2m^*v_0}{p_{\rm TF}^2}\right)^2 \tag{2.16}$$

is the value of $\sigma_{\mathbf{v}}$ at a vanishing p. The function $\sigma_{\mathbf{v}}(p)$ is plotted against pseudomomentum in Fig. 2.3. It is seen to take its highest value as p vanishes. Figure 2.3 also shows the momentum-relaxation rate,

$$\frac{1}{\tau_{\rm v}(p)} = N_{\rm D}\left(\frac{p}{m^*}\right)\sigma_{\rm v}(p) \tag{2.17}$$

as a function of pseudomomentum. That rate is seen to take its highest value $\approx N_{\rm D}\sigma_{\rm TF}\,p_{\rm TF}/2m^*$ as $p \approx p_{\rm TF}/2$; that is to say, the de Broglie wavelength of the matter wave, $\hbar/p$, is on a par with the range $\lambda_{\rm TF}$ of the potential.

Since the electron gas is degenerate, p is $p_{\rm F} = m^*v_{\rm F}$, and $\sigma_{\mathbf{v}}(p_{\rm F}) = 1.32\ {\rm nm}^2$ according to (2.15). This is fourfold the experimental value 0.33 $\rm nm^2$. Now the Born approximation holds in the high-energy range $p \gg p_{\rm TF}$ while here $p = p_{\rm F}$ exceeds $p_{\rm TF}$ by only 20%. The approximation overestimates the scattering rate of a screened Coulomb potential by a factor of about 4, as compared with the rate derived numerically from a phase-shift analysis [1, 22, 66]. Reducing the Born cross section by a factor of 4 brings the theoretical value in good agreement with the experimental one. We consider the agreement as very satisfactory in view of the scatter in the experimental data, the rough geometrical approximation made in computing the wire resistivities from the bulk formula (2.3), and the absence of any adjustable parameter.

To conclude this section, the Ohmic behaviour of the Si nanowires reported by Weber et al. can be explained on the basis of the physics relevant at the microscale. The physics is semiclassical as stated in Section 2.1; in particular, the motion of conduction electrons is guided by Bloch waves, which, however, do not interfere after they

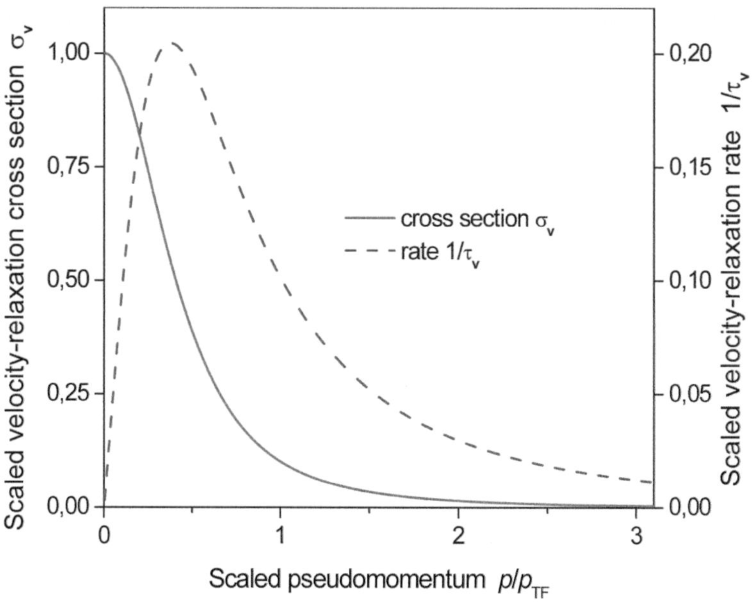

Figure 2.3 Velocity-relaxation cross section (σ_v) and scattering rate ($1/\tau_v$) of a screened Coulomb potential, plotted against pseudomomentum p. The pseudomomentum scale p_{TF} is $\hbar/\lambda_{TF}$, where λ_{TF} is the range of the potential. The cross section has been scaled by its value σ_{TF} at a vanishing energy $p^2/2m^* \to 0$. The scattering rate has been scaled by $N_D(p_{TF}/m^*)\sigma_{TF}$.

are reflected or transmitted by the scatterers which alter the crystal periodicity. The main finding of [83] is now textbook material [63]. In Section 2.6, another experiment on conduction at the nanoscale will be interpreted which shows a more important departure from classical physics. Before dealing with that experiment, a few theoretical considerations which motivated the experiment are reviewed in Section 2.5.

2.5 Can a Length of Perfect Conductor Have a Resistance?

In 1981, Landauer asked the question, 'Can a length of perfect conductor have a resistance?' [41], where 'perfect' meant that over the given length the conduction electrons travel ballistically, i.e. without undergoing a collision with the lattice of atoms. A warning is due to dispel a frequent confusion. Many papers refer to an earlier work of Landauer published in 1957. In a 1996 comment [44], Landauer remarks that his 1957 paper 'is most often cited in connection with the now widely used expressions which relate conductance to the overall transmissive behaviour of a sample which transmits carriers in a quantum mechanically coherent and elastic fashion. That result is, however, not contained in the 1957 paper'. Landauer's 1981 calculation of the electric current I in response to a small voltage U across the terminals to a perfect metallic wire threw up the linear relation (2.2) with an electrical conductance given by

$$G = \bar{\mathcal{T}} \left(\frac{2e^2}{h} \right), \tag{2.18}$$

where $\bar{\mathcal{T}}$ is an average wave-mechanical transmission coefficient of electrons between the terminals. A transmission coefficient is expected because a classically allowed motion has a probability possibly less than unity in quantum mechanics [40]. This was already noted in the physics of vacuum diodes where electrons travel ballistically through a vacuum gap [33, 57]. The transmission coefficient of the gap cannot be predicted reliably as it is affected by surface roughness and chemical contamination of the electrodes; it may take an experimental value $\approx 1/3$ [14] [3] and it is sometimes unmentioned [87]. In 1984, Sinkkonen et al. investigated Landauer's question for a semiconducting wire where the occupancy of electron states is much less than unity [72]. Then, the formula for G contains an extra occupancy factor. The calculated conductance decreases as L^{-1} at lengths L longer than the electronic mean free path because of collisions occurring in the wire; this is consistent with the 'Pouillet law' (2.3). For L of a few atomic layers, an increase of G is expected as transmission is enhanced by tunnelling through classically forbidden states lying between the terminals.

[3] In commercial vacuum diodes the cathode is heated while the anode is not, which gives rise to current rectification. The current–voltage characteristic is linear in the low-voltage range [14]. The slope of the characteristic depends on the difference in electrode temperatures, but Fig. 4 of [16] shows that a reduction in the cathode temperature does not entail a marked variation of the slope.

The first experiments answering Landauer's question were performed in 1988 on degenerate GaAs wires at $T = 4$ K, where the electronic mean free path can reach several micrometres [78, 85]. The wires were electrostatic constrictions of the quasi-two-dimensional electron gas lying on the GaAs side of a GaAs–GaAlAs heterojunction; see the next section. The observed wire conductance was finite and exhibited quantization in approximate multiples of $2e^2/h$. The latter feature and other ones had not been anticipated by theory, as was remarked by van Houten et al. [77] and by Landauer [43]. The quantization of conductance could be related to the fact that a constriction of the electron flow acts as an electron wave guide with discrete transverse eigenmodes [84], nowadays termed conduction channels. Each one entails a conductance approximately equal to $2e^2/h$. The quantization of conductance in a metal wire had been hinted at in 1987 when Gimzewski and Möller observed a discontinuous change in the current flowing between the tip of a scanning tunnelling microscope and a silver substrate [30]. Later on, the phenomenon was observed by several groups. In a table-top experiment the opening of the contact of an electromechanical relay switch showed discrete steps of decreasing conductance [60]. The conductance steps may be integer mutiples of $2e^2/h$, corresponding to $\bar{\mathcal{T}} = 1$, but non-integer steps are observed, too. Theoretically speaking, smooth potential barriers behave almost classically with a transmission coefficient equal to unity or zero [40]. Most experimenters prefer to report 'classical' observations such that $\bar{\mathcal{T}} = 1$ or 0. Metallic nanocontacts are less well-characterized and reproducible than semiconductor nanostructures, and later work is reviewed in [2, 37].

Ohm's law $I = GU$ cannot be separated from the Joule–Lenz heating effect. The battery which maintains a steady voltage across a conductor supplies—isothermally—a steady electric power $UI = GU^2$ eventually manifested as heat. This is an irreversible process associated with a dissipation quantitatively defined as the rate of entropy creation; that rate is GU^2/T if temperature is homogeneous. As remarked by Landauer, 'there is a central problem with most of the more totally quantum-mechanical formulations. They use the quantum mechanics of conservative systems, but then they go on and try to explain dissipative effects. They also need to invoke the quantum mechanics of closed systems [...]. But a resistor has leads, and particles must be able to go in and out of those. All of this requires cheating' [42]. Because $G \propto \bar{\mathcal{T}}$, the theoretical picture of conduction in a ballistic wire reached in 1997 is summed up in the statement 'Conductance is transmission' [45]. But in 1999 Imry and Landauer [35] remind us that 'the dissipation and the irreversibility are in the reservoirs; carriers returning to them from the sample eventually suffer inelastic collisions'. Again it is remarked that dissipation cannot arise in a purely quantum-mechanical account of ballistic conduction rooted in matter-wave transmission and reflection: 'Quantum theory, as described by the Schrödinger equation, is a theory of conservative systems, and does not allow for dissipation. [...] Some supplementary handwaving is needed to calculate a dissipative effect such as conductance, for a sample with boundaries where electrons enter and leave'.

To address this and other questions, in 2001 de Picciotto et al. performed an experiment on GaAs wires whose very high electron mobility attested to the ballistic nature of transport over a length of 6 μm [26]. A linear I–U plot was found and a four-point probe measurement allowed to map the electrochemical-potential profile along the wire. This experiment is presented and theoretically analysed in the next section.

2.6 The Resistance of a Ballistic Wire

2.6.1 The Four-Terminal Measurement of Resistance

In [26] a thin layer of GaAs lies between two thick parallel layers of AlGaAs; see Fig. 2.4. The offset of the conduction bands of GaAs and AlGaAs is 0.24 eV. For electrons donated by dopants located in AlGaAs, the GaAs layer acts as a potential well of depth 0.24 eV. Because this depth largely exceeds kT at $T = 4$ K, the donated electrons are confined between the two AlGaAs layers, called barriers, and they make up a quasi-two-dimensional electron gas. In the GaAs layer, an electron moves as a matter wave whose transverse wave-vector k_z is quantized owing to the thinness of the well, while the components k_x and k_y are free to vary continuously. On growing an additional doped AlGaAs layer onto the edge of the AlGaAs–GaAs–AlGaAs stack, the wave propagation is restricted to one single dimension x because the positively charged ionized dopants present in the additional layer confine the GaAs conduction electrons near that layer. We thus have a guided matter wave travelling along the x axis but stationary in the two other dimensions, i.e. k_y is quantized besides k_z. The carrier's transverse motion is quantized into discrete modes indexed by (i, j) while the longitudinal motion along x is free [84]. The eigenvalues of the carrier's energy lie in subbands indexed by (i, j) and they can be expressed as

$$E_{i,j}(p_x) = E_{i,j} + \frac{p_x^2}{2m^*}, \tag{2.19}$$

where $E_{i,j}$ is the energy at the bottom of the subband (i, j) and $p_x = \hbar k_x$ may vary continuously.

For a matter wave to propagate freely along x, scattering should be avoided. This is achieved by (i) avoiding structural imperfections such as impurities, which is why remote dopants located outside the GaAs layer are used; and (ii) taking a temperature low enough that lattice vibrations are absent. Regarding the latter point, lattice vibrations are quantized in the form of phonons with an equilibrium occupancy of the Bose–Einstein type. That occupancy vanishes with temperature, and classically speaking the lattice is frozen. Spontaneous emission of a phonon by a conduction electron is, however, possible at any temperature at a finite rate [7, 66]. Electron–phonon scattering occurs at a rate $1/\tau_{e-ph}$ associated with a length $v_g \tau_{e-ph}$ called the electron–phonon mean free path, where the group velocity $v_g = \partial E_{i,j}/\partial p_x$ is taken at the filling level of the subband (i, j). If the free path is longer than the wire's length L, transport along the wire is ballistic. A 6 μm-long wire has been fabricated where the large electrical mobility measured, $e\mu = 4 \times 10^6$ cm^{2}V^{-1}s^{-1}, indicates a mean free time $\tau = m^*\mu$ equal to 1.4×10^{-10} s, and a mean free path equal to or exceeding the wire's length.

Figure 2.4, drawn after Fig. 1a of [26], shows three tungsten gates. Upon application of a gate voltage negative with respect to the source-to-drain channel, the electron gas underneath the gates is depleted. Varying the gate voltage provides a convenient tool to control the number of occupied modes in the wire. A small bias across source and drain is observed to give rise to a linear response

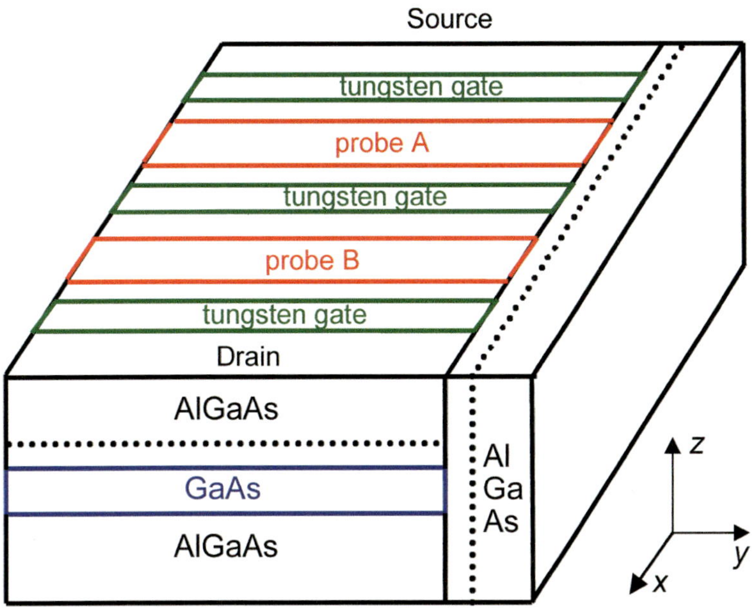

Figure 2.4 Schematic view of the structure used to create a ballistic GaAs wire, after Fig. 1a of [26]. A thin undoped GaAs layer is sandwiched between two thick parallel AlGaAs layers. A third AlGaAs layer is grown onto the edge. The GaAs layer is populated by electrons originating from sheets of donor impurities, indicated by dots, located in two AlGaAs layers. Electrons are electrostatically confined in the upper right corner of the GaAs layer, and they make up a conducting wire of a nanoscopic cross section where they can move ballistically along x. The two-terminal measurement of resistance involves the electrochemical-potential drop between source and drain serving as the current leads. The four-terminal measurement involves the electrochemical-potential drop between two probes, A and B, placed along the pathway of the electrons. Three tungsten strips serve as the gate electrode controlling the conductance of the source-to-drain channel.

of current with a resistance depending on the gate voltage. For a gate voltage lying below -4.5 V, the resistance is infinite. For a voltage lying between -4.5 and -3.8 V, Fig. 1b of [26] shows a quantized resistance $\approx$ 19 kΩ across the two current leads. This indicates the existence of one conduction channel (transverse mode). The resistance value is greater than $h/2e^2 \approx 13$ kΩ because of 'non-ideal coupling between the cleaved-edge-overgrowth wire and its two-dimensional source and drain contacts'; see [25] for more details. That is to say, part of the electron wave is reflected as it leaves the source contact to enter the ballistic wire, and likewise at the drain contact. From Landauer's formula (2.18), a non-ideal transmission coefficient $\bar{\mathcal{T}} \approx 13/19 \approx 70\%$ is inferred. For a less negative gate voltage, Fig. 1b of [26] shows conductance steps smaller than $h/2e^2$, which again indicate transmission coefficients less than unity. In the language of classical wave physics, the wire is a long uniform constriction ending in tapered structures which, like an acoustic horn, permit a wave in the constriction to merge into the wider leads with some reflection.

In the central region of the wire, the strips A and B serve as voltage probes. They may cause additional back-scattering of the matter wave as they distort the electrostatic field. This is included in the observed transmission coefficient $\approx 70\%$ which the authors principally ascribe to non-ideal coupling of the ballistic wire to the source and drain regions. More importantly, a voltmeter lowers the voltage actually present in a circuit if the voltmeter impedance does not largely exceed the resistance of the circuit under measurement. In the experiment, the impedance associated with each probe was measured to be 250 kΩ so that the voltmeter invasiveness is about $19/500 \approx 4\%$ [26].

Figure 2 of [26] shows that, while the two-terminal resistance moves through the characteristic quantized-resistance steps, in contrast the four-terminal resistance measured by the probes A and B hovers around zero resistance. Thus, the voltage across the probes located along the pathway of the electrons vanishes. As temperature is well-defined and homogeneous, this means that the electrochemical-potential profile is flat along the wire. While Ohm's law (2.2) is obeyed with $R \approx h/2e^2$ for one occupied mode, that law is *not* associated with an electrochemical potential profile sloping down from source toward drain. The voltage, and thence the resistance, should be ascribed to the source and drain which carry current into and out of the ballistic wire.

In Fig. 3 of [26] at zero magnetic field, the internal voltage of the ballistic wire is measured to be half the drain voltage reckoned with respect to the source taken to be at ground. That is to say, the electrochemical potential $\tilde{g}_{\mathrm{w}}$ in the wire lies midway between the electrochemical potentials, $\tilde{g}_s$ and $\tilde{g}_d$, in the source and drain terminals. This finding is consistent with the symmetry of the device in the exchange of source and drain which changes $+I$ into $-I$.

2.6.2 Thermodynamic Interpretation of the Four-Terminal Resistance

To properly address heating in an electric resistor, one should explicitly consider the flow of heat and the possibility of an inhomogeneous temperature. The simplest framework to do so is linear non-equilibrium thermodynamics [5, 27, 56]. It agrees with the Boltzmann kinetic theory under a weak deviation from equilibrium [38, 73]. Since equilibrium in a resistor is achieved under no voltage and no temperature difference across the terminals, non-equilibrium thermodynamics is a valid framework under weak voltage and weak temperature difference. As recognized in Section 2.2, the voltage across two points is the difference of their electrochemical potentials, $\tilde{g}$, per carrier charge. In non-equilibrium thermodynamics $\tilde{g}$ and T are defined locally and their gradients give rise to an electron current density, $\mathbf{j}_N$, and a heat current density, $T\mathbf{j}_S$, where $\mathbf{j}_S$ is the entropy current density [27, 38]. The current densities are reckoned with respect to the lattice. They are related to the gradients through linear relations which may be written in several equivalent ways. The one most convenient to our purpose is

$$\mathbf{j}_N = n\mu[\nabla(-\tilde{g}) - (s + s^*)\nabla T], \tag{2.20}$$

$$T\mathbf{j}_S = -\kappa\nabla T + T(s + s^*)\mathbf{j}_N, \tag{2.21}$$

where κ is Fourier's thermal conductivity, $s = -(\partial a/\partial T)_n$ is the entropy per carrier, a is A/N and s^* is an entropy of transport, following the nomenclature and notation of [56]. While s is a thermostatic quantity, the extra term s^* is not; its kinetic nature is disclosed in a simple microscopic model where s^* is seen to be related to the energy dependency of the carrier scattering rate [5, 27]. While s depends on how entropy is gauged, s^* does not [17]. In the linear-response relations above, the first term on the right-hand side expresses the fact that electrons (heat) flow(s) down the electrochemical (temperature) gradient. The second term is the Seebeck (Peltier) contribution to the electron (heat) current density. If temperature is homogeneous the electron current density reduces to $n\mu\nabla(-\tilde{g})$. Otherwise relation (2.20) includes the Seebeck thermoelectric effect: the flow of carriers between two points is due not only to the voltage across the points but also to the temperature difference.

The local production of entropy per unit time and unit volume, reckoned in watt per K and per m^3, is div $\mathbf{j}_\mathrm{S}$. It is called the dissipation (per unit volume) and will be denoted by $\dot{\Sigma}$. It is calculated to be [17, 27, 56]

$$\dot{\Sigma} = \frac{1}{T}[\mathbf{j}_\mathrm{N} \cdot \nabla(-\tilde{g}) + \mathbf{j}_\mathrm{S} \cdot \nabla(-T)]. \tag{2.22}$$

On bringing relations (2.20) and (2.21) into (2.22), it is found that

$$\dot{\Sigma} = \frac{1}{\mu}\left(\frac{\mathbf{j}_\mathrm{S}^2}{nT}\right) + \kappa\left(\frac{\nabla T}{T}\right)^2. \tag{2.23}$$

The positivity of dissipation is due to that of the friction coefficient $1/\mu$ and the thermal conductivity κ. Irreversibility in a steady state is manifested in the dissipation. It is caused by the friction undergone by the carriers dragged off by the driving force and/or the flow of heat from hot to cold locations. Dissipation due to a thermal leak down a temperature gradient can damp a macroscopic motion in the same way as friction [15, 88]. It is seen in (2.23) that the Seebeck and Peltier effects do not contribute to the dissipation.

2.6.3 Microscopic Interpretation of the Four-Terminal Resistance

The thermodynamical framework holds if the carriers have well-defined values of their intensive thermodynamic variables, $\tilde{g}$ and T, at every location. This condition is met in ordinary conductors where the scattering rate of conduction electrons is high. Taking Cu at 1 K as an example, the electrical conductivity $e^2\mu n \approx 5 \times 10^{10}\ \Omega^{-1}\mathrm{m}^{-1}$ [49] and the electron density $n \approx 10^{29}\ \mathrm{m}^{-3}$ yield $\mu \approx 2 \times 10^{19}$ s/kg. From the Drude formula, the rate of momentum relaxation $1/\tau$ is estimated to be $5 \times 10^{10}\ \mathrm{s}^{-1}$. Since an electron exchanges energy with the lattice at a fast rate (though slower than $1/\tau$ owing to the weak fractional energy change per scattering event), electrons quickly reach a state of thermodynamic equilibrium characterized by a Fermi–Dirac occupancy of the states at the lattice temperature with a chemical potential g depending on the electron density n and temperature T. In a steady state specified by finite current densities, the deviation from equilibrium decreases with decreasing current densities. To the first order in $\mathbf{j}_N$ and $\mathbf{j}_S$, the steady-state occupancy of an electron state,

$$f(\mathbf{p}, \mathbf{r}) = f_0(E_k(\mathbf{p}), \mathbf{r}) + f_1(\mathbf{p}, \mathbf{r}), \tag{2.24}$$

is the sum of the equilibrium occupation function f_0, which is even in $\mathbf{p}$, and of a small contribution f_1 odd in $\mathbf{p}$ so that the current densities do not vanish. In (2.24), $E_k(\mathbf{p}) = E_c + \mathbf{p}^2/2m^*$ for electrons of effective mass m^*. For transverse confinement along y and z,

$$E_k(\mathbf{p}) = E_{i,j} + \frac{p_x^2}{2m^*}, \tag{2.25}$$

with $E_{i,j}$ taking discrete values. The equilibrium contribution f_0 is a Fermi–Dirac function of the total (mechanical) energy $E = E_k(\mathbf{p}) + \bar{e}V(\mathbf{r})$ with position-dependent $\tilde{g}(\mathbf{r})$ and $T(\mathbf{r})$. This is the so-called state of local equilibrium where the deviation from equilibrium f_1 is much smaller than the equilibrium function f_0 [12, 38, 73].

To simplify the theoretical interpretation, we consider a one-dimensional geometry roughly approximating the experimental one. Granted local equilibrium in the wire terminals lying at $x < 0$ and $x > L$, we ask the question whether conduction electrons have well-defined values of $\tilde{g}$ and T in the ballistic region $0 < x < L$. This amounts to asking whether the zero-order state occupancy has a Fermi–Dirac pattern in that range. The answer can be obtained from kinetic theory where the occupancy $f(\mathbf{p}, \mathbf{r}, t)$ at time t obeys a Vlasov-like equation,

$$\frac{\partial f}{\partial t} + \mathbf{v}_g(\mathbf{p}) \cdot \nabla f + \nabla(-\bar{e}V) \cdot \left(\frac{\partial f}{\partial \mathbf{p}}\right) = 0, \tag{2.26}$$

in which $\mathbf{v}_g(\mathbf{p}) \equiv \partial E_k/\partial \mathbf{p}$ is the group velocity of the Bloch wave, and the electric force $\nabla(-\bar{e}V)$ undergone by a carrier is determined self-consistently from the Poisson equation. In one dimension x, the latter is[4]

$$\frac{\partial^2 V}{\partial x^2} + \frac{\bar{e}n(x,t)}{\varepsilon} = 0, \tag{2.27}$$

where ε is the static permittivity and

$$n(x,t) = \int f(\mathbf{p}, x, t)\frac{\mathrm{d}^3\mathbf{p}}{h^3}.(2.28) \tag{2.28}$$

Equation (2.26) is a Boltzmann equation whose right-hand side vanishes owing to the absence of collisions [4, 32, 38]. Such an equation holds in the gap of a vacuum diode where the electron undergoes an electric force without undergoing collisions. Equation (2.26) is semiclassical in that the velocity in vacuum $\mathbf{p}/m_0$ is replaced by the group velocity determined by the electronic band structure of the crystal. The right-hand side of (2.26) is approximated to zero because, on the GaAs side of an AlGaAs–GaAs heterojunction, the scattering rate $1/\tau = 1/m^*\mu$ is very low as the electron mobility is very high.

A time-independent, energy-dependent solution $f_0(E_\mathrm{k}(\mathbf{p}), \mathbf{r})$ in the range $0 < x < L$ should satisfy the condition

$$\mathbf{v}_g(\mathbf{p}) \cdot [\nabla f_0 + \left(\frac{\partial f_0}{\partial E_\mathrm{k}}\right)_\mathbf{r} \nabla(-\bar{e}V)] = 0, \tag{2.29}$$

at any $\mathbf{p}$ and $\mathbf{r}$, so that the term in square brackets should vanish. Changing the energy variable E_k into the total energy E so that $f_0(E_k, \mathbf{r}) \equiv F(E, \mathbf{r})$, the term in square brackets is ∇F. Its vanishing means that F is a position-independent function $F(E)$. This reflects the conservation of the total energy of a conduction electron in the ballistic region. In addition, the function F should fulfil continuity conditions at the boundaries $x = 0$ and L. Granted local equilibrium in the lead at $x < 0$, F at the boundary $x = 0$ has to be a Fermi–Dirac function with the parameters $\tilde{g}(0)$ and T of the lead at $x = 0$:

$$F(E) = \left[\exp\left(\frac{E - \tilde{g}(0)}{kT(0)}\right) + 1\right]^{-1}. \tag{2.30}$$

Therefore, in the ballistic region f_0 is a Fermi–Dirac function of E with position-independent parameters, namely $\tilde{g}_\mathrm{w} = \tilde{g}(0)$ and $T_\mathrm{w} =$

[4]The Poisson equation, which strictly holds in electrostatics, is approximately true for a slow time dependence owing to the high value of the speed of light.

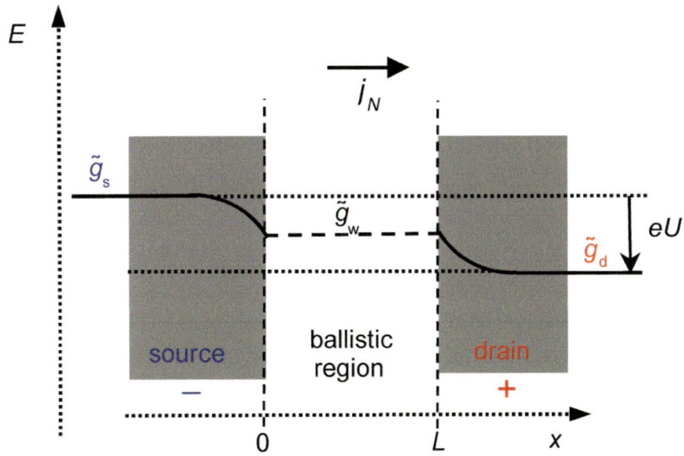

Figure 2.5 Schematic energy diagram of the source–ballistic-wire–drain structure, showing the spatial variation of the electrochemical potential $\tilde{g}(x)$ of the electrons in the leads and in the ballistic region, under a bias $U = (\tilde{g}_s - \tilde{g}_d)/e$ causing an electron current density j_N; the bias is exaggerated in the figure. The electrochemical potential $\tilde{g}_w$ is independent of position along the ballistic wire. The slightly negative slope $d\tilde{g}/dx$ far before $x = 0$ and far beyond $x = L$ is approximated to zero in the figure owing to the large electrical conductivity in the leads.

$T(0)$. By the same token, because of local equilibrium in the lead at $x > L$, we also have $\tilde{g}_w = \tilde{g}(L)$ and $T_w = T(L)$. The ballistic region thus acts as an electrical short-circuit equalizing the electrochemical potentials at its ends, as shown in Fig. 2.5. The region also acts as a thermal short-circuit equalizing the temperatures at its ends. The first point has been experimentally observed in [26], but the second point has not because the wire and the leads were held at the same temperature. The second point can be put to test in an experiment imposing distinct temperatures to the leads and probing intermediate temperatures.

Three other points are worth noting.

(i) The flat electrochemical-potential profile in the ballistic region is reminiscent of the behaviour of the junction of a metal with a high-mobility semiconductor (Schottky diode). In the semiconductor, the electrochemical-potential profile is flat under direct bias or weak reverse bias. The junction is then governed by the ballistic, instead of diffusion, theory. The former is often designated as 'thermionic theory' in analogy to vacuum diodes [31, 65].

(ii) The right-hand side of Eq. (2.26) is taken as zero while actually it is of the order of f_0/τ. The right-hand side may be neglected if it is much smaller than either term on the left-hand side, for example $\mathbf{v}_g(\mathbf{p}) \cdot \nabla f_0 \approx v_g f_0/L$. As $v_g\tau$ is the electronic mean free path, this is just the condition $v_g\tau \gg L$ expressing the ballistic nature of transport over length L.

(iii) The quantum coherence of transport in the ballistic region gives rise to the transmission and reflection of an electron wave travelling forward (backward) from $x = 0^-$ (L^+) to $x = L^+$ (0^-), with pseudo-wave-vector $+k_x$ $(-k_x)$. The transmission and reflection coefficients take into account the multiple reflections and transmissions undergone by the matter wave at the boundaries, just like in a Fabry–Pérot interferometer in optics. The coefficients connect the anisotropic contributions f_1 on both sides of each boundary in such a way that current is conserved in the steady state. This is Ben Abdallah's model of quantum (coherent) transport in a layer connected to terminals where transport is semiclassical (incoherent) [6]. This model, just like ours, rests upon a Vlasov-like kinetic equation in the quantum (ballistic) region.

In the ballistic region the electron gas has well-defined intensive thermodynamic variables, and they are uniform. The driving forces $\nabla(-\tilde{g})$ and $\nabla(-T)$ vanish, yet the electron current density is finite. That a current can flow under no driving force is akin to Newton's First Law in mechanics where a particle can have a finite velocity under no force. Thermodynamically speaking, transport is non-dissipative in that no entropy is produced while particles and heat are flowing. The vanishing of $\dot{\Sigma}$, which is due to that of $\nabla\tilde{g}$ and ∇T, is in line with the conservation of *mechanical* energy mentioned above. The absence of dissipation reflects the reversible nature of the transport phenomenon in the ballistic region. For, if $f(\mathbf{p}, \mathbf{r}, t)$ is a solution of Eq. (2.26), the time-reversed function $f(-\mathbf{p}, \mathbf{r}, -t)$ is also a solution with a velocity $\mathbf{v}_g(-\mathbf{p}) = -\mathbf{v}_g(\mathbf{p})$.

In contradistinction, in the terminals where transport is described by a Boltzmann kinetic equation including a right-hand side, time-reversal invariance is broken by the scattering integral which is unchanged upon $t \to -t$. This is the way irreversibility enters Boltzmann's H theorem [4, 38]. Irreversibility is easier to grasp in the time-dependent regime dealt with in kinetic theory than in the steady regime dealt with in non-equilibrium thermodynamics. In the latter irreversibility is quantitatively assessed in an abstract way as the rate of entropy production $\dot{\Sigma}$. It vanishes in the ballistic region but takes positive values in the ordinary conductors making up the terminals. The Joule heating, which is the isothermal contribution to dissipation, occurs in the leads where $1/\mu$ and $\nabla\tilde{g}$ are non-zero[5]. Microscopically speaking, in an electron-lattice scattering event the interchanged energy is spread out at random among the many degrees of freedom (vibrational modes) of the lattice, hence irreversibility and the production of entropy. The uniformity of temperature, not shown in Fig. 2.5, is compatible with a flow of heat because of the Peltier term in relation (2.21).

2.6.4 Thermal Conduction at the Nanoscale

From non-equilibrium thermodynamics it is expected that a non-uniform temperature field $T(x)$ should vary in the same way as function $\tilde{g}(x)$ shown in Fig. 2.5 if the transport of heat in the range $0 < x < L$ were non-dissipative. That is to say, $T(x)$ should exhibit a vanishing gradient in that range, and finite gradients in the contacts at $x < 0$ and $x > L$. This comes about because the dissipation in a phonon gas traversed by an entropy current density $\mathbf{j}'_S$ is given by $\mathbf{j}'_S \cdot \nabla(-T)/T$ instead of (2.22), since phonons are uncharged particles without a chemical potential as their number is not conserved. Thus, dissipationless transport in the range $0 < x < L$ implies a flat tem-

[5]In [24] and, [47] the electrochemical potential has well-defined values far before $x = 0$ and far after $x = L$, but *not* near $x = 0^-$ and $x = L^+$ where $\tilde{g}(0^-)$ and $\tilde{g}(L^+)$ are taken as double-valued [24, 26]. In keeping with Ockham's rule, we follow conventional non-equilibrium thermodynamics in considering that, in a reservoir of particles, $\tilde{g}$ is single-valued at any given location. If the reservoir releases or collects particles, there appears a correction $f_1 \propto \nabla\tilde{g}$ to function f_0 and the latter keeps a local-equilibrium pattern characterized by a single-valued electrochemical potential at any location until $x = 0^-$ and $x = L^+$, and a single-valued temperature as well.

perature profile in that range. As a result, the thermal conductance Γ of a non-dissipative wire should be independent of its length L, analogous to the electrical conductance G being independent of L. Indeed a thermal conductance independent of length has been observed in GaP nanowires of 25 nm diameter up to a length $L = 15$ μm [76]. A thermal resistance Γ^{-1} independent of, instead of proportional to, the length had been theoretically anticipated in a slab of a highly rarefied gas where the mean free path of molecules exceeds the width of the slab [50]. In the solid state, a thermal resistance independent of L is associated with a ballistic transport of phonons over the length of the wire, making Fourier's local formula $\kappa' \nabla(-T)$ for the heat current density invalid. On doubling the diameter of the GaP nanowire, a transition from ballistic to collisional, alternatively called diffusive, transport of phonons is observed. The plot of the thermal resistance against L becomes linear in a GaP wire of 50 nm diameter, similar to the 'Pouillet law' (2.3) for the electrical resistance.

The local lattice temperature in GaP wires has been independently probed by Raman thermometry [76]. The measured temperature profile along the 50 nm GaP wire is linear, in agreement with the picture of diffusive transport. The measured gradient is consistent with the applied difference of temperatures, indicating no significant contact resistance near to the ends of the wire. In contrast, in the ballistic 25 nm wire, symmetric temperature jumps are observed at the contacts, and the constant temperature observed along this wire lies mid-way between the contact temperatures. This parallels the findings of [26] regarding the electrochemical potential in the GaAs nanowire. It is remarkable that Ohm, following Fourier's analytical theory of heat, established his law by noting that electric current could be substituted for heat flow and potential for temperature [59]. The parallelism survives to the nanoscale.

2.7 Concluding Remarks

When thermodynamic equilibrium is negligibly perturbed by a small current, the electrical resistance of a solid conductor is the power to release in order to drive a unit electric current across the conductor, under isothermal conditions. This was our starting point to interpret resistances measured in semiconductor nanostructures where a current is flowing in a pipe of nanoscopic cross section. In the Si wires of [83], the resistance is observed to be proportional to the wire's length and it is a bulk property. The release of electric power along the wire has been ascribed to screened Coulomb scattering with the ionized dopants lying within the wire and providing the charge carriers. As shown in quantitative fashion in Appendix 2.A, the drift velocity is essentially controlled by elastic scattering off the ionized dopants while the electric power is released to the vibrational modes of the lattice, as manifested in Joule heating, in order to maintain the carriers near thermal equilibrium. The resistance is then proportional to the length of the wire according to the 'Pouillet law' (2.3).

In contrast, in the GaAs nanowire of [26] the resistance is unrelated to the length in so far as it is shorter than the mean free path. This has been ascribed to the absence of events scattering the charge carriers along the wire as the dopants providing the carriers are located outside the conducting wire. The separation of dopants and carriers entails a very high mobility, $e\mu = 4 \times 10^6$ cm^{2}V^{-1}s^{-1} in the GaAs wire, compared to $e\mu \approx 15$ cm^{2}V^{-1}s^{-1} in the Si wires. Because the GaAs wire length is shorter than the electronic mean free path, transport within this wire is non-dissipative or almost so. The dissipation principally occurs in the source and drain terminals where mobility is low and the scattering rate is high. Accordingly, the law of addition of series resistances derived from the 'Pouillet law' is invalid within the ballistic region, just as in a vacuum [14].

We also analysed an experimental study of GaP wires of nanoscopic cross sections subjected to a temperature difference across the ends. For a given wire length, thermal conduction can be dissipative owing to a finite temperature gradient in the wire, or not. The thermodynamic behaviour is related to diffusive or ballistic transport of the heat carriers at the nanoscopic level, in a way similar to charge carriers in electrically conducting wires.

Early theoretical investigations emphasized quantum features expected to play a prominent role in electronic conduction at the nanoscale. The technology of semiconductor nanostructures allowed to make highly controlled wires of nanoscopic cross sections so as to disclose those features. Together with addressing the quantum features, Landauer stressed that dissipation should be paid attention to although Clausius' entropy is absent in the usual framework of quantum mechanics. Quantization of conductance was observed in nanostructures although it had not been anticipated by theory. In this chapter, we followed Landauer in paying attention to dissipation as experimentally mapped in a GaAs nanostructure. In accordance with Ockham's rule, we did not invoke extraneous quantum features specific to a nanoscopic realm lying beneath near-equilibrium thermodynamics. We only made use of the quantum notion of a group velocity derived from Bloch's electronic band structure and of the reflection/transmission coefficients of electron waves through the source-wire-drain structure. A semiclassical Vlasov equation was able to account for the behaviour observed in the ballistic wire. Mathematical and ongoing physical research help to understand how the more totally quantum-mechanical description envisaged in the early theoretical investigations can reduce to semiclassical kinetics. Some wave-mechanical features are erased by scattering off static obstacles or the thermally agitated lattice because of static or thermal disorder. This is sketched in Appendix 2.B for the interested reader although the theoretical picture is not complete.

Appendix 2.A. The Fokker–Planck Account of Ohm's Local Law and the Heating Effect

Let $\underline{n}(E, t)$, reckoned in $\mathrm{m}^{-3}\mathrm{J}^{-1}$ in SI unit ($\mathrm{cm}^{-3}\mathrm{eV}^{-1}$ in practical unit), denote the spectral carrier density whose integral over energies E in the conduction band is the carrier density n,

$$\int_{0}^{+\infty} \underline{n}(E, t)\mathrm{d}E = n. \tag{2.A.1}$$

We are considering only homogeneous, i.e. position-independent issues although the Fokker–Planck framework is able to handle inhomogeneous transport as well [8, 12]. In (2.A.1) the conduction-band bottom, denoted by E_c in Section 2.4, is taken as the zero of energies; but it should not be forgotten that this bottom is temperature-dependent if reckoned with respect to the vacuum level outside the solid. Denoting by $N(E)$ the density of quantum states per unit volume and per unit energy range, we write

$$\underline{n}(E,t) = N(E) f_0(E,t), \tag{2.A.2}$$

where the occupancy $f_0(E,\ t)$ of a state of energy E at time t lies between 0 and 1. Actually the occupancy depends on the pseudomomentum $\mathbf{p}$ of the Bloch state; but we are considering low fields causing little disturbance of the isotropic thermal distribution of pseudomomenta. That is to say, states $\mathbf{p}$ in a given energy shell $E(\mathbf{p}) = E$ are almost equally occupied. Then, the average velocity v_d is much smaller than the instantaneous velocity $|\mathbf{v}_g|$; this is the drift regime which is such that $v_\mathrm{d}(E) \ll v_\mathrm{g}(E)$ at given energy E.

Because the motion of carriers along the energy axis is caused by weakly inelastic scattering events[6] manifested as small energy steps, the spectral carrier density should obey an equation of local conservation in energy space,

$$\frac{\partial \underline{n}}{\partial t} + \frac{\partial J_\mathrm{m}}{\partial E} = 0, \tag{2.A.3}$$

involving a current density in energy space J_m, reckoned in $\mathrm{s}^{-1}\mathrm{m}^{-3}$ in SI unit. That current density is written on a drift-diffusion pattern,

$$J_\mathrm{m}(E) = W_\mathrm{m}(E)\underline{n} - \frac{\partial}{\partial E}[D_\mathrm{m}(E)\underline{n}]. \tag{2.A.4}$$

In (2.A.4) the drift coefficient $W_\mathrm{m}(E)$, reckoned in J/s or eV/s, reflects the average motion along the axis of energies; $W_\mathrm{m}(E)$ means a gain (loss) of energy if it is positive (negative). For a non-degenerate electron gas, $W_\mathrm{m}(E) < 0$ if $E \gg kT$ while $W_\mathrm{m}(E) > 0$ if $E \ll kT$. This evinces the electron's tendency to equilibrate at the temperature T of the environment acting as a thermostat. The

[6]Only acoustic phonons exist in Si, and the fractional energy change per scattering event $\approx \pm v_\mathrm{g}/c_\mathrm{s}$ is small, with c_s denoting the speed of sound. Optical phonons causing significantly inelastic events exist in GaAs, but at $T = 4$ K their scattering rate with electrons is negligible [84].

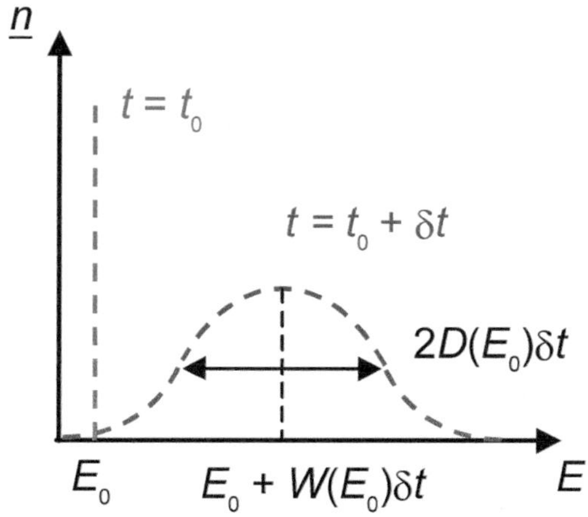

Figure 2.6 In a Fokker–Planck equation with current density $W(E)\underline{n}$ $-\partial[D(E)\underline{n}]/\partial E$, a monoenergetic spectral carrier density $\underline{n}(E, t = t_0) = n\,\delta(E - E_0)$ at time t_0 shifts and spreads into a Gaussian function centred at $E_0 + W(E_0)\,\delta t$ with a variance $2D(E_0)\,\delta t$ at the later time $t = t_0 + \delta t$. The Fokker–Planck equation links up the infinitesimal evolutions.

second contribution to $J_{\mathrm{m}}(E)$ reflects a superimposed fluctuational (unbiased) motion along the axis of energies due to the stochastic nature of the electron-environment scatterings. In other words, in (2.A.4) the drift term embodies the average motion whereas the diffusion term embodies the spreading of an ensemble of carriers about the average position. This is why the latter kind of motion is not specified by a velocity, but by a diffusivity $D_{\mathrm{m}}(E)$ reckoned in J^2/s or eV^2/s. This is shown in Fig. 2.6. Including a finite number of higher-order derivatives in the current density (2.A.4) may better approximate the evolution at very short times, but it violates the positivity of $\underline{n}(E, t)$ [11, 52].

The current density in energy space embodying interchange of energy with the force field is cast on the same pattern,

$$J_{\mathrm{F}}(E) = W_{\mathrm{F}}(E)\underline{n} - \frac{\partial}{\partial E}[D_{\mathrm{F}}(E)\underline{n}], \tag{2.A.5}$$

where the coefficient $W_{\mathrm{F}}(E)$ [resp. $D_{\mathrm{F}}(E)$] is a velocity (resp. diffusivity) in energy space. Because a variation in position under a force F entails a variation in energy, the coefficients are simply related to the drift velocity $v_d(E)$ and the diffusivity $D(E)$ in position space, namely

$$W_F(E) = Fv_d(E),\ D_F(E) = F^2 D(E). \tag{2.A.6}$$

Now in kinetic theory $D(E)$ is given by

$$D(E) = \frac{1}{3} v_g(E)\lambda(E), \tag{2.A.7}$$

where $\lambda(E) \equiv v_g(E)\tau(E)$ is the mean free path of an electron of energy E and $\tau(E)$ is the relaxation time of its instantaneous (group) velocity $\mathbf{v}_g$. That time is such that $\mathbf{v}_g/\tau(E)$ is the average deceleration, due to scattering events, of electrons of initial velocity $\mathbf{v}_g$. Both elastic and inelastic scatterings relax velocity, whence

$$\frac{1}{\tau(E)} = \frac{1}{\tau_{el}(E)} + \frac{1}{\tau_{in}(E)}. \tag{2.A.8}$$

For simplicity, all scatterings are taken to be isotropic. Under a small force F an electron of energy E drifts at a velocity $v_d(E) = \mu(E)F$ where $\mu(E)$ is the mobility at energy E. The entropyless nature of a force field entails a connection between energy-dependent mobility and diffusivity, namely [9, 39]

$$N(E)\mu(E) = \frac{d[N(E)D(E)]}{dE}. \tag{2.A.9}$$

Under the joint effect of scattering events and the applied force, the evolution of $\underline{n}$ is governed by equation (2.A.3) where $J_m(E)$ is replaced by the *total* current density in energy space, namely $J(E) \equiv J_m(E) + J_F(E)$. In a steady state $\partial J/\partial E$ vanishes so that $J(E) = J(0^+)$. Now $J(E)$ is a continuous function of energy and $J(0^-) = 0$ as no electron can go to or come from negative energies. Therefore, $J(E)$ vanishes at all energies. Were the current density approximated by the drift term $[W_m(E) + W_F(E)]\ \underline{n}(E)$ alone, we would obtain an *algebraic* equation whose solution is the energy at which gain from the force is exactly balanced by loss to the medium. Including the fluctuational (diffusion) term provides a *differential* equation whose solution $\underline{n}(E)$ is the steady-state energy distribution function under the force F [10].

The rate of energy loss to the lattice (production of heat) is, per unit volume,

$$\Pi_{\text{Joule}} = \left| \int_0^{+\infty} J_{\text{m}}(E)\text{d}E \right|. \tag{2.A.10}$$

At steady state, we may rewrite (2.A.10) as the rate of gain from the force field per unit volume,

$$\Pi_{\text{Joule}} = \int_0^{+\infty} J_{\text{F}}(E)\text{d}E. \tag{2.A.11}$$

In (2.A.10) [resp. (2.A.11)] the fluctuational term in the current density J_{m} (resp. J_{F}) does not contribute to the integral. Since we are concerned with low-field, near-equilibrium transport such that $f_0(E)$ is close to the equilibrium occupancy, we may write

$$f_0(E) = f_{\text{e}}(E) + \text{O}(F^2), \tag{2.A.12}$$

$$f_{\text{e}}(E) = f_{\text{FD}}\left(\frac{E - g}{kT}\right) \text{ and } f_{\text{FD}}(y) \equiv (e^y + 1)^{-1}. \tag{2.A.13}$$

The discrepancy $f_0(E) - f_{\text{e}}(E)$ is of second order in the force because it is unchanged upon force reversal. Letting $f_0(E) = f_{\text{e}}(E)$ in (2.A.10) throws up $\Pi_{\text{Joule}} = 0$ since in equilibrium the electron-lattice interchange of energy is balanced, i.e. $J_{\text{m}}(E) = 0$. Calculating a non-zero Π_{Joule} from (2.A.10) requires the knowledge of the non-equilibrium correction $\text{O}(F^2)$ to $f_{\text{e}}(E)$. In contrast, from (2.A.11) a non-zero value is calculated on replacing $f_0(E)$ with the equilibrium occupancy,

$$\Pi_{\text{Joule}} = F^2 \int_0^{+\infty} N(E)\mu(E) f_{\text{e}}(E)\text{d}E + \text{O}(F^4). \tag{2.A.14}$$

Upon integration by parts, owing to the mobility-diffusivity connection (2.A.9), we obtain

$$\Pi_{\text{Joule}} = F^2 \int_0^{+\infty} N(E)D(E)\left(-\frac{\partial f_{\text{e}}}{\partial E}\right)\text{d}E + \text{O}(F^4). \tag{2.A.15}$$

Consider a non-degenerate gas such that $\partial f_e/\partial E \approx -f_e/kT$. From (2.A.15) we get

$$\Pi_{\text{Joule}} \approx n\frac{\langle D(E)\rangle_e}{kT}F^2, \tag{2.A.16}$$

where

$$\langle \ldots \rangle_e \equiv \frac{\int_0^{+\infty} N(E)f_e(E)(..)\mathrm{d}E}{\int_0^{+\infty} N(E)f_e(E)\mathrm{d}E} \tag{2.A.17}$$

is the canonical average of equilibrium statistical mechanics. Expression (2.A.11) gives the alternative expression $\Pi_{\text{Joule}} = n\langle \mu(E)\rangle_e F^2$ whereby the Nernst–Townsend relation linking energy-averaged mobility and diffusivity, namely $\langle \mu(E)\rangle_e = \langle D(E)\rangle_e/kT$, ensues [9, 81]. Consider now a degenerate gas such that $\partial f_e/\partial E$ takes significant values in a narrow range $\approx kT$ about the Fermi level $g = E_c + E_F$. Letting $-\partial f_e/\partial E \approx \delta(E-g)$ in (2.A.15) yields

$$\Pi_{\text{Joule}} \approx N(E_F)\ D(E_F)\ F^2. \tag{2.A.18}$$

The production of heat, or equivalently the gain of energy from the force field, is due to electrons of energies lying about the Fermi level. Above $E \approx g + kT$ the states are empty, and below $E \approx g - kT$ electrons cannot exit their state owing to the Pauli principle (remember that only elastic or weakly inelastic transitions are possible).

If elastic scattering dominates, i.e. $\tau_{el} \ll \tau_{in}$, the diffusion coefficient (2.A.7) is given by

$$D(E) \approx \frac{1}{3}v_g(E)\lambda_{el}(E)\left[1 - \frac{\tau_{el}(E)}{\tau_{in}(E)}\right]. \tag{2.A.19}$$

Then, the diffusion coefficient of carriers is essentially governed by the elastic scattering mechanism with static impurities, and so too is the production of heat as given by (2.A.15). If we look at (2.A.10), where only the drift term $W_m(E)\underline{n}(E)$ contributes to the integral, we can see that the inelastic mechanism is indispensable since $W_m(E) \propto 1/\lambda_{in}$ (for energy-independent mean free path); and the Joule power would vanish in the purely elastic limit $\lambda_{in} \to +\infty$. But this limit is unphysical as it makes a steady state impossible because of impossible energy balance. No heating occurs under no field, but in the purely elastic limit $\lambda_{in} \to +\infty$ the electrons and the lattice cannot interchange energy. Then, electrons cannot thermalize at the lattice temperature [21], which again is unphysical.

Appendix 2.B. Beyond Ockham's Rule

We travel not for trafficking alone;
By hotter winds our fiery hearts are fanned:

For lust of knowing what should not be known,
We take the Golden Road to Samarkand.
—James Elroy Flecker

Early theoretical analyses of short-scale transport stressed the quantum-mechanical coherence of ballistic transport [41, 44, 47]. Accordingly, Ferry remarked that the observation by Weber et al. 'that Ohm's law remains valid, even at very low temperatures, [is] a surprising result that reveals classical behaviour in the quantum regime' [29]. Following Ockham's rule the observation has been accounted for herein within the semiclassical framework making use of a classical-looking phase-space occupation function $f(\mathbf{p}, \mathbf{r})$. Quantumness is present (i) dynamically through the Bloch states solving for a time-independent one-electron Schrödinger equation including the periodic crystal potential, and through the scattering cross section off a screened Coulomb potential; and (ii) statistically through the exclusion principle. A wave-mechanical feature present in the semiclassical description is the fact that ballistic motion occurs at a group velocity determined by the electronic band structure. A more totally quantum-mechanical account of electron transport would start from a wave function written as a linear combination of delocalized Bloch states of given pseudomomentum $\mathbf{p}$ or of Wannier states localized about a given position $\mathbf{r}$. In such an account, the use of a phase-space occupation function $f(\mathbf{p}, \mathbf{r})$ is incompatible with the indeterminacy principle. As noted in [53], 'in his correspondence with Moyal, Dirac was suspicious of the phase-space formulation of quantum mechanics. He believed that the uncertainty principle made it impossible to talk about functions that were well defined for position and momentum simultaneously'. In Bardeen's cautious words, 'it is presumed that the space variation of $f(\mathbf{p}, \mathbf{r})$ is sufficiently slow so that no difficulties arise from the indeterminacy principle' [5]. Despite Dirac's reluctance, connecting the quantum account with that of semiclassical kinetic theory based upon a phase-space function governed by a Vlasov- or Boltzmann-like equation has been achieved. In free space, the connecting tool is the Wigner transform of the time-dependent Schrödinger wave function [53, 86]. In a crystal where a strong periodic potential is present, the connecting tool is the Wannier–Bloch transform of the envelope of the wave function [19, 54, 55]. The Wigner or Wannier–Bloch transform is real- instead of complex-valued, but it enfolds the coherence of the quantum dynamics. The transform obeys a quantum Liouville equation formally similar to Vlasov's kinetic equation. It is possible to expand the Wannier–Bloch transform as a series whose dominant term is the occupation function solving for Vlasov's equation. The connection cannot be thought of in oversimple terms, however, for two reasons. First, a Wannier–Bloch transform is a *quasi*-probability density in phase space which in most cases takes negative values in some subvolumes of phase space, and therefore this function lacks the physical meaning of a genuine probability density. Rather, this function is a mathematical tool enabling the calculation of phase-space averages of certain physical observables [53, 86]. Secondly, even though the dominant term in the expansion is non-negative and thence may be viewed as a probability density, it is not an arbitrary function as happens in classical statistical physics, since that term is restricted by the uncertainty principle.

The transition from the quantum to the semiclassical description is rooted in the fact that wave-mechanical features get smeared upon scattering off static obstacles or thermal agitation of the lattice because of the static [20] or thermal [21] disorder. By disorder, it is meant that an electron is scattered by a *stochastic* energyscape which has many realizations over which the Wannier–Bloch transform is to be averaged inasmuch as we are not concerned with the motion of one electron but by the transport of many electrons. Now ensemble averaging is equivalent to space averaging provided that large enough spatial scales be involved. Upon approaching the correlation length of the disordered energyscape, the equivalence breaks down and sample-dependent conductance fluctuations show up. They are observed to be sensitive to applied magnetic fields which rotate the phase of a wave function and therefore affect the coherence enfolded in the unaveraged Wannier–Bloch transform. This is why part of the experimental literature on nanoscopic conduction is concerned with the magnetic-field dependency of conductance, which is seen as a signature of disorder [82]. With respect to a major issue of this chapter, two remarks are in order.

(i) Wave-mechanical reflection and transmission in the GaAs nanostructure survive upon averaging because their coefficients arise from the *deterministic*, as opposed to stochastic, potential-energy profile of the source-wire-drain structure, just like the group velocity arises from the deterministic crystal potential.
(ii) The fact that the Wannier–Bloch transform within the ballistic GaAs nanowire can be approximated by its semiclassical limit is not due to disorder in the nanowire, but to the disorder present in the contacts with the wire. This disorder entails semiclassical phase-space functions in the contacts, and the semiclassicality spreads from the contacts into the ballistic wire through the boundary conditions [6].

References

1. Abrahams, E. (1954). Electron-electron scattering in alkali metals, *Phys. Rev.* **95**: 839–840.
2. Agraït, N., Yeyati, A. L., and van Ruitenbeek, J. M. (2003). Quantum properties of atomic-sized conductors, *Phys. Reports* **177**: 81–279.
3. Andrews, D. L. (2009). Very little is new in nanoscience, *Eur. J. Phys.* **30**: S1–S2.
4. Balescu, R. (1975). *Equilibrium and Non-Equilibrium Statistical Mechanics*, New York, Wiley, pp. 413–421.
5. Bardeen, J. (1967). Conduction: metals and semiconductors, *Handbook of Physics* 2nd edn, edited by E. U. Condon and H. Odishaw, New York, McGraw-Hill, pp 4-72–4-101.
6. Ben Abdallah, N. (1998). A hybrid kinetic–quantum model for stationary electron transport, *J. Stat. Phys.* **90**: 627–662.
7. Brenann, K. F. (1999). *The Physics of Semiconductors*, Cambridge, Cambridge University Press.

8. Bringuier, E. (1997). Fokker–Planck approach to nonlocal high-field transport, *Phys. Rev. B* **56**: 5328–5331.
9. Bringuier, E. (1998). Anatomy of the drift-diffusion relationship, *Phil. Mag. B* **77**: 959–964.
10. Bringuier, E. (1998). Use of the Fokker-Planck equation in high-field transport problems, *Am. J. Phys.* **66**: 995–1002.
11. Bringuier, E. (1999). The current equation in strong electric fields, *Phil. Mag. B* **79**: 1659–1672.
12. Bringuier, E. (2000). Nonequilibrium statistical mechanics of drifting particles, *Phys. Rev. E* **61**: 6351–6358.
13. Bringuier, E. (2005). *Electrocinétique: Transport de l'électricité dans les milieux matériels*, Paris, CNRS Editions, chapters 13–15.
14. Bringuier, E. (2013). The electrical resistance of vacuum, *Eur. J. Phys.* **34**: 931–952.
15. Bringuier, E. (2015). The frictionless damping of a piston in thermodynamics, *Eur. J. Phys.* **36**: 055024.
16. Bringuier, E. (2016). The dissipation in Landauer's ballistic resistor, <hal-01423188>.
17. Bringuier, E. (2018). The thermodynamical foundation of electronic conduction in solids, *Eur. J. Phys.* **39**: 025101.
18. Bringuier, E. (2019). The Boltzmann equation and relaxation-time approximation for electron transport in solids, *Eur. J. Phys.* **40:** 025103.
19. Bringuier, E. (2021). The basis of the semiclassical description of electron transport in solids, *Eur. J. Phys.* **42**: 025502.
20. Bringuier, E. (2022). Disorder-induced quantum-to-classical transition, or how the world becomes classical, *Eur. Phys. J. Web Conf.* **263**: 01011.
21. Bringuier, E. (2024). Order to disorder transformation in an Ohmic resistor: Quantum theory of Joule heating, *Eur. Phys. J. Web Conf.* **300**, 01010.
22. Chattopadhyay, D., and Queisser, H. J. (1981). Electron scattering by ionized impurities in semiconductors, *Rev. Mod. Phys.* **53**: 745–768.
23. Das, M. P., and Green, F. (2012). Comments on "Ohm's law survives to the atomic scale" by Weber et al. **doi** 10.48550/arXiv.1204.4893
24. Datta, S. (1995). *Electronic Transport in Mesoscopic Systems*, Cambridge, Cambridge University Press, Sec. 2.3.
25. de Picciotto, R., Stormer, H. L., Yacoby, A., Pfeiffer, L. N., Baldwin, K. W., and West, K. W. (2000). 2D–1D coupling in cleaved edge overgrowth, *Phys. Rev. Lett.* **85**: 1730–1733.

26. de Picciotto, R., Stormer, H. L., Pfeiffer, L. N., Baldwin, K. W., and West, K. W. (2001). Four-terminal resistance of a ballistic quantum wire, *Nature* **411**: 51–54.
27. Domenicali, C. A. (1954). Irreversible thermodynamics of thermoelectricity, *Rev. Mod. Phys.* **26**: 237–275.
28. Drude, P. (1900). Zur Elektronentheorie der Metalle. I Teil, *Ann. Phys. (Leipzig)* **306**: 566–613; Zur Elektronentheorie der Metalle. II Teil, *Ann. Phys. (Leipzig)* **308**: 369–402; Zur Elektronentheorie der Metalle. Berichtigung, *Ann. Phys. (Leipzig)* **312**: 687–692.
29. Ferry D. K. (2012). Ohm's law in a quantum world, *Science* **335:** 45–46.
30. Gimzewski, J. K., and Möller, R. (1987). Transition from the tunneling regime to point contact studied using scanning tunneling microscopy, *Phys. Rev. B* **36**: 1284–1287.
31. Grundmann, M. (2010). *The Physics of Semiconductors: An introduction including Nanophysics and Applications* 2nd edn, Berlin, Springer, pp. 532–539.
32. Hénon, M. (1982). Vlasov's equation? *Astron. Astrophys.* **114**: 211–212.
33. Herring, C., and Nichols, M. H. (1949). Thermionic emission, *Rev. Mod. Phys.* **21**: 185–270.
34. Horowitz, P., and Hill, W. (2015). *The Art of Electronics* 3rd edn, Cambridge, MA, Cambridge University Press.
35. Imry, Y., and Landauer, R. (1999). Conductance viewed as transmission, *Rev. Mod. Phys.* **71**: S306–S312.
36. Joule, J. P. (1841). On the heat evolved by metallic conductors of electricity, and in the cells of a battery during electrolysis, *Phil. Mag.* **XIX**: 260–277.
37. Klavsyuk, L., and Saletsky, A. M. (2015). Formation and properties of metallic atomic contacts, *Usp. Fiz. Nauk* **185**: 1009–1030 (in Russian); *Physics–Uspekhi* **58**: 933–951 (English translation).
38. Kreuzer, H. J. (1981). *Nonequilibrium Thermodynamics and its Statistical Foundations*, Oxford, Clarendon Press, pp. 177–188.
39. Kurosawa, T. (1965). Notes on the theory of hot electrons in semiconductors, *J. Phys. Soc. Japan* **20**: 937–942.
40. Landau, L. D., and Lifshitz, E. M. (1965), *Quantum Mechanics: Non-relativistic Theory* 2nd edn, Oxford, Pergamon, sections 25 and 125.
41. Landauer, R. (1981). Can a length of perfect conductor have a resistance?, *Phys. Lett.* **85A**: 91–93.

42. Landauer, R. (1989). Nanostructure physics: fashion or depth?, *Nanostructure Physics and Fabrication*, edited by M. A. Reed and W. P. Kirk, New York, Academic, pp. 17–30.
43. Landauer, R. (1989). Conductance determined by transmission: probes and quantised constriction resistance, *J. Phys.: Condens. Matter* **1**: 8099–8110.
44. Landauer, R. (1996). Comment on 'Spatial variation of currents and fields due to localized scatterers in metallic conduction', *J. Math. Phys.* **37**: 5259.
45. Landauer, R. (1997). Conductance is transmission, *Nanowires*, Dordrecht, Kluwer, pp. 1–7.
46. Lenz, E. (1843). Ueber die Gesetze der Wärme-Entwickelung durch den galvanischen Strom, *Bulletin de la Classe physico-mathématique de l'Académie impériale des Sciences de Saint-Pétersbourg* **I**: 209–253; Ueber die Gesetze der Wärmeentwickelung durch den galvanischen Strom, *Bulletin de la Classe physico-mathématique de l'Académie impériale des Sciences de Saint-Pétersbourg* **II**: 161–188.
47. Lesovik, G. B., and Sadovskyy, I. A. (2011). Scattering matrix approach to the description of quantum electron transport, *Usp. Fiz. Nauk* **181**: 1041–1096 (in Russian); *Physics–Uspekhi* **54**: 1007–1059 (English translation).
48. Levich, B. G. (1973). *Theoretical Physics: An advanced text,* vol. **4** *Quantum Statistics and Physical Kinetics*, Amsterdam, North Holland, p. 241.
49. Lide, D. R. (editor) (1995). *CRC Handbook of Chemistry and Physics* 76th edn, Boca Raton, FL, CRC Press, p. 12–46.
50. Lifshitz, E. M., and Pitaevski, L. P. (1981). *Physical Kinetics*, Oxford, Pergamon, Sec. 15.
51. Lorentz, H. A. (1952). *The Theory of Electrons* 2nd edn, New York, Dover, Sec. 29.
52. Marcinkiewicz, J. (1939). Sur une propriété de la loi de Gauß, *Math. Z.* **44**: 612–618.
53. Mariño, M. (2022). *Advanced Topics in Quantum Mechanics*, Cambridge, Cambridge University Press, chapter 3.
54. Markowich P. A., Ringhofer C. and Schmeiser C. (1990). *Semiconductor Equations*, Vienna, Springer, chapter 1.
55. McIrvine, E. C., and Overhauser, A. W. (1959). New quantum-mechanical representation, *Phys. Rev.* **115**: 1531–1536.

56. Nakashima, K., and Takeyama, N. (1992). Steady-state thermodynamics for crossed transport of heat and matter, *J. Phys. Soc. Japan* **61**: 2754–2762.

57. Nottingham, W. B. (1956). Thermionic emission, *Handbuch der Physik* **XXI**, edited by S. Flügge, Berlin, Springer, pp. 1–178.

58. Ockham, W. (1957). *Philosophical Writings*, Edinburgh, Thomas Nelson and Sons.

59. Ohm, G. S. (1827). *Die galvanische Kette, mathematisch bearbeitet*, Berlin, T. H. Riemann; (1860). *Théorie mathématique des courants électriques*, Paris, Hachette; (1891). *The galvanic circuit investigated mathematically*, New York, Van Nostrand.

60. Ott, F., and Lunney, J. (1998). Quantum conduction: A step-by-step guide, *Europhysics News* **29** January/February, pp. 13–16; Retracing quantum steps, *ibid.* March/April, p. 55.

61. Palermo, C., and Torres, J. (2020). *Manuel d'électricité* 2nd edn, Paris, Dunod, p. 9.

62. Peierls, R. E. (1991). *More Surprises in Theoretical Physics*, Princeton, NJ, Princeton University Press, pp. 52–55.

63. Premaratne, M, and Agrawal, G. P. (2021). *Theoretical Foundations of Nanoscale Quantum Devices*, Cambridge, Cambridge University Press.

64. Reif-Acherman, S. (2016). Georg Simon Ohm and the first comprehensive theory of electrical conductivity in metals, *Proc. IEEE* **104**: 198–209.

65. Rhoderick, E. H. (1972). Comments on the conduction mechanism in Schottky diodes, *J. Phys. D: Appl. Phys.* **5**: 1920–1929; Rhoderick, E. H., and Williams, R. H. (1988). *Metal–Semiconductor Contacts* 2nd edn, Oxford, Clarendon Press, pp. 90–109.

66. Ridley, B. K. (1999). *Quantum Processes in Semiconductors* 4th edn, Oxford, Clarendon Press.

67. Rueß, F. J., Oberbeck, L., Butcher, M. J. et al. (2005). The use of etched registration markers to make four-terminal electrical contacts to STM-patterned nanostructures, *Nanotechnology* **16**: 2446–2449.

68. Rueß, F. J., Goh, K. E. J., Butcher, M. J. et al. (2007). Narrow, highly P-doped, planar wires in silicon created by scanning probe microscopy, *Nanotechnology* **18**: 044023.

69. Ryskin, G. (1997). Simple procedure for correcting equations of evolution: Application to Markov processes, *Phys. Rev. E* **56**: 5123–5127.

70. Sagan, C. (1995). *The Demon-Haunted World: Science as a candle in the dark*, New York, Random House.

71. Schlup W. A. (1969). Exact solution of the Boltzmann equation in the relaxation approximation, *Z. Phys. B* **10**: 116–43.
72. Sinkkonen, J., Eränen, S., and Stubb, T. (1984). Linear conductance of short semiconductor structures, *Phys. Rev. B* **30**: 4813–4815.
73. Smith, A. C., Janak, J. F., and Adler, R. B. (1967). *Electronic Conduction in Solids*, New York, McGraw-Hill.
74. Soto, R. (2016). *Kinetic Theory and Transport Phenomena*, Oxford, Oxford University Press, p. 177.
75. Uehling, E. A., and Uhlenbeck, G. E. (1933). Transport phenomena in Einstein–Bose and Fermi–Dirac gases. I, *Phys. Rev.* **43**: 552–561.
76. Vakulov, D., Gireesan, S., Swinkels, M. Y., Chavez, R., Vogelaar, T., Torres, P. et al. (2020). Ballistic phonons in ultrathin nanowires, *Nano Lett.* **20**: 2703–2709.
77. van Houten, H., Beenakker, C. W. J., and van Wees, B. J. (1992). Quantum point contacts, *Nanostructured Systems*, Semiconductors and Semimetals vol. **35**, edited by M. Reed, Boston, Academic.
78. van Wees, B. J., van Houten, H., Beenakker, C. W. J., Williamson, J. G., Kouwenhoven, L. P., van der Marel, D. et al. (1988). Quantized conductance of point contacts in a two-dimensional electron gas, *Phys. Rev. Lett.* **60**: 848–850.
79. Varney, R. N., and Fisher, L. H. (1980). Electromotive force: Volta's forgotten concept, *Am. J. Phys.* **48**: 405–408.
80. Volta, A. (1801). De l'électricité dite galvanique, *Annales de Chimie* **XL**: 225–256.
81. Wannier, G. H. (1966). *Statistical Physics*, New York, Wiley, pp. 442, 479, 498.
82. Webb R. A. and Washburn S. (1988). Quantum interference fluctuations in disordered metals, *Phys. Today* **41** (12): 46–53.
83. Weber, B., Mahapatra, S., Ryu, H., et al. (2012). Ohm's law survives to the atomic scale, *Science* **335:** 64–67.
84. Weisbuch, C. (1987). Fundamental properties of III–V semiconductor two-dimensional quantized structures: The basis for optical and electronic device applications, *Applications of multiquantum wells, selective doping, and superlattices*, Semiconductors and Semimetals vol. **24**, edited by R. Dingle, San Diego, CA, Academic, pp. 1–133.
85. Wharam, D. A., Thornton, T. J., Newbury, R., Pepper, M., Ahmed, H., Frost, J. E. F. et al. (1988). One-dimensional transport and the quantisation of the ballistic resistance, *J. Phys. C: Solid State Physics* **21**: L209–L214.

86. Wigner, E. P. (1971). Quantum-mechanical distribution functions revisited, *Perspectives in Quantum Theory: Essays in Honor of Alfred Landé,* edited by W. Yourgrau and A. van der Merwe, Cambridge, MA, M.I.T. Press, pp. 25–36.
87. Yamamoto, S. (2006). Fundamental physics of vacuum electron sources, *Rep. Prog. Phys.* **69**: 181–232.
88. Zener, C. (1937). Internal friction in solids: I. Theory of internal friction in reeds, *Phys. Rev.* **52**: 230–235; (1938); Internal friction in solids: II. General theory of thermoelastic internal friction, *Phys. Rev.* **53**: 90–100.

Chapter 3

Physics of 2D Magnets and Magnetic Thin Films: Surface Structure and Phase Transitions, Criticality and Skyrmions

H. T. Diep
Laboratoire de Physique Théorique et Modélisation,
CY Cergy Paris Université, CNRS, UMR 8089,
2, Avenue Adolphe Chauvin, 95302 Cergy-Pontoise Cedex, France
diep@cyu.fr

Recently, there is an increasing renewed interest in 2D magnetism such as Van der Waals magnets. The physics of 2D magnetism and ultra-thin magnetic films has a long history. This chapter is a review devoted to some fundamental theoretical properties of 2D magnets and magnetic thin films including frustrated systems and topological spin textures. These properties allow to understand macroscopic behaviors experimentally observed in thin films and superlattices where the surface and the interface play a crucial role. The chapter begins with a review on 2D magnets, their spin structures and phase transitions. Next, the case of thin films is considered. The theory of surface spin waves is discussed in various situations with and without surface reconstruction of spin ordering. Various interactions are taken into account: surface interaction

2D and 3D Nanostructures: Structure, Properties, and Applications
Edited by Jean-Claude Serge Levy

ISBN 978-981-5129-97-7 (Hardcover), 978-1-003-74871-7 (eBook)
www.jennystanford.com

DOI: 10.1201/97810037487-3

different from the bulk one, competing interactions, Dzyaloshinskii–Moriya interaction. Surface phase transitions are shown in some particularly striking cases. Finally, some cases of topological spin textures called "skyrmions" are reviewed. All the results shown in this chapter have been published in various research papers cited in the text. Therefore, we will discuss some important results but avoid to enter complicated methods. Instead, the reader is referred to original papers for detailed demonstrations.

3.1 Introduction

Investigations on properties of nanoscale magnetic systems have been intensively carried out in the last five decades. This is because there was a search for smaller and smaller systems with high capacities of memory stockage. Magnetic thin films and nanodots are among the objects the most studied experimentally, numerically and theoretically [1–3]. In the 1980s, the discovery of giant magneto-resistance (GMR) has led to a generation of high-density hard disks [6, 7]. Intensive researches continue to look for new effects, new phenomena which can lead to even more efficient, energy-saving electronic applications. Spintronics is one of the future technologies which is based on spins or groups of spin. A single spin with up and down orientations represents the smallest entity for a computer binary bit or logic gates in principle, but how to control the orientation of such a spin? Recent research focuses on groups of a few dozens of spins having the double-fold states such as nanodots and skyrmions which have left or right chirality. With a magnetic field, for example, one can change its chirality. In multiferroic materials, the chirality of skyrmions can be controlled by an electric field.

Since the discovery of the effect of the frustration in magnetic materials, an enormous number of investigations has been realized, theoretically, experimentally and numerically. The reader is referred to Refs. [4, 5, 8] for reviews on various topics.

In this chapter, we show some examples of 2D magnetic mono- or bi-layers where striking surface effects have been discovered. We show surface states, surface phase transitions and other properties that may have applications in spintronics. The criticality of thin films is presented. We show that this is an important subject which may allow us to understand critical exponents experimentally observed in thin films. We also review important works since 2003 on what is called "skyrmion" (topologically protected spin structure containing a dozen of spins). We recall theoretical and experimental works on "skyrmion crystal" which is a "superstructure" of skyrmions periodically arranged in a lattice structure.

Since the aim of this chapter is to review some striking aspects of the surface magnetism, we shall show main results and discuss on their physical implications. We shall not show complicated technical details in order to make the reading easy. We will refer the reader to original papers for the mathematical and computing techniques.

The chapter is organized as follows. In Section 3.2, we briefly present the notion of frustration which allows to understand the following sections. In Section 3.3, we consider the case of thin films where striking surface properties are reviewed. We discuss here both non-frustrated and frustrated films. In Section 3.4, we show results on the criticality of thin films. In Section 3.5, we describe skyrmions resulted from the Dzyaloshinskii–Moriya interaction. Concluding remarks are given in Section 3.6.

3.2 Frustration

The frustration is caused by the competition between various interactions in the system. It can be caused by only one type of interaction in the case where the lattice structure does not allow to fully satisfy every interaction bond: this is the case of a triangular lattice with an antiferromagnetic interaction between nearest neighbors (NN), or the face-centered cubic (FCC) lattice and the hexagonal-closed-packed (HCP) lattice, with antiferromagnetic NN interaction. These cases are called "geometry frustration".

The effects of the frustration are numerous. We can mention a few of them (the reader is referred to chapter 1 of Ref. [8]): (i) the GS degeneracy is very high, in many cases the GS degeneracy is infinite, (ii) in the case of vector spins (XY and Heisenberg spins), the GS spin configuration is non-collinear, unlike the collinear configuration in ferromagnets and antiferromagnets. A famous example is the 120-degree spin structure in the antiferromagnetic triangular lattice and the helimagnetic structure which has been determined a long time ago [9, 10] (iii) the nature of the phase transition is often difficult to determine. To our knowledge, the phase transition in most (if not all) frustrated systems in 3D is of the first-order transition. Well-known examples are the cases of HCP antiferromagnets [11, 12], the FCC antiferromagnets [13] and the antiferromagnetic stacked triangular lattice [14–18].

Let us examine some simple cases in the following.

3.2.1 Definition

We show below some popular cases for readers who are not familiar with frustrated spin systems.

In the pairwise Ising and Heisenberg model, the interaction energy between two spins $\mathbf{S}_i$ and $\mathbf{S}_j$ is given by $E = -J\left(\mathbf{S}_i \cdot \mathbf{S}_j\right)$ where J is the exchange integral. If J is ferromagnetic (positive) then the minimum of E is $-J$ corresponding to the state in which $\mathbf{S}_i$ is parallel to $\mathbf{S}_j$. If J is antiferromagnetic (negative), the minimum of E corresponds to the state where $\mathbf{S}_i$ is antiparallel to $\mathbf{S}_j$. For a crystal where the NN interaction is ferromagnetic, the GS of the system corresponds to the spin configuration where all spins are parallel. This is true for any lattice structure. However, if J is antiferromagnetic, the spin configuration of the GS depends on the lattice structure: (i) for lattices containing no elementary triangles, such as square lattice, simple cubic lattices, ... the GS is the configuration in which each spin is antiparallel to its neighbors, namely every bond is fully satisfied (lowest energy); (ii) for lattices containing elementary triangles such as the triangular lattice, the fcc lattice and the hcp lattice, it is impossible to construct a GS where all bonds have the lowest energy J (see Fig. 3.1): The GS does not correspond to the minimum of the interaction of every spin pair. In this case, one says that the system is frustrated.

We consider another spin system, which can be called frustrated system because each bond is only partially satisfied: this is the case with different kinds of conflicting interactions and the GS does not correspond to the minimum of each kind of interaction.

Let us examine a chain of spins where the NN interaction J_1 is ferromagnetic while the next NN (NNN) interaction J_2 is antiferromagnetic. As long as $|J_2| \ll J_1$, the GS is ferromagnetic: every NN bond is then satisfied but the NNN ones are not. Of course, when $|J_2|$ exceeds a critical value, the ferromagnetic GS is no longer valid (see the helimagnet example below): both the NN and NNN bonds are not fully satisfied.

The translational invariance allows in many cases to consider only an elementary cell of the lattice in search for the system GS. The lattice cell is in general a polygon formed by faces hereafter called "plaquettes". For example, the elementary cell of the simple cubic (SC) lattice is a cube with six square plaquettes, the elementary cell of the FCC lattice is a tetrahedron formed by four triangular plaquettes. Let $J_{i,j}$ be the interaction between two NN spins of the plaquette. According to the definition of Toulouse [19], the plaquette is frustrated if the parameter P defined below is negative

$$P = \prod_{\langle i,j \rangle} \text{sign}(J_{i,j}), \tag{3.1}$$

where the product is performed over all $J_{i,j}$ around the plaquette. Two examples of frustrated plaquettes are shown in Fig. 3.1: a triangle with three antiferromagnetic bonds and a square with three ferromagnetic bonds and one antiferromagnetic bond. P is negative in both cases. One sees that if one tries to put Ising spins on those plaquettes, at least one of the bonds around the plaquette will not be satisfied. For vector spins, we show below that in the lowest energy state, each bond is only partially satisfied.

Let us touch upon the GS degeneracy. One notes that for the triangular plaquette, the degeneracy is three, and for the square plaquette it is four, in addition to the degeneracy associated with global spin reversal. Therefore, the degeneracy of an infinite lattice composed of such plaquettes is infinite, in contrast to the unfrustrated case.

At this stage, it is worth noting that in the above discussion we have assumed the interaction between two spins to be of the form $E = -J\left(\mathbf{S}_i \cdot \mathbf{S}_j\right)$. However, the frustration takes place also in systems with other types of interactions such as the Dzyaloshinskii–Moriya interaction $E = -J\left|\left(\mathbf{S}_i \wedge \mathbf{S}_j\right)\right|$: a spin system is frustrated whenever the minimum of the system energy does not correspond

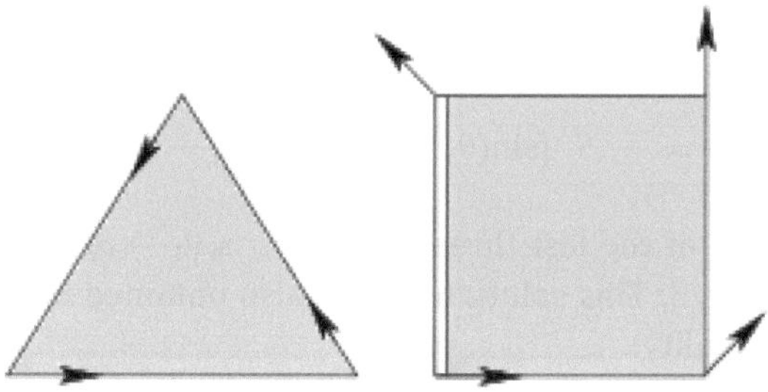

Figure 3.2 Non collinear spin configuration of frustrated triangular and square plaquettes with XY spins: ferro- and antiferromagnetic interactions J and $-J$ are indicated by thin and double lines, respectively.

non collinear. One shows that the helical configuration displayed in Fig. 3.3 is obtained by minimizing the interaction energy:

$$E = -J_1 \sum_i \mathbf{S}_i \cdot \mathbf{S}_{i+1} + |J_2| \sum_i \mathbf{S}_i \cdot \mathbf{S}_{i+2}$$

$$= S^2 \left[-J_1 \cos\theta + |J_2| \cos(2\theta)\right] \sum_i 1$$

$$\frac{\partial E}{\partial \theta} = S^2 \left[J_1 \sin\theta - 2|J_2| \sin(2\theta)\right] \sum_i 1 = 0$$

$$= S^2 \left[J_1 \sin\theta - 4|J_2| \sin\theta \cos\theta\right] \sum_i 1 = 0, \qquad (3.4)$$

where one has supposed that the angle between NN spins is θ.

The two solutions are

$$\sin\theta = 0 \longrightarrow \theta = 0 \ \text{(ferromagnetic solution)}$$

and

$$\cos\theta = \frac{J_1}{4|J_2|} \longrightarrow \theta = \pm \arccos\left(\frac{J_1}{4|J_2|}\right). \qquad (3.5)$$

The last solution is possible if $-1 \leq \cos\theta \leq 1$, i.e. $J_1/(4|J_2|) \leq 1$ or $|J_2|/J_1 \geq 1/4 \equiv \varepsilon_c$.

As before, there are two degenerate configurations: clockwise and counter-clockwise.

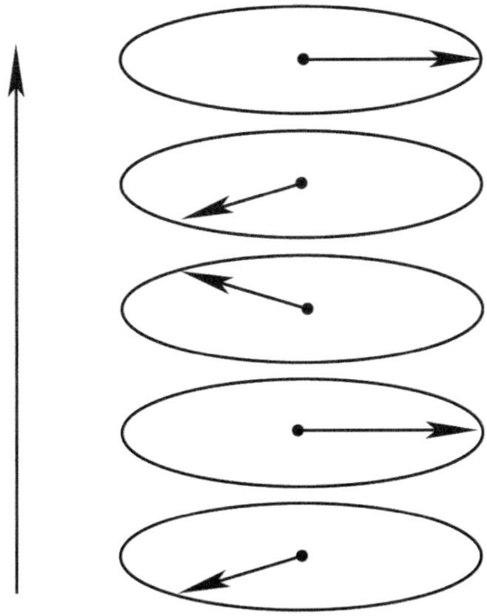

Figure 3.3 Helical configuration when $\varepsilon = |J_2|/J_1 > \varepsilon_c = 1/4$ ($J_1 > 0$, $J_2 < 0$).

One can define a chiral order parameter for each plaquette. In the case of a triangular plaquette, the chiral parameter is given by

$$\kappa = \frac{2}{3\sqrt{3}} \left[\mathbf{S}_1 \wedge \mathbf{S}_2 + \mathbf{S}_2 \wedge \mathbf{S}_3 + \mathbf{S}_3 \wedge \mathbf{S}_1 \right], \tag{3.6}$$

where coefficient $2/\left(3\sqrt{3}\right)$ was introduced so that the $\pm 2\pi/3$ degeneracy corresponds to $\kappa_i = \pm 1$.

We can form a triangular lattice using plaquettes as shown in Fig. 3.4. The GS corresponds to the state where all triangle plaquettes of the same orientation have the same chirality: plaquettes $\triangle$ have positive chirality ($\kappa = 1$) and plaquettes $\triangledown$ have negative chirality ($\kappa = -1$). Mapping the chiralities into the Ising spins, we have a perfect antiferromagnetic Ising order. This order is broken at a phase transition temperature where κ vanishes.

Let us enumerate two frequently encountered frustrated spin systems where the NN interaction is antiferromagnetic: the FCC lattice and the HCP lattice. These two lattices are formed by stacking tetrahedra with four triangular faces. The frustration due

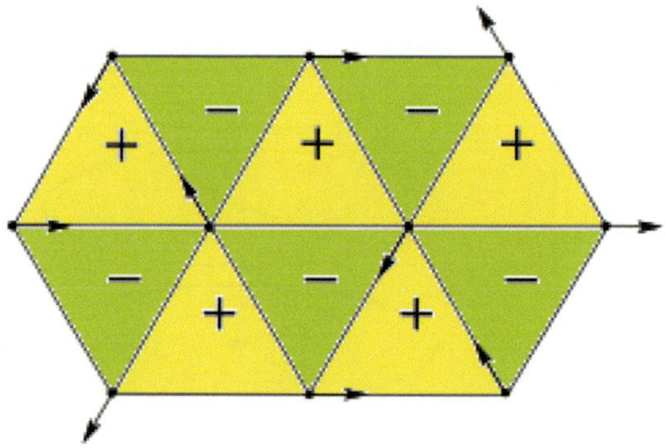

Figure 3.4 Antiferromagnetic triangular lattice with XY spins. The positive and negative chiralities are indicated by $+$ and $-$.

to the lattice structure such as in these cases is called "geometry frustration".

3.3 Surface States: Surface Spin Structures and Surface Spin Waves

Let us take for example a thin film. The spins on the surfaces have less neighbors than those in the bulk. This causes many effects: the electronic orbitals, the exchange interactions, the spin configuration, the lattice constants, . . . are more or less modified. As consequences, the electronic structures and the spin-wave spectrum may have what we call "surface states". These states lead to macroscopic properties of the magnet different from the bulk ones such as the surface magnetization and the surface phase transition. In the following, we will show the link between the existence of a surface and those striking surface properties.

3.3.1 Surface Spin Structures

In films with ferromagnetic interaction between NN, the GS spin configuration is ferromagnetic everywhere. The lack of neighbors of surface spins suffices in some cases to generate surface spin waves such as in the body-centered cubic (BCC) semi-infinite films with

(001) surface orientation (see the calculation using Green's function in [21]). In the case where there are competing interactions, namely ferromagnetic and antiferromagnetic interactions simultaneously acting on a spin, there are often a surface reconstruction which shows a surface spin ordering different from the bulk spin ordering. We give some examples in the following.

Example 3.1. Helimagnetic spin structure in a thin film. This system has been studied in Ref. [22]. We have determined the surface spin structure and used the Green's function method for non-collinear spin configuration [23] to calculate spin-wave spectrum. The magnetization of each film layer has been calculated as a function of temperature (T), using the spin-wave spectrum. This is briefly shown below. Low-temperature properties in helimagnets such as spin waves [24–27] and heat capacity [28] have been extensively investigated. Helimagnets belong to a class of frustrated vector-spin systems. Note that some works have been devoted to helimagnets: surface spin structures [29], MC simulations [30], magnetic field effects on the phase diagram in Ho [31] and a few experiments [32, 33]. Helical magnets present potential applications in spintronics with predictions of spin-dependent electron transport in these magnetic materials [34–36].

We consider hereafter the simplest model for a film: the helical ordering is along one direction, namely the c-axis perpendicular to the film surface.

Let us examine a thin film of BCC lattice of N_z layers, with two symmetrical surfaces perpendicular to the c-axis, for simplicity. The Hamiltonian reads

$$\mathcal{H} = \mathcal{H}_e + \mathcal{H}_a \tag{3.7}$$

where the isotropic exchange part is given by

$$\mathcal{H}_e = -\sum_{\langle i,j\rangle} J_{i,j}\mathbf{S}_i \cdot \mathbf{S}_j \tag{3.8}$$

$J_{i,j}$ being the interaction between two quantum Heisenberg spins $\mathbf{S}_i$ and $\mathbf{S}_j$ occupying the lattice sites i and j. The anisotropic part is chosen as

$$\mathcal{H}_a = -\sum_{<i,j>} I_{i,j} S_i^z S_j^z \cos\theta_{ij} \tag{3.9}$$

where $I_{i,j}$ is the anisotropic interaction along the in-plane local spin-quantization axes z of $\mathbf{S}_i$ and $\mathbf{S}_j$, supposed to be positive, small compared to J_1, and limited to NN on the c-axis. Let us mention that according to the theorem of Mermin and Wagner [37] continuous isotropic spin models such as XY and Heisenberg spins do not have long-range ordering at finite temperatures in two dimensions. Since we are dealing with the Heisenberg model in a thin film, it is useful to add an anisotropic interaction to create a long-range ordering and a phase transition at finite temperatures.

To generate a bulk helimagnetic structure, the simplest way is to take a ferromagnetic interaction between NNs, say J_1 (>0), and an antiferromagnetic interaction between NNNs, $J_2 < 0$. It is obvious that if $|J_2|$ is smaller than a critical value $|J_2^c|$, the classical GS spin configuration is ferromagnetic [25, 26]. Since our purpose is to investigate the helimagnetic structure near the surface and surface effects, let us consider the case of a helimagnetic structure only in the c-direction perpendicular to the film surface. In such a case, we assume a non-zero J_2 only on the c-axis. This assumption simplifies formulas but does not change the physics of the problem since including the uniform helical angles in two other directions parallel to the surface will not introduce additional surface effects. Note that the bulk case of the above quantum spin model has been studied by the Green's function method [27].

To calculate the classical GS surface spin configuration, we write down the expression of the energy of spins along the c-axis, starting from the surface:

$$\begin{aligned} E = & -Z_1 J_1 \cos(\theta_1 - \theta_2) - Z_1 J_1[\cos(\theta_2 - \theta_1) \\ & + \cos(\theta_2 - \theta_3)] + \ldots \\ & - J_2 \cos(\theta_1 - \theta_3) - J_2 \cos(\theta_2 - \theta_4) \\ & - J_2[\cos(\theta_3 - \theta_1) + \cos(\theta_3 - \theta_5)] + \ldots \end{aligned} \tag{3.10}$$

where $Z_1 = 4$ is the number of NNs in a neighboring layer, θ_i denotes the angle of a spin in the i-th layer made with the Cartesian x axis of the layer. The interaction energy between two NN spins in the two adjacent layers i and j depends only on the difference $\alpha_i \equiv \theta_i - \theta_{i+1}$. The GS configuration corresponds to the minimum of E. We have to solve the set of equations:

$$\frac{\partial E}{\partial \alpha_i} = 0, \quad \text{for } i = 1, N_z - 1 \tag{3.11}$$

Explicitly, we have

$$\frac{\partial E}{\partial \alpha_1} = 8J_1 \sin\alpha_1 + 2J_2 \sin(\alpha_1 + \alpha_2) = 0 \tag{3.12}$$

$$\frac{\partial E}{\partial \alpha_2} = 8J_1 \sin\alpha_2 + 2J_2 \sin(\alpha_1 + \alpha_2) + 2J_2 \sin(\alpha_2 + \alpha_3) = 0 \tag{3.13}$$

$$\frac{\partial E}{\partial \alpha_3} = 8J_1 \sin\alpha_3 + 2J_2 \sin(\alpha_2 + \alpha_3) + 2J_2 \sin(\alpha_3 + \alpha_4) = 0 \tag{3.14}$$

$$\frac{\partial E}{\partial \alpha_4} = \ldots$$

where we have expressed the angle between two NNNs as follows: $\theta_1 - \theta_3 = \theta_1 - \theta_2 + \theta_2 - \theta_3 = \alpha_1 + \alpha_2$, etc. In the bulk case, putting all angles α_i in Eq. (3.13) equal to α we get $\cos\alpha = -J_1/J_2$ as expected. For the spin configuration near the surface, let us consider in the first step only three parameters α_1 (between the surface and the second layer), α_2 and α_3. We take $\alpha_n = \alpha$ from $n = 4$ inward up to $n = N_z/2$, the other half being symmetric. Solving the first two equations, we obtain

$$\tan\alpha_2 = -\frac{2J_2(\sin\alpha_3 + \sin\alpha_1)}{8J_1 + 2J_2(\cos\alpha_3 + \cos\alpha_1)} \tag{3.15}$$

The iterative numerical procedure is as follows: (i) replacing α_3 by $\alpha = \arccos(-J_1/J_2)$ and solving (3.12) and (3.15) to obtain α_1 and α_2, (ii) replacing these values into (3.14) to calculate α_3, (iii) using this value of α_3 to solve again (3.12) and (3.15) to obtain new values of α_1 and α_2, (iv) repeating step (ii) and (iii) until the convergence is reached within a desired precision. In the second step, we use α_1, α_2 and α_3 to calculate by iteration α_4, assuming a bulk value for α_5. In the third step, we use α_i $(i = 1 - 4)$ to calculate α_5 and so on. The results calculated for various J_2/J_1 for a film of $N_z = 8$ layers are shown in Table 3.1.

Some remarks are in order: (i) result shown is obtained by iteration with errors less than 10^{-4} degrees, (ii) strong angle variations are observed near the surface with oscillation for strong

Table 3.1 Values of $\cos\theta_{n,n+1} = \alpha_n$ between two adjacent layers are shown for various values of J_2/J_1

J_2/J_1	$\cos\theta_{1,2}$	$\cos\theta_{2,3}$	$\cos\theta_{3,4}$	$\cos\theta_{4,5}$	α(bulk)
−1.2	0.985(9.79°)	0.908(24.73°)	0.855(31.15°)	0.843(32.54°)	33.56°
−1.4	0.955(17.07°)	0.767(39.92°)	0.716(44.28°)	0.714(44.41°)	44.42°
−1.6	0.924(22.52°)	0.633(50.73°)	0.624(51.38°)	0.625(51.30°)	51.32°
−1.8	0.894(26.66°)	0.514(59.04°)	0.564(55.66°)	0.552(56.48°)	56.25°
−2.0	0.867(29.84°)	0.411(65.76°)	0.525(58.31°)	0.487(60.85°)	60°

Note: Only angles of the first half of the 8-layer film are shown: other angles are, by symmetry, $\cos\theta_{7,8}=\cos\theta_{1,2}$, $\cos\theta_{6,7}=\cos\theta_{2,3}$, $\cos\theta_{5,6}=\cos\theta_{3,4}$. The values in parentheses are angles in degrees. The last column shows the value of the angle in the bulk case (infinite thickness). For presentation, angles are shown with two digits.

J_2, (iii) the angles at the film center are close to the bulk value α (last column), meaning that the surface reconstruction affects just a few atomic layers, (iv) the bulk helical order is stable just a few atomic layers away from the surface even for films thicker than $N_z = 8$ (see below). This helical stability has been experimentally observed in holmium films [38].

Note that using the numerical steepest descent method described in Ref. [39], we find the same result. In Ref. [22], we have used the Green's function method for non-collinear spin configurations. The calculations are lengthy, but the obtained spin-wave spectrum allows us to calculate the layer magnetization. An example is shown in Fig. 3.5. Note that the surface magnetization is very low at finite T with respect to interior layers.

Another example of surface spin configuration different from the bulk one is shown in Example 3.2 below.

3.3.2 Surface Spin Waves

When a crystal is infinite in all directions, spin waves are excited due to the interactions between spins on the lattice. A spin-wave is characterized by a wave vector $\mathbf{k}$ which has three components in three lattice directions. The spin-wave energy $E = \hbar\omega$ depends on $\mathbf{k}$, ω being the spin-wave frequency. The spin-wave spectrum is given by $E = \hbar\omega(\mathbf{k})$ in the k-space defined inside the first Brillouin zone. Now, if the crystal is semi-infinite or a thin film, the values of $\mathbf{k}$ in the direction perpendicular to the surface are no more determined by

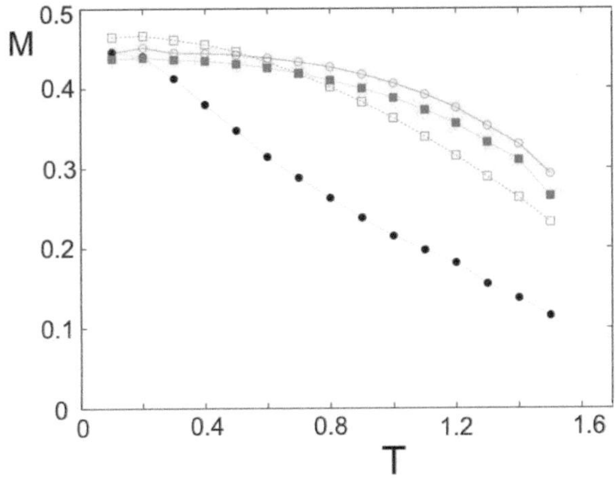

Figure 3.5 Layer magnetizations as functions of T for the surface interaction $J_1^s = 0.3$ with $J_2/J_1 = -2$, $d = 0.1$ and $N_z = 16$. Black circles, blue void squares, magenta squares and red void circles are for first, second, third and fourth layers, respectively.

the periodic condition of the lattice, namely the Fourier transform cannot be done in that direction. As a consequence, surface spin waves can be observed: these waves propagate parallelely to the surface but are damped when propagating toward inside the crystal (or film). They are called "surface-localized spin waves". We have studied in the past the existence of surface spin wave in a BCC semi-infinite crystal [21, 40] and in ferrimagnetic semi-infinite crystal [41]. The reader is referred to those references for details.

3.3.3 Surface Phase Transitions

In the presence of a surface, interaction parameters between spins on the surface and spins on the beneath layer may be different from the bulk interaction parameters. As a consequence, when the temperature T increases, one may observe a phase transition at the surface distinct from that of the bulk. We show below some examples of this phenomenon:

Example 3.2. We consider a thin film made up by stacking N_z planes of triangular lattice of $N \times N$ lattice sites.

The Hamiltonian is given by

$$\mathcal{H} = -\sum_{\langle i,j \rangle} J_{i,j}\mathbf{S}_i \cdot \mathbf{S}_j - \sum_{<i,j>} I_{i,j} S_i^z S_j^z \tag{3.16}$$

where $\mathbf{S}_i$ is the Heisenberg spin at the lattice site i, $\sum_{\langle i,j \rangle}$ indicates the sum over the NN spin pairs $\mathbf{S}_i$ and $\mathbf{S}_j$. The last term, which will be taken to be very small, is needed to make the film with a finite thickness to have a phase transition at a finite temperature in the case where all exchange interactions $J_{i,j}$ are ferromagnetic. This guarantees the existence of a phase transition at finite temperature, since it is known that a strictly two-dimensional system with an isotropic continuous spin model (XY and Heisenberg models) does not have long-range ordering at finite temperature [37].

Interaction between two NN surface spins is equal to J_s. Interaction between layers and interaction between NN in interior layers are supposed to be ferromagnetic and all equal to $J = 1$ for simplicity. The two surfaces of the film are frustrated if J_s is antiferromagnetic ($J_s < 0$).

With the above model we have shown [39] that the ground state is non-collinear near the surface: the three spins on a triangle on the surface form an 'umbrella' with an angle α between them and an angle β between a surface spin and its beneath neighbor (see Figs. 3.6 and 3.7). This non-planar structure is due to the interaction of the spins on the beneath layer, just like an external applied field in the z direction. Of course, when J_s is larger than J_s^c one has the collinear ferromagnetic GS as expected: the frustration is not strong enough to resist the ferromagnetic interaction from the beneath layer.

We show in Fig. 3.7 cos(α) and cos(β) as functions of J_s. The critical value J_s^c is found between -0.18 and -0.19. This value can be calculated analytically by assuming the 'umbrella structure' [39]. We obtain

$$\cos\beta = -\frac{J + I}{9J_s + 6I_s}. \tag{3.17}$$

For given values of I_s and I, we see that the solution (3.17) exists for $J_s \leq J_s^c$ where the critical value J_s^c is determined by $-1 \leq \cos\beta \leq 1$. For $I = -I_s = 0.1$, $J_s^c \approx -0.1889J$.

Using MC simulations, we showed that in the case of highly frustrated surface, there is a surface phase transition occurring at

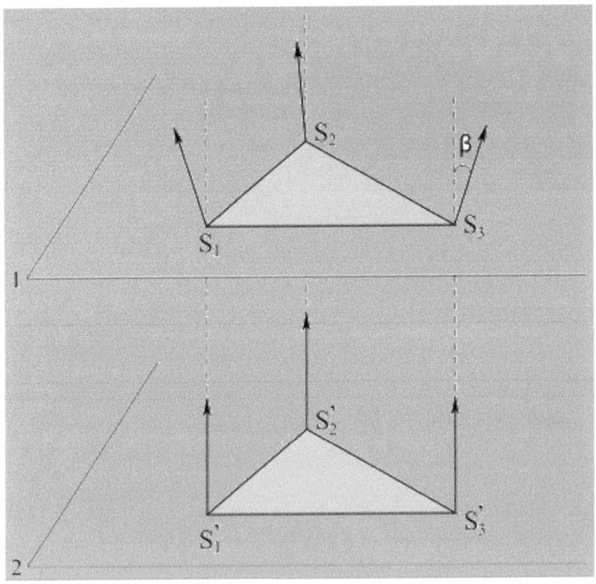

Figure 3.6 Non-collinear surface spin configuration. Angles between spins on layer 1 are all equal (noted α), while angles between vertical spins are β.

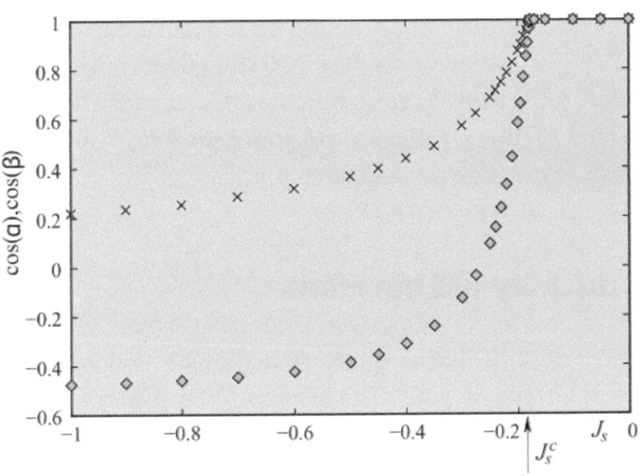

Figure 3.7 Non-collinear surface spin configuration. Angles between spins on layer 1 are all equal (noted α), while angles between vertical spins are β.

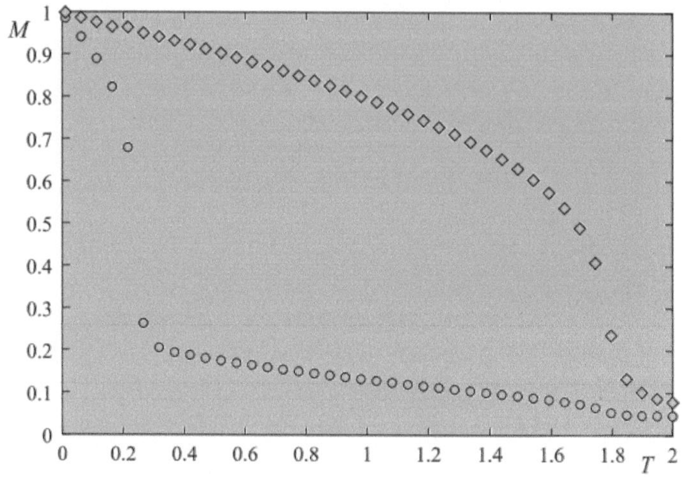

Figure 3.8 Magnetizations of layer 1 (circles) and layer 2 (diamonds) versus temperature T in unit of J/k_B for $J_s = -0.5$ with $I = -I_s = 0.1$.

low T as seen in Fig. 3.8: The surface magnetization (M_1) falls sharply at $T_1 \simeq 0.25J/k_B$ while the magnetization of the beneath layer (M_2) falls at $T_2 \simeq 1.8J/k_B$. Note that between T_1 and T_2, M_1 is not zero because of M_2 acting on it as an external field. Note that in Ref. [39], we have also studied the quantum counterpart of the model (3.16) using Green's function. The results show quantum fluctuations of the layer magnetizations at low T but the main features as found above at high T.

3.4 Criticality of Thin Films

The theory of criticality of phase transitions was demonstrated in the case of infinite crystals. The beauty of the theory resides in the fact that, despite an infinite number of systems, the phase transitions belong to a small number of "universality class". The university class depends on a few numbers of parameters: the interaction nature (short range, long range, ...), the symmetry of the order parameter (spins of Ising, XY, Heisenberg, Potts, ...) and the space dimension (2D, 3D, ...). Each universality class is defined by 6 critical exponents. When the transition is of second order, one can define in the vicinity of T_c the following critical exponents :

$$C_V = A\left|\frac{T - T_c}{T_c}\right|^{-\alpha} \tag{3.18}$$

$$\overline{M} = B\left[\frac{T_c - T}{T_c}\right]^{\beta} \tag{3.19}$$

$$\chi = C\left|\frac{T - T_c}{T_c}\right|^{-\gamma} \tag{3.20}$$

$$\xi \propto \left[\frac{T - T_c}{T_c}\right]^{-\nu} \tag{3.21}$$

$$\overline{M} = H^{1/\delta} \tag{3.22}$$

Note that the same α is defined for $T > T_c$ and $T < T_c$ but the coefficient A is different for each side of T_c. This is also the case for γ. However, β is defined only for $T < T_c$ because $\overline{M} = 0$ for $T \geq T_c$. The definition of δ is valid only at $T = T_c$ when the system is under an applied magnetic field of amplitude H. Finally, at $T = T_c$ one defines exponent η of the correlation function by

$$G(r) \propto \frac{1}{r^{d-2+\eta}} \tag{3.23}$$

There is another exponent called "dynamic exponent" z defined via the relaxation time τ of the spin system for $T \geq T_c$:

$$\tau \propto \xi^z \propto \left[\frac{1}{T - T_c}\right]^{z\nu} \tag{3.24}$$

There are thus six critical exponents α, β, γ, δ, ν and η. However, there are four relations between them: [42–45]

$$\alpha + 2\beta + \gamma = 2 \tag{3.25}$$

$$\alpha + \beta(1 + \delta) = 2 \tag{3.26}$$

$$\alpha = 2 - d\nu \tag{3.27}$$

$$\gamma = \nu(2 - \eta) \tag{3.28}$$

Therefore, there are only two of them that are to be computed. The last two relations are called "hyperscaling relations".

Note that when a phase transition is of first order, the first derivatives of the free energy such as the internal energy and the order parameter (magnetization, for example) are discontinuous at the phase transition, one cannot define the critical exponents given above. The phase transition is not "critical".

Now, if we have a thin film where the thickness is finite, does the theory of criticality apply? The answer is difficult as we see by the work presented below. Also, when a transition is of first order in 3D, does it remain so in a film with a finite thickness? We show examples of the two mentioned cases below.

3.4.1 Critical Exponents of Thin Films

In Ref. [46], we have studied the question whether or not the phase transition known in the bulk state changes its nature when the system is made as a thin film. We have considered the case of a bulk second-order transition. For that purpose, we studied the case of an Ising ferromagnet. The criticality of a film with uniform interaction has been studied by Capehart and Fisher as a function of the film thickness using a scaling analysis [47] and by MC simulations [48, 49]. The results by Capehart and Fisher indicated that as long as the film thickness is finite the phase transition is strictly that of the 2D Ising universality class. However, they showed that at a temperature away from the transition temperature $T_c(N_z)$, the system can behave as a 3D one when the spin-spin correlation length $\xi(T)$ is much smaller than the film thickness, i.e. $\xi(T)/N_z \ll 1$. As T gets very close to $T_c(N_z)$, $\xi(T)/N_z \to 1$, the system undergoes a crossover to 2D criticality. We will return to this work for comparison with our results shown below.

We have used the high-performance multiple histogram technique which is known to reproduce with very high accuracy the critical exponents of second order phase transitions [50–53]. The reader is referred to Ref. [46] for technical details, especially on the finite-size scaling relations. We have shown that under a thin film shape, i.e. with a finite thickness, the transition shows "effective critical exponents" whose values are between 2D and 3D universality classes. As an example, we show the exponent ν as a function of the film thickness in Fig. 3.9 (other exponents can be found in Ref. [46]).

The deviation of ν from the 2D value when N_z increases is due, as discussed earlier, to the crossover to 3D. Other exponents suffer the

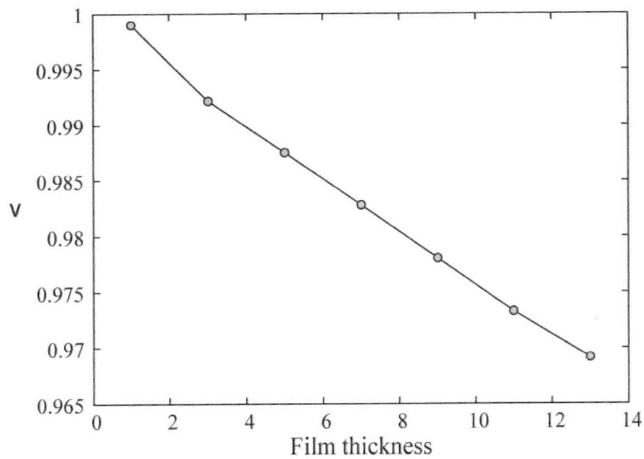

Figure 3.9 Effective exponent ν versus N_z.

same deviations. Note that these results are in excellent agreement with the exact results $\nu_{2D} = 1$ (and $\gamma_{2D} = 1.75$ not shown here). The very high precision of our method is thus verified in the rather modest range of the system sizes $L = 20 - 80$ used in the present work. Note that the result of Ref. [48] gave $\nu = 0.96 \pm 0.05$ for $N_z = 1$ which is very far from the exact value.

Let us show the prediction of Capehart and Fisher [47] on the critical temperature as a function of N_z. Defining the critical-point shift as

$$\varepsilon(N_z) = [T_c(L = \infty, N_z) - T_c(3D)] / T_c(3D) \tag{3.29}$$

they showed that

$$\varepsilon(N_z) \approx \frac{b}{N_z^{1/\nu}} [1 + a/N_z] \tag{3.30}$$

where $\nu = 0.6289$ (3D value). Using $T_c(3D) = 4.51$, we fit the above formula with $T_c(L = \infty, N_z)$ obtained from our MC simulations (see Ref. [46]), we should find, as long as the thickness is finite, the 2D universality class. This has been done, the scaling of Capehart and Fisher is validated by our results: the MC results and the fitted curve are shown in Fig. 3.10. Note that if we do not use the correction factor $[1 + a/N_z]$, the fit is not good for small N_z. The prediction of Capehart and Fisher is thus very well verified.

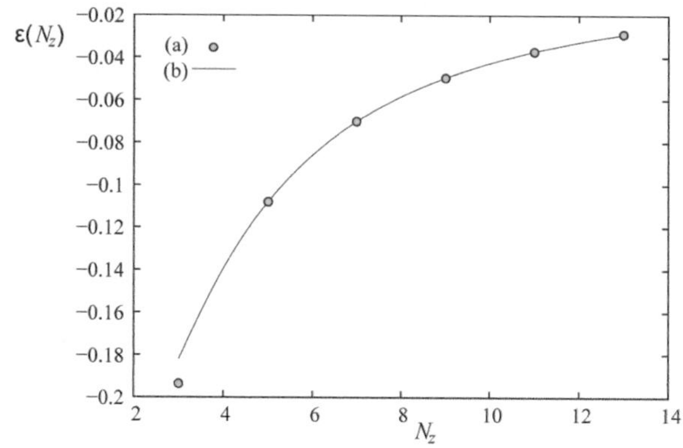

Figure 3.10 Critical temperature at infinite L as a function of the film thickness. Points are MC results, continuous line is the prediction of Capehart and Fisher. The agreement is excellent.

3.4.2 Cross-Over from First- to Second-Order Transition in Thin Films

We have shown in Ref. [54] the crossover of the phase transition from first to second order in the frustrated Ising FCC AF film. This crossover occurs when the film thickness $N_z = 2$ is smaller than a value between 2 and 4 FCC lattice cells. These results are obtained with the highly performing Wang-Landau flat histogram technique which allows to determine a weak first-order transition with efficiency [55–59]. This algorithm is very efficient for classical statistical spin models. The algorithm uses a random walk in energy space in order to obtain an accurate estimate for the density of states $\rho(E)$ which is defined as the number of spin configurations for any given energy E.

For $N_z = 2$, we found that in a range of temperature the surface spins stay ordered while interior spins are disordered. We interpret this as an effect of the frustration reduction: due to the lack of neighbors, the surface spins are less frustrated than the interior spins. As a consequence, interior spins are disordered at a lower temperature. This has been verified by the Green's function calculation [54].

The second-order transition for $N_z = 2$ is governed by the surface disordering and is characterized by critical exponents whose values are deviated from those of the 2D Ising universality class. We believe that this deviation results from the effect of the disordered interior spins which act as "correlated" random fields on the surface spins. We do not know if the critical exponents found here belong to a new universality class or they are just "effective critical exponents" which could be scaled in one way or another into the 2D Ising universality class. Anyway, these exponents seem to obey a weak universality [60]. An answer to this question is still desirable.

3.5 Skyrmions in a Mono-Layer Film

A skyrmion is a vortex-like spin configuration which has a double chirality. It has a variety of origin: we can mention spin systems with Dzyaloshinskii–Moriya (DM) interaction [61, 62], highly frustrated systems [63–65], systems with combined competing interactions, ... Due to the robustness of skyrmions and their small sizes (a dozen of atoms), there are many potential applications in spintronics [66]. This explains an enormous number of publications on skyrmions every single day. We shall show some examples in the following.

Let us mention the important papers appeared since 2003, from Ref. [67] to Ref. [91]. The reader is invited to look at the titles of these references to have an idea on what has been done. Surely, I miss a lot of important papers.

Hereafter, we consider for simplicity the two-dimensional case where the spins are on a square lattice in the xy plane of linear dimension N. We are interested in the stability of the skyrmion crystal generated in a system of spins interacting with each other via a DM interaction and a symmetric isotropic Heisenberg exchange interaction in an applied field perpendicular to the xy plane. All interactions are limited to NN. The results below have been published in Ref. [91]. The full Hamiltonian is given by

$$\mathcal{H} = -J\sum_{\langle ij\rangle}\vec{S}_i.\vec{S}_j + D\overrightarrow{z}\cdot\sum_i\overrightarrow{S_i}\wedge(\overrightarrow{S}_{i+x} + \overrightarrow{S}_{i+y}) \\ -H\sum_i S_i^z \tag{3.31}$$

where the DM interaction vector is taken along the z axis and the exchange interaction are taken between NN on both x and y directions. Rewriting it in a convenient form, we have

$$\begin{aligned}\mathcal{H} &= -J\sum_{\langle ij\rangle}\vec{S}_i.\vec{S}_j + D\sum_i[S_i^yS_{i+x}^z - S_i^zS_{i+x}^y - S_i^xS_{i+y}^z \\ &+S_i^zS_{i+y}^x] - H\sum_i S_i^z \\ &= -J\sum_{\langle ij\rangle}\vec{S}_i.\vec{S}_j + D\sum_i[S_i^y(S_{i+x}^z - S_{i-x}^z) \\ &-S_i^z(S_{i+x}^y - S_{i-x}^y) - S_i^x(S_{i+y}^z - S_{i-y}^z) \\ &+S_i^z(S_{i+y}^x - S_{i-y}^x)] - H\sum_i S_i^z \end{aligned} \tag{3.32}$$

For the i-th spin, one can write

$$\mathcal{H}_i = -S_i^xH_i^x - S_i^yH_i^y - S_i^zH_i^z \tag{3.33}$$

where the local-field components are given by

$$H_i^x = J\sum_{NN}S_j^x + D(S_{i+y}^z - S_{i-y}^z) \tag{3.34}$$

$$H_i^y = J\sum_{NN}S_j^y - D(S_{i+x}^z - S_{i-x}^z) \tag{3.35}$$

$$\begin{aligned}H_i^z &= J\sum_{NN}S_j^z + D(S_{i+x}^y - S_{i-x}^y) - D(S_{i+y}^x - S_{i-y}^x) \\ &+H\end{aligned} \tag{3.36}$$

To determine the ground state (GS), we minimize the energy of each spin, one after another. This can be numerically achieved as the following. At each spin, we calculate its local-field components acting on it from its NN using the above equations. Next we align the spin in its local field, i.e. taking $S_i^x = H_i^x/\sqrt{H_i^x**2 + H_i^y**2 + H_i^z**2}$, etc. The denominator is the modulus of the local field. In doing so, the spin modulus is normalized to be 1. As seen from Eq. (3.33), the energy of the spin $\vec{S}_i$ is minimum. We take another spin and repeat the same procedure until all spins are visited. This is one iteration of the steepest-descent method which has been used in the previous section. The reader is referred to Ref. [91] for details.

We show in Fig. 3.11 an example calculated with $D/J = 1$ and $H/J = 0.5$. Fig. 3.11 shows the 2D view of the GS where one sees

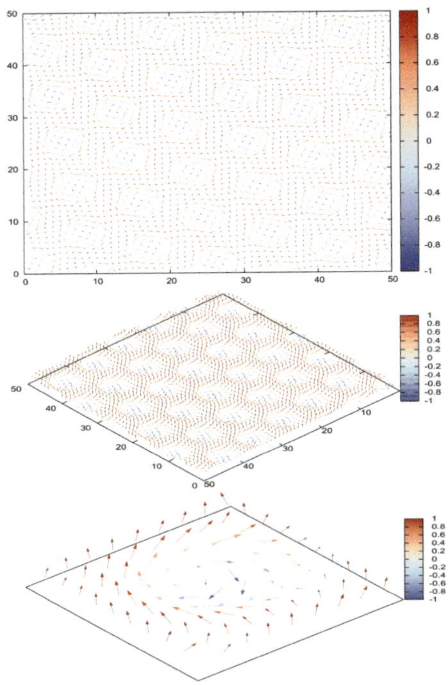

Figure 3.11 Ground state for $D/J = 1$ and $H/J = 0.5$, a crystal of skyrmions of triangular structure is observed. Top: Skyrmion crystal viewed in the xy plane, Middle: a 3D view, Bottom: zoom of the structure of a single skyrmion. The value of S_z is indicated on the color scale. See text for comments.

that skyrmions form a crystal with a triangular arrangement. Figure 3.11b shows a 3D view and Fig. 3.11c shows the vortex-like spin structure of a single skyrmion where one sees that the spin at the center is pointing down, and the spins at the skyrmion boundary are pointing in the +z direction.

We have shown in Ref. [91] that the skyrmion crystal is stable at finite temperature. To see that skyrmion stability, we have defined a new order parameter as follows:

$$M(T) = \frac{1}{N^2(t_a - t_0)} \sum_i | \sum_{t=t_0}^{t_a} \vec{S}_i(T, t) \cdot S_i^0(T = 0)| \qquad (3.37)$$

where $\vec{S}_i(T, t)$ is the i-th spin at the time t, at temperature T, and $\vec{S}_i(T = 0)$ is its state in the GS. The order parameter $M(T)$ is close to 1 at very low T where each spin is only weakly deviated from its state in the GS. $M(T)$ is zero when every spin strongly fluctuates in the paramagnetic state. The above definition of $M(T)$ is similar to the Edwards-Anderson order parameter used to measure the degree of freezing in spin glasses [92]: we follow each spin with time evolving and take the spatial average at the end.

We show in Fig. 3.12 the order parameter M versus T (red data points) as well as the average z spin component (blue data points) calculated by the projection procedure for the total time $t = 10^5 + 10^6$ MC steps per spin. As seen, both two curves indicate a phase transition at $T_c \simeq 0.26J/k_B$. The fact that M does not vanish above T_c is due to the effect of the applied field. As said above, each skyrmion has a center with spins of negative z components (the most negative at the center), the spins turn progressively to positive z components while going away from the center.

The stability of skyrmions at high temperatures is very important for making devices. Nowadays, skyrmions have been observed at room temperatures in various materials (see, for example, Refs. [71, 73]).

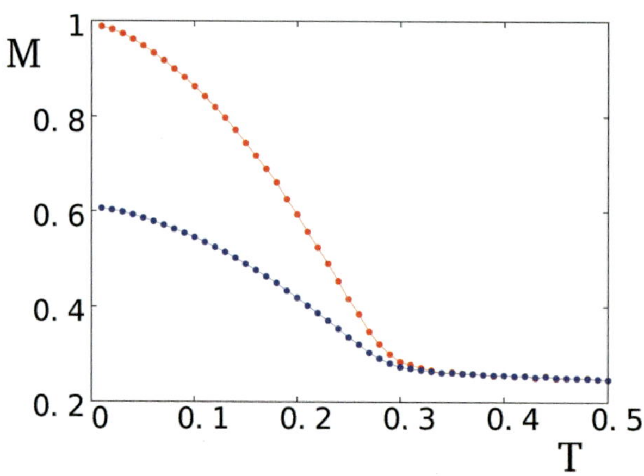

Figure 3.12 Order parameter defined in Eq. (3.37) versus T (red), for $H = 0.5$ and $N = 1800$, averaged during $t_a = 10^5$ MC steps per spin after an equilibrating time $t_0 = 10^5$ MC steps. The projection of the $S_z(T)$ on S_z^0 of the ground state are shown in blue. See text for comments.

To conclude this section, note that we have also studied a number of issues concerning applications of skyrmions. To keep the length of this chapter reasonable, we do not develop these works of us here, we just mention a few of them in Ref. [93] to Ref. [101]. The reader interested in the subject of skyrmions may have a look at these papers.

3.6 Concluding Remarks

This chapter gives an overview on some important properties of magnetic thin films obtained mostly from statistical physics and computer simulations by the author and his collaborators. On each item many references have been cited in addition to those cited in our original papers. We have discussed the effect of surface spin waves on macroscopic thermodynamic properties in thin films such as the low surface magnetization and the surface phase transition. Several examples have been shown. The criticality of the phase transition in thin films has also been analyzed with high-performance numerical simulation methods such as the multi-histogram technique and the Wang-Landau algorithm. The reader is referred to our cited original papers for the details of these complicated numerical methods and also the analytical Green's function theory.

Many topics have not been reviewed above, they include spin transport in magnetic materials, topological Hall effects [102–104] and numerous experimental works on real materials.

Acknowledgments

This chapter is a review based on subjects that the author has investigated mostly with his doctorate students over the years. The author is thankful to them, in particular R. Quartu, V.-T. Ngo, D.-T. Hoang, X.-T. Pham Phu, S. El Hog and I. Sharafullin, for fruitful collaborations.

References

1. A. Zangwill, *Physics at Surfaces*, Cambridge University Press (1988).
2. J.A.C. Bland and B. Heinrich (editors), *Ultrathin Magnetic Structures*, vol. I and II, Springer-Verlag (1994).
3. H. T. Diep, *Theory of Magnetism - Application to Surface Physics*, World Scientific, Singapore (2014).
4. K. Binder, in *Phase Transitions and Critical Phenomena*, ed. by C. Domb, J.L. Lebowitz (Academic, London, 1983) vol. 8.
5. H.W. Diehl, in *Phase Transitions and Critical Phenomena*, ed. by C. Domb, J.L. Lebowitz (Academic, London, 1986) vol. 10 (see also H.W. Diehl, Int. J. Mod. Phys. B **11**, 3503 (1997)).
6. M. N. Baibich, J. M. Broto, A. Fert, A., F. Nguyen Van Dau, F. Petroff, P. Etienne, G. Creuzet, A. Friederich, and J. Chazelas, *Phys. Rev. Lett.* **61**, 2472–2475 (1988).
7. G. Binash, P. Grünberg, F. Saurenbach, and W. Zinn, *Phys. Rev. B* **39**, 4828 (1989).
8. H. T. Diep, Ed., *Frustrated Spin Systems*, 3rd Edition, World Scientific, Singapore (2020).
9. A. Yoshimori, A New Type of Antiferromagnetic Structure in the Rutile Type Crystal, *J. Phys. Soc. Jpn* **14**, 807 (1959).
10. J. Villain, La structure des substances magnetiques, *Phys. Chem. Solids* **11**, 303 (1959).
11. H. T. Diep, First-order transition in the hexagonal-close-packed lattice with vector spins, *Phys. Rev. B* **45**, 2863 (1992).
12. D.-T. Hoang and H. T. Diep, Hexagonal-close-packed lattice: Ground state and phase transition, *Phys. Rev. E* **85**, 041107 (2012) ; http://arxiv.org/abs/1112.5724.
13. H. T. Diep and H. Kawamura, First-order transition in the FCC Heisenberg antiferromagnet, *Phys. Rev. B* **40**, 7019 (1989).
14. B. Delamotte, D. Mouhanna, and M. Tissier, Nonperturbative renormalization group approach to frustrated magnets, *Phys. Rev. B* **69**, 134413 (2004).
15. B. Delamotte, M. Dudka, Y. Holovatch, and D. Mouhanna, About the relevance of the fixed dimension perturbative approach to frustrated magnets in two and three dimensions, *Phys. Rev. B* **82**, 104432 (2010).
16. V.-T. Ngo and H. T. Diep, Phase transition in Heisenberg stacked triangular antiferromagnets: End of a controversy, *Phys. Rev. E* **78**, 031119 (2008).

17. V.-T. Ngo and H. T. Diep, Stacked triangular XY antiferromagnets: End of a controversial issue on the phase transition, *Journal of Applied Physics* **103**, 07C712 (2008).
18. M. Reehorst, S. Rychkov, B. Sirois and B. C. van Rees, Bootstrapping frustrated magnets: The fate of the chiral O(N) × O(2) universality class, arXiv:2405.19411v2.
19. G. Toulouse, Spin glass with non-random interactions, *Commun. Phys.* **2**, 115 (1977).
20. B. Berge, H. T. Diep, A. Ghazali and P. Lallemand, Phase transitions in two-dimensional uniformly frustrated XY spin systems, *Phys. Rev. B* **34**, 3177 (1986).
21. H. T. Diep, J.C.S. Levy and O. Nagai, Effect of surface spin-waves and surface anisotropy in magnetic thin films at finite temperature, *Physics Status Solidi (b)* **93**, 351 (1979).
22. H. T. Diep, Quantum theory of helimagnetic thin films, *Phys. Rev. B* **91**, 014436 (2015), http://arxiv.org/abs/1412.0184.
23. R. Quartu and H. T. Diep, Partial order in frustrated quantum spin systems, *Phys. Rev. B* **55**, 2975 (1997).
24. I. Harada and K. Motizuki, Effect of magnon-magnon interaction on spin wave dispersion and magnon sideband in MnS, *J. Phys. Soc. Jpn.* **32**, 927 (1972).
25. E. Rastelli, L. Reatto and A. Tassi, Quantum fluctuations in helimagnets, *J. Phys. C* **18**, 353 (1985).
26. H. T. Diep, Low-temperature properties of quantum Heisenberg helimagnets, *Phys. Rev. B* **40**, 741 (1989).
27. R. Quartu and H. T. Diep, Phase diagram of body-centered tetragonal Helimagnets, *J. Magn. Magn. Mater.* **182**, 38 (1998).
28. S. M. Stishov, A. E. Petrova, S. Khasanov, G. Kh. Panova, A. A. Shikov, J. C. Lashley, D. Wu, and T. A. Lograsso, Magnetic phase transition in the itinerant helimagnet MnSi: Thermodynamic and transport properties, *Phys. Rev. B* **76**, 052405 (2007).
29. V. D. Mello, C. V. Chianca, Ana L. Danta, and A. S. Carriç, Magnetic surface phase of thin helimagnetic films, *Phys. Rev. B* **67**, 012401 (2003).
30. F. Cinti, A. Cuccoli, and A. Rettori, Exotic magnetic structures in ultrathin helimagnetic holmium films, *Phys. Rev. B* **78**, 020402(R) (2008).
31. L. J. Rodrigues, V. D. Mello, D. H. A. L. Anselmo and M. S. Vasconcelos, Magnetic structures in ultra-thin Holmium films: Influence of external magnetic field, *J. Mag. Mag. Mater.* **377**, 24 (2014).

32. E. A. Karhu, S. Kahwaji, M. D. Robertson, H. Fritzsche, B. J. Kirby, C. F. Majkrzak, and T. L. Monchesky, Helical magnetic order in MnSi thin films, *Phys. Rev. B* **84**, 060404(R) (2011).
33. E. A. Karhu, U. K. Röβler, A. N. Bogdanov, S. Kahwaji, B. J. Kirby, H. Fritzsche, M. D. Robertson, C. F. Majkrzak, and T. L. Monchesky, Chiral modulation and reorientation effects in MnSi thin films, *Phys. Rev. B* **85**, 094429 (2012).
34. J. Heurich, J. König, and A. H. MacDonald, Persistent spin currents in helimagnets, *Phys. Rev. B* **68**, 064406 (2003).
35. O. Wessely, B. Skubic, and L. Nordström, Spin-transfer torque in helical spin-density waves, *Phys. Rev. B***79**, 104433 (2009).
36. F. Jonietz, S. Mühlbauer, C. Pfleiderer, A. Neubauer, W. Münzer, A. Bauer, T. Adams, R. Georgii, P. Böni, R. A. Duine, K. Everschor, M. Garst, and A. Risch, Spin transfer torques in MnSi at ultralow current densities, *Science* **330**, 1648 (2010).
37. N. D. Mermin and H. Wagner, Absence of ferromagnetism or antiferromagnetism in one- or two-dimensional isotropic Heisenberg models, *Phys. Rev. Lett.* **17**, 1133 (1966). Erratum *Phys. Rev. Lett.* **17**, 1307 (1966).
38. V. Leiner, D. Labergerie, R. Siebrecht, Ch. Sutter, and H. Zabel, Investigation of the magnetism in thin single Ho(00.1) films via neutron reflectivity measurements, *Physica B* **283**, 167 (2000).
39. V.-T. Ngo and H. T. Diep, Effects of frustrated surface in Heisenberg thin films, *Phys. Rev. B* **75**, 035412 (2007).
40. Diep-The-Hung, Temperature-dependent surface magnetisation and critical temperature of ferromagnetic thin films, *Phys. Stat. Solidi (b)* **103**, 809 (1981).
41. Diep-The-Hung, I. Harada and O. Nagai, Theory of surface spin-wave in a semi-infinite ferrimagnet, *Physics Letters A* **53**, 157 (1975).
42. K. G. Wilson, Renormalization Group and Critical Phenomena. I. Renormalization group and the Kadanoff scaling picture, *Phys. Rev. B* **4**, 3174 (1971).
43. D. J. Amit, *Field Theory, The Renormalization Group and Critical Phenomena*, World Scientific, Singapore (1984).
44. J. Cardy, *Scaling and Renormalization in Statistical Physics*, Cambridge University Press, London (1996).
45. J. Zinn-Justin, *Quantum Field Theory and Critical Phenomena*, 4th ed., Oxford Univ. Press (2002).

46. X. T. Pham Phu, V. Thanh Ngo and H. T. Diep, Critical behavior of magnetic thin films, *Surf. Science* **603**, 109 (2009).
47. T. W. Capehart and M. E. Fisher, Susceptibility scaling functions for ferromagnetic Ising films, *Phys. Rev. B* **13**, 5021 (1976).
48. P. Schilbe, S. Siebentritt and K. H. Rieder, Monte Carlo calculations on the dimensional crossover of thin Ising films, *Phys. Lett. A* **216**, 20 (1996).
49. M. Caselle and M. Hasenbusch, Deconfinement transition and dimensional cross-over in the 3D gauge Ising model, *Nucl.Phys. B* **470**, 435 (1996).
50. A. M. Ferrenberg and R. H. Swendsen, *Phys. Rev. Lett.* **61**, 2635 (1988).
51. A. M. Ferrenberg and R. H. Swendsen, Optimized Monte Carlo data analysis, *Phys. Rev. Lett.* **63**, 1195 (1989).
52. A. M. Ferrenberg and D. P. Landau, Critical behavior of the three-dimensional Ising model: A high-resolution Monte Carlo study, *Phys. Rev. B***44**, 5081(1991).
53. A. Bunker, B. D. Gaulin, and C. Kallin, Multiple-histogram Monte Carlo study of the Ising antiferromagnet on a stacked triangular lattice, *Phys. Rev. B* **48**, 15861 (1993).
54. X. T. Pham Phu, V. Thanh Ngo and H. T. Diep, Crossover from first to second order transition in frustrated Ising antiferromagnetic films, *Phys. Rev. E* **79**, 061106 (2009) ; arXiv:0902.4310.
55. F. Wang and D. P. Landau, Efficient, Multiple-range random walk algorithm to calculate the density of states, *Phys. Rev. Lett.* **86**, 2050 (2001).
56. F. Wang and D. P. Landau, Determining the density of states for classical statistical models, *Phys. Rev. E* **64**, 056101 (2001).
57. G. Brown and T.C. Schulhess, Wang–Landau estimation of magnetic properties for the Heisenberg model, *J. Appl. Phys.* **97**, 10E303 (2005).
58. B. J. Schulz, K. Binder, M. Müller, and D. P. Landau, Avoiding boundary effects in Wang–Landau sampling, *Phys. Rev. E* **67**, 067102 (2003).
59. A. Malakis, S. S. Martinos, I. A. Hadjiagapiou, N. G. Fytas, and P. Kalozoumis, Entropic sampling via Wang–Landau random walks in dominant energy subspaces, *Phys. Rev. E* **72**, 066120 (2005).
60. Masuo Suzuki, New universality of critical exponents, *Prog. Theor. Phys.* **51**, 1992 (1974).
61. I. E. Dzyaloshinskii, Thermodynamical theory of “weak” ferromagnetism in antiferromagnetic substances, *Sov. Phys. JETP* **5**, 1259 (1957).

89. M. Heide, G. Bihlmayer, and S. Blügel, Dzyaloshinskii–Moriya interaction accounting for the orientation of magnetic domains in ultrathin films: Fe/W(110), *Phys. Rev. B* **78**, 140403(R) (2008).

90. S. Rohart. and A. Thiaville, Skyrmion confinement in ultrathin film nanostructures in the presence of Dzyaloshinskii–Moriya interaction, *Phys. Rev. B* **88**, 184422 (2013).

91. S. El Hog, A. Bailly-Reyre and H. T. Diep, Stability and phase transition of skyrmion crystals generated by Dzyaloshinskii–Moriya interaction, *J. Mag. Mag. Mater.* **455**, 32–38 (2018).

92. M. Mézard, M. Parisi and M. Virasoro, *Spin Glass Theory and Beyond An Introduction to the Replica Method and Its Applications*, World Scientific, Singapore (1986).

93. I. F. Sharafullin, M. Kh. Kharrasov and H. T. Diep, Dzyaloshinskii–Moriya interaction in magneto-ferroelectric superlattices: Spin waves and skyrmions, *Phys. Rev. B* **99**, 214420 (2019).

94. J. Xia, X. Zhang, M. Ezawa, O. A. Tretiakov, Z. Hou, W. Wang, G. Zhao, X. Liu, H. T. Diep and Y. Zhou, Current-driven skyrmionium in a frustrated magnetic system, *Appl. Phys. Lett.* **117**, 012403 (2020).

95. X. Zhang, J. Xia, M. Ezawa, O. A. Tretiakov, H. T. Diep, G. Zhao, X. Liu and Y. Zhou, A frustrated bimeronium: Static structure and dynamics, *Appl. Phys. Lett.* **118**, 052411 (2021).

96. J. Xia, X. Zhang, O. A. Tretiakov, H. T. Diep, J. Yang, G. Zhao, M. Ezawa, Y. Zhou, X, Liu, Bifurcation of a topological skyrmion string, *Phys. Rev. B* **105**, 214402 (2022).

97. X. Zhang, J. Xia, O. A. Tretiakov, H. T. Diep, G. Zhao, J. Yang, Y. Zhou, M. Ezawa, X. Liu, Current-induced helicity switching of frustrated skyrmions on a square-grid obstacle pattern, *J. Magn. Soc. Jpn.* **47**, 20–27 (2023).

98. Sahbi El Hog, Fumitake Kato, Hiroshi Koibuchi, Hung T. Diep, Finsler geometry modeling and Monte Carlo study of skyrmion shape deformation by uniaxial stress, *Phys. Rev. B* **104**, 024402 (2021).

99. Sahbi El Hog, Fumitake Kato, Satoshi Hongo, Hiroshi Koibuchi, Gildas Diguet, Tetsuya Uchimoto, Hung T. Diep, The stability of 3D skyrmions under mechanical stress studied via Monte Carlo calculations, *Results in Physics* **38**, 105578 (2022).

100. G. Diguet, B. Ducharne, S. El Hog, F. Kato, H. Koibuchi, T. Uchimoto, H.T. Diep, Monte Carlo studies of skyrmion stabilization under geometric confinement and uniaxial strains, *J. Magn. Magn. Mater.* **579**, 170819 (2023).

101. G. Diguet, B. Ducharne, S. El Hog, F. Kato, H. Koibuchi, T. Uchimoto and H. T. Diep, Monte Carlo studies on geometrically confined skyrmions in nanodots: Stability and morphology under radial stresses, *Comput. Mater. Sci.* bf 243, 113137 (2024).

102. T. Kurumaji, T. Nakajima, M. Hirschberger, A. Kikkawa, Y. Yamasaki, H. Sagayama, H. Nakao, Y. Taguchi, T. Arima and Y. Tokura, Skyrmion lattice with a giant topological Hall effect in a frustrated triangular-lattice magnet, *Science* **365**, Issue 6456 pp. 914–918 (2019).

103. B. Göbel, A. Mook, J. Henk and I. Mertig, Antiferromagnetic skyrmion crystals: Generation, topological Hall and topological spin Hall effect, arXiv:1707.05267v1 (2017).

104. K. Hamamoto, M. Ezawa and N. Nagaosa, Quantized topological Hall effect in skyrmion crystal, *Phys. Rev. B* **92**, 115417 (2015).

Chapter 4

Emulating Spin Ice Physics Using Two-Dimensional Arrays of Magnetic Nanostructures

Nicolas Rougemaille and Johann Coraux
Université Grenoble Alpes, CNRS,
Grenoble INP, Institut NEEL,
38000 Grenoble, France
nicolas.rougemaille@neel.cnrs.fr

At the turn of the millennium, micro- and nanofabrication techniques allowed the scientific community to tackle a wide range of questions in frustrated magnetism through the prism of nanoscience. Lithographically patterned arrays of interacting magnetic nanostructures, in particular, emerged as a powerful experimental platform for probing intriguing and fascinating phenomena often associated to highly frustrated magnets. This chapter provides historical and scientific contexts to introduce the reader to spin ice physics, while trying to explain why a research community strongly rooted in nanomagnetism and spintronics recently started to design and study artificial frustrated magnetic architectures.

2D and 3D Nanostructures: Structure, Properties, and Applications
Edited by Jean-Claude Serge Levy

ISBN 978-981-5129-97-7 (Hardcover), 978-1-003-74871-7 (eBook)
www.jennystanford.com

DOI: 10.1201/97810037487-4

4.1 Preamble

When brought to sufficiently low temperatures, matter generally orders, as in the case of a crystalline solid. Some systems remain disordered, though, in the manner of a liquid, even at the lowest temperatures achievable experimentally. Entropy, a thermodynamic quantity that characterizes the degree of disorder, is then very useful to describe the low-temperature properties of matter. While most materials have their entropy approaching a constant value as temperature tends to absolute zero, others instead preserve much of the gas-like entropy they had at high temperatures. Nevertheless, these two disordered phases at high and low temperatures are not equivalent: a liquid and a gas have distinct properties. In liquids, the disorder is said to be correlated as correlations between degrees of freedom can be strong, at least at short distances, even though the system strongly fluctuates. This is not the case in an ideal gas, for which these correlations are zero on average, at all distances.

One might wonder how a material can fail to order at low temperatures. This can happen when the degrees of freedom (atom positions, local charge state, direction of the magnetic moment, etc.) feel competing interactions. As we shall see, these competing interactions sometimes lead to the propagation of contradicting information, preventing the existence of a unique configuration for the considered degree of freedom that minimizes the system energy. This situation favoring conflicting information can have different origins, but is often referred to as frustration. What may seem surprising at first glance is that frustration can manifest on certain degrees of freedom only. For instance, a magnetic compound may be well crystallized at low temperatures, but never magnetically ordered, even at thermodynamic equilibrium. In other words, the atomic structure of the compound can be crystallized, whereas the magnetic moments carried by the atoms forming the crystal cannot order. In frustrated magnetism, this magnetically disordered state is known as a spin liquid. Let us stress the term thermodynamic equilibrium: we are not talking here about out-of-equilibrium systems, like spin glasses, but magnetic states of matter characterized by a liquid-like behavior, even at extremely low temperatures. There are natural materials that appear either solid or liquid at low temperatures depending on the physical quantity that is probed. This chapter illustrates that one can fabricate, artificially, a meta-material exhibiting all the properties of a spin liquid, and which is advantageously amenable to direct imaging of its correlated disorder in space and time, at the scale of the elementary constituent. More specifically, we will show why lithographically patterned arrays of interacting magnetic nanostructures prove to be an interesting platform in which to visualize intriguing behaviors of matter that are difficult to access otherwise. In other words, we will show how nanosciences, combined with magnetic imaging techniques, enabled these last years to tackle scientific questions traditionally addressed in natural, bulk magnetic compounds, and to emulate a class of lattice spin models whose origins date back to the early twentieth century. We will provide the reader with some historical and scientific background to understand why a community exploring nanomagnetism and spintronics is now also involved in frustrated magnetism.

There are numerous journal articles and book chapters describing how arrays of interacting magnetic nanostructures have shed new light on spin liquid physics. This literature describes how the initial goal to manufacture a frustrated meta-material triggered new research directions, occasionally with applicative aims, such as bio-inspired computing, connecting distinct subdisciplines of magnetism. This chapter will refer to some of these works for the reader wishing to delve deeper into specific notions and concepts. However, it is primarily aimed at readers seeking a general introduction to artificial spin ice physics, without necessarily having any prior knowledge in frustrated magnetism. We hope it will provide a complementary viewpoint to the already existing literature.

This chapter is divided into four parts. Section 4.2 provides the foundations and motivations at the origin of numerous studies in frustrated magnetism on either natural or artificial materials. In Section 4.3, we show why arrays of magnetic nanostructures are a convenient platform for imaging artificial spin liquids and for probing their properties at the scale of the elementary constituent (i.e., the individual spin degree of freedom). Two emblematic cases are presented in Section 4.4. The final section briefly mentions other research avenues the community is exploring.

4.2 A Brief Historical and Scientific Overview

4.2.1 Ice, Spin Ice, Frustration and the Third Law of Thermodynamics

When we learn thermodynamics, we soon hear about the three laws: the conservation of energy, the increase of entropy (arrow of time), and the third one, dealing with the state of matter at absolute zero. Although the first and second laws are generally well-admitted, the formulation of the third one, associated to the concept of entropy, does not always appear, even today, as unique or universal. This situation is somehow related to the fact that one usually assumes that matter orders as absolute zero is approached, although this is not always true, as we mentioned above. This might be the reason why the third law states, at least in its commonly taught formulation, that "the entropy of a pure crystalline substance tends to a constant, which can be taken to be zero, as the temperature approaches absolute zero." This formulation of the third law was given by Planck [1], and is used for most substances and compounds considered in physics or chemistry. However, other formulations were proposed, and in fact vigorously debated, in the beginning of the 20th century, notably by Nernst, Planck and Einstein [2–4], and later also by Lewis [5].

When scientists developed cryogenics in the early 1920s, some started to challenge these formulations experimentally. Among the works done at that time, Giauque's measurement of the heat capacity of ice was certainly a milestone in the third law controversy [6]. Indeed, in its usual form (the one stable under atmospheric conditions), ice's entropy (expressed per water molecule) does not vanish at low temperatures, contradicting Planck's formulation. In fact, the number of possible arrangements of molecules in ice, i.e., the level of disorder in a crystal at thermodynamic equilibrium, increases faster than the system size! This back-then surprising behavior was vividly debated in the scientific community of the time, and that was Linus Pauling in 1935 who proposed the now admitted interpretation of the zero point entropy of ice [7]. His explanation lies in the properties of the O–H chemical bond, which is shorter than half of the O–O distance. As a consequence, hydrogens can jump between two energetically equivalent positions along the axis defined by two neighboring oxygens. Each hydrogen atom can then be close or far from a given oxygen, and hydrogen positions in ice still fluctuate greatly at low temperatures. In other words, although oxygen atoms form a perfect crystal, there is an essentially infinite number of hydrogen configurations minimizing the system's energy, even at low temperatures, making ice a very unusual solid.

What is surprising in the behavior of the hydrogen atoms is that their positional disorder is not random, although strongly fluctuating (they are not randomly distributed on either of the two equivalent sites between two oxygens). As the stoichiometry of the water molecule requires two hydrogens to form a bond with each oxygen atom, two of the four hydrogen atoms surrounding an oxygen are close to it (covalent bond), whereas the other two are further away (hydrogen bond). Said differently, each oxygen atom has two short bonds with two hydrogen atoms and two longer bonds with the other two, as schematized in Fig. 4.1. If the arrangement of hydrogen atoms can fluctuate greatly at low temperatures, it still has to satisfy this local constraint everywhere and at any time. This is known as the ice rule [8] or the 2-in/2-out constraint. Let us stress that although strong, this local constraint leaves an extensive number of possible hydrogen arrangements in the ground state. The disorder of hydrogen atoms in ice is correlated.

Sixty years after Pauling's description, a class of magnetic oxide compounds has been found to have a crystal structure made of corner-sharing tetrahedra [9], exactly like ordinary ice. In these compounds, oxygen atoms are also located at the center of the tetrahedra and magnetic rare earth elements sit on their corners. But unlike the hydrogen atoms in ice, the rare earth atoms have a fixed position and carry a magnetic moment that is oriented along the O–O axis. Interestingly, the magnetic interaction between neighboring rare earth atoms is ferromagnetic, and because of the crystal geometry together with the directions of the magnetic moment axes, pairwise spin interactions cannot be all satisfied

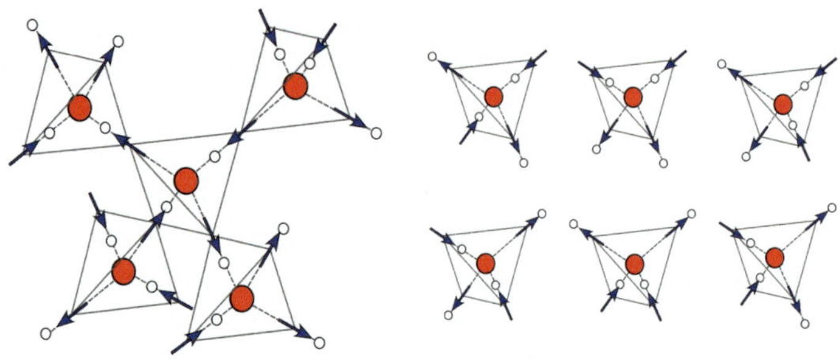

Figure 4.1 (Left) Schematics of the atoms' arrangement in ordinary ice. The crystal structure consists of a three-dimensional lattice of corner-sharing tetrahedra. The center of each tetrahedron is occupied by an oxygen atom (in red), whereas hydrogen atoms (in white) sit on its four corners, which define the midpoint of the O–O distance. The hydrogen atoms are distributed in such a way that two are bond to an oxygen atom, whereas the other two are bond to the neighboring O atoms. This can be illustrated by showing the displacement vector (represented by blue arrows) of the hydrogen atoms along the O–O axis. (Right) Since each O is surrounded by four H atoms, which can be either close or far, 6 arrangements are possible per tetrahedron. Assuming N tetrahedra in the ice crystal, Pauling estimated the number of possible ground state configurations to be $(3/2)^N$, an exponentially growing number of atom arrangements. Reproduced with permission from Ref. 71.

simultaneously. The system is said to be geometrically frustrated. Minimizing the magnetic energy requires two moments to point towards the center of the associated tetrahedron, whereas the other two point outwards. This local constraint on the magnetic moment is reminiscent of the hydrogen disorder in ice and is also called the ice rule (see Fig. 4.2). Similarly to the structural disorder in ice, the magnetic ground state in these oxide compounds is extensively degenerate [9, 10]. These compounds, naturally named *spin ices*, are another class of systems contradicting Planck's formulation of the third law.

In fact, and this is important for the rest of this chapter, this concept of geometrical frustration is not specific to bulk (three-dimensional) rare earth compounds, and can also be relevant in two dimensions. The Ising model on a triangular lattice with antiferromagnetic nearest-neighbor interactions is the archetype

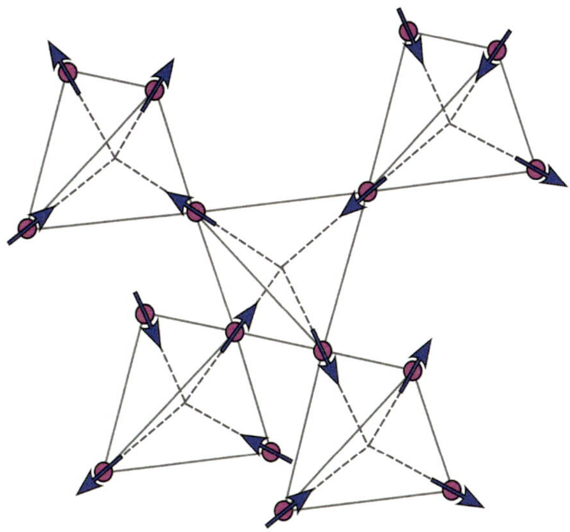

Figure 4.2 Schematics of a spin ice compound. The rare earth element appears in purple, and the Ising magnetic moment carried by each of these atoms is represented by a blue arrow. The crystal structure is the same as the one of ice and consists of a three-dimensional lattice of corner-sharing tetrahedra. Note that the magnetic moments are aligned along the axis bridging two neighboring tetrahedron centers, where oxygen atoms seat (not shown here). There is thus a direct correspondence between the displacement vectors showing the hydrogen arrangements in ice and the magnetic configurations in spin ice compounds. Reproduced with permission from Ref. 71.

of a two-dimensional frustrated spin system [11]. A variant of this model is to construct a lattice of equilateral triangles sharing their corners rather than their edges. This is known as a kagome lattice [12, 13]. In either case, the odd number of corners of the triangular unit makes it impossible to alternate the Ising spin direction between nearest neighbors (see Fig. 4.3). The ground state is then obtained when each triangle contains one, and only one bond coupling two neighboring Ising variables of the same value. The key result here is that there is an extensive number of Ising configurations satisfying this local constraint, which is the counterpart of the ice rule, but in two dimensions. We emphasize

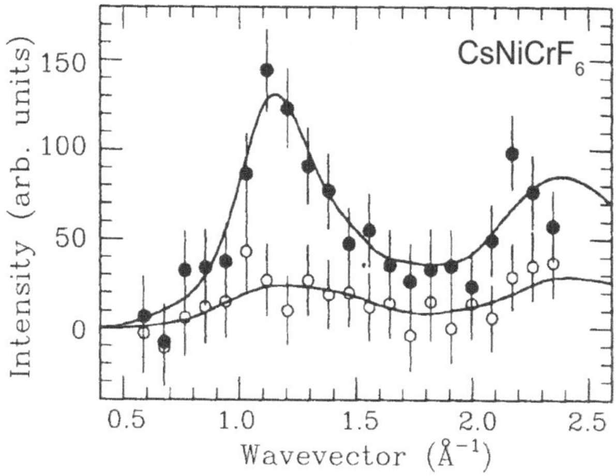

Figure 4.4 Magnetic scattering cross section for a single crystal of the magnetic $CsNiCrF_6$ compound, measured through neutron experiments performed at 4 K (full circles) and 30 K (open circles). The curves reveal a liquid-like behavior of the magnetic moments within the $CsNiCrF_6$ crystal, hence the name spin liquid to describe the disordered magnetic ground state of geometrically frustrated lattices. Reproduced with permission from Ref. 73.

then captures the physics of spin ice excitations, which behave as point-like, magnetically charged quasiparticles. These defects being the source or sink of a magnetic flux, they are classical analogues of magnetic monopoles [18, 19], and spin ices have been coined Coulomb phases [16, 17].

To summarize, spin ice compounds are spin liquids, but specific spin liquids: they are Coulomb spin liquids. This is so because an emergent electrostatics describes their low-energy physics nicely, and the properties of their excitations more specifically. This electrostatics also accounts for the power-law decaying spin-spin correlations in spin ices. Indeed, although one might naively think that correlations decay exponentially with the separation distance, they actually have the spatial dependence of a dipole-dipole interaction at long distances. In fact, the algebraic decay of the magnetic correlations in spin ices generally signals a Coulomb phase physics. Let us stress that the concept of Coulomb phase has general significance in physics and does not restrict to spin ices (Coulomb phases are also found in nonmagnetic systems) [16].

4.2.3 From Natural Spin Ice Compounds to Artificial Spin Ice Matter

Over the past 25 years, spin ices have proved to be a rich condensed matter platform in which to study exotic phenomena. The magnetic properties in these compounds are generally determined through very low-temperature (sometimes sub-100 mK) measurements of thermodynamic quantities, such as the magnetic susceptibility or specific heat, and via neutron scattering techniques. There is, however, no direct means to visualize the correlated disorder of spin ices at the scale of the elementary constituent (directly observing the microscopic details within a liquid is not possible either). Local probing of individual spins would indeed be a challenge as a measurement at the atomic level will likely alter the state that one intends to capture.

A second class of shortcomings when exploring spin ice compounds is that one can only study what nature provides or permits. Crystals simply do not grow in any kind of conceivable structure or with any kind of magnetic ion, which prevents access to physical effects governed by unusual lattice symmetry or geometry, for example. If the lattice geometry together with the nature of the spin degree of freedom could be controlled at will in a convenient experimental framework, one would end up with an almost infinite playground in which to investigate frustrated magnetism.

This sounds like a utopia, but this is essentially what artificial systems can do by providing two-dimensional equivalents of frustrated bulk compounds. In fact, at about the same time as the spin ice story was unfolding [9], several groups worldwide reported artificially made realizations of magnetically frustrated systems [20–22]. But the breakthrough came a couple of years later when two groups independently created *artificial spin ice* systems by patterning magnetic thin films via electron beam lithography techniques, yielding two-dimensional networks of tightly packed nanomagnets, each behaving as a classical pseudo-spin with a well-defined uniaxial anisotropy and two possible orientations (i.e.,

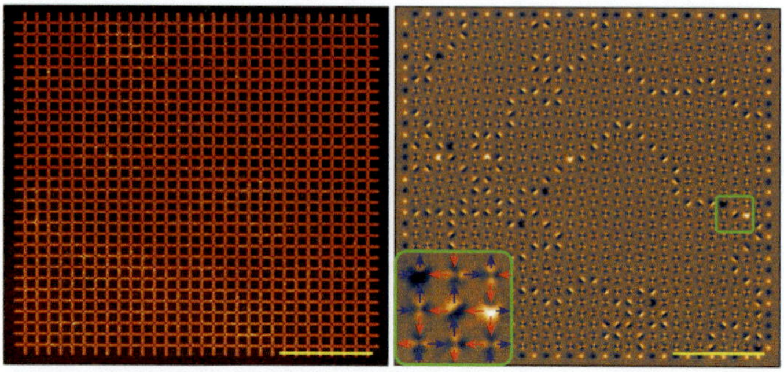

Figure 4.5 Example of room-temperature atomic force microscopy (AFM) (left) and magnetic force microscopy (MFM) (right) images of a square array of nanomagnets. The AFM data reveals the topography of the array, with magnets of $500\times100\times25$ nm^3 dimensions. The MFM data is free of topographic information, and reveals the orientation of the Ising-like magnets. In the bottom left corner of the right panel, the green frame gives a detail of the image, focusing on a 3×3 vertex region and providing the correspondence between the MFM contrast and the (in-plane) spin orientation. Overall, the MFM data shows extended patches of similar ice-rule-complying vertices (with similar kind of magnetic contrast) separated by lines of other kinds of ice-rule-complying vertices. The brightest or darkest vertices (one of each is observed within the green frame) correspond to vertices that violate the ice rule, where three spins point in and one points out (black) and vice versa (white). The scale bar represents 5 μm in both images. Reproduced with permission from Ref. 71.

an Ising variable), in mutual interaction with the neighboring nanomagnets [23–25].

Studying these magnetic 2D meta-materials, a new viewpoint on frustrated magnetism becomes accessible. Firstly, the magnetic state of each and every individual Ising-like nanomagnet can be imaged in a routine fashion using standard magnetic imaging techniques. Real space (see Fig. 4.5)—sometimes even real-time observations—accordingly reveals the orientation of all magnetic units within extended lattices. In addition, whereas spin ice compounds exhibit exotic properties at cryogenic temperatures only, artificial spin ices can be designed to obtain such properties at room temperature (in fact, the working temperature is tunable). Finally, although they were initially made to mimic the unconventional low-temperature physics of spin ices, their flexibility and ease of fabrication allowed to test and revisit many predictions from statistical mechanics, but also to investigate many-body phenomena in systems having no natural counterparts [26–28]. This is certainly why artificial spin ices received considerable attention recently and triggered a wealth of studies at the frontier between condensed matter physics, nanomagnetism and statistical thermodynamics.

4.3 Why Emulating Spin Ice Physics with Arrays of Interacting Nanomagnets?

As we have just seen, arrays of magnetic nanostructures allow the investigation of spin ice physics under the prism of magnetic imaging. One can thus measure all quantities that can be deduced from a spin configuration, both locally and globally. Magnetization, spin-spin correlations as a function of distance or the density of defects violating the ice-rule constraint are typical observables that can be determined experimentally and compared to theoretical predictions. Let us stress once again that these quantities can be visualized in real space and it is therefore possible, for example, to know where magnetic defects are located, whether their spatial distribution is homogeneous, whether their density varies between successive experiments or over time, and so on. We can therefore understand why arrays of magnetic nanostructures rapidly become a platform of choice for emulating spin ice models, and more generally for probing the physics of spin liquids and their excitations.

The democratization of nanofabrication techniques over the last few decades likely contributed to the dissemination of arrays of magnetic nanostructures in frustrated magnetism. Indeed, most nanoscience research centers nowadays host comprehensive micro/nanofabrication platforms and multi-technique characterization tools, including imaging, of the thus-produced samples. In fact, while current efforts in nanomagnetism and spintronics exploit increasingly sophisticated nanofabrication techniques, the design of artificial spin systems is based on processes that are essentially well mastered. In modern spintronics, it is not uncommon to see magnetic stacks composed of 10 different layers, or magnetic architectures at the nanoscale; in contrast, frustrated nanomagnetism essentially relies on arrays made of permalloy or other alloys that pose generally no major fabrication difficulties. Moreover, the characteristic sizes of the nanomagnets are often closer to the micron than to the nanometer, and magnetic elements with typical lateral dimensions of 500×150 nm^2 are usually perfectly suited. These dimensions are roughly an order of magnitude larger than the spatial resolutions of most magnetic imaging techniques (except for optical techniques) and nearly two orders of magnitude larger than the limiting resolutions in electron lithography. Today, nanofabrication techniques even allow to design *three-dimensional* arrays that can, at least to some extent, still be imaged at the scale of the individual spin [29]. We should also mention that magnetically sensitive microscopy techniques have made spectacular advances in recent years, and can now image complex magnetization patterns that can be resolved vectorially, even in volume [30].

The above-mentioned dimensions have an important practical consequence: the Curie temperature of the nanomagnets is usually that of the bulk material, and is therefore substantially larger than room temperature for common elements or alloys. In other words, arrays of magnetic nanostructures are ferromagnetic at room temperature, and there is no need to resort to cryogenic environments, often the source of experimental difficulties in magnetic imaging. Most works on nanomagnetic arrays are therefore carried out using techniques such as magnetic force microscopy, which is widely used in laboratories. This does not exclude the use of more sophisticated techniques such as transmission electron microscopy (Lorentz or holography) or photoemission electron microscopy at synchrotron facilities, to name a few [31]. But the ability to image magnetic configurations "on the table" and at room temperature is one of the main advantages of artificial spin ice structures. In fact, the situation is even more favorable. Nanomagnets can be made thermally active [32–35] and the characteristic time to reverse the magnetization in a given nanomagnet can be adjusted to the time of the experimental measurement. This can be achieved by reducing the volume of the nanomagnets, in particular their thickness, to lower their blocking temperature to the desired level. For example, nanomagnets thinned down to 3 nm become thermally active around room temperature, with a typical reversal time of the order of a second or a minute [34]. Playing with the composition of magnetic alloys, the Curie temperature can be also adjusted [35–37]. The dynamics of the system under study can then be adapted to the instrument, rather than the other way around, providing a versatility that is difficult to achieve with other approaches.

Maybe surprisingly, the first frustrated artificial systems were not magnetic in essence, but superconducting [20–22]. The idea was to fabricate assemblies of superconducting current loops in electrically isolated or connected rings, forming arrays of interacting induced magnetic moments, pointing perpendicular to the lattice plane. Imaging these magnetic moments with sufficient spatial resolution required cryogenic instrumentation operating at very low temperatures (<4 K). Despite the difficulties inherent in superconducting materials, promising results were obtained in the late 1990s, notably on two-dimensional kagome and triangular arrays (see Fig. 4.6 for the kagome array). Compared to the relatively straightforward fabrication and study of the (now standard) arrays of nanomagnets, these pioneering studies were a true experimental tour de force. One shortcoming was, besides, the low amplitude of the magnetic moments induced in the superconducting rings, implying weak coupling between neighboring rings, even between nearest neighbors. This was a strong limitation to correlate the system, which precluded the observation of clear manifestations of frustrated magnetism. On the contrary, the magnetostatic interactions between nanomagnets are strong, making disordered and highly correlated states at reach.

The use of frustrated magnetic nanostructures was also promoted by a significant progress in the modeling and prediction of micromagnetic effects. Simulating magnetization textures and their dynamics, possibly under stimuli (temperature, magnetic field, electric current, etc.), is today facilitated by the widespread availability of high-performance numerical tools [38–39]. In particular, the way magnetization reverses in a nanostructure under the effect of an external applied field or temperature is usually well understood, whether the nanostructure is isolated or in a lattice.

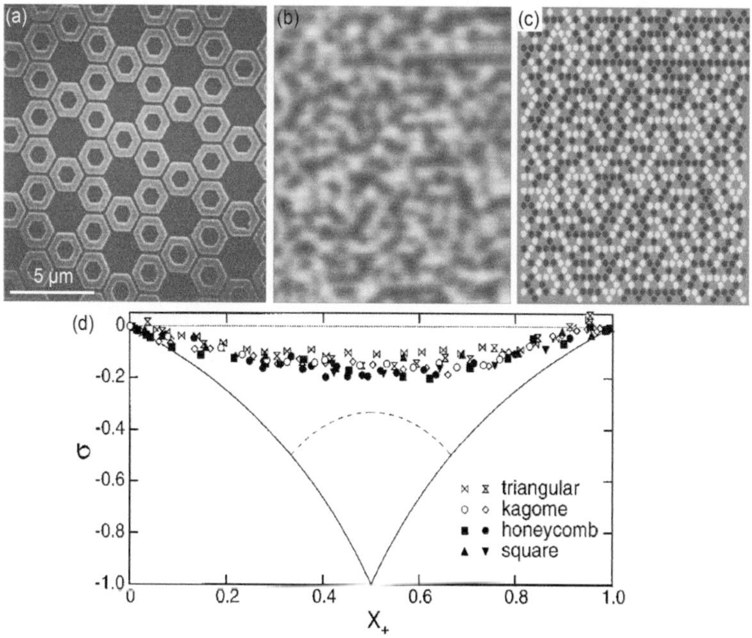

Figure 4.6 (a) Scanning electron micrograph of an array of aluminum rings arranged on a kagome lattice. The width and thickness of each ring are 400 nm and 240 nm, respectively. (b) Scanning Hall probe microscopy image, taken at 650 mK, and detecting the magnetic moments induced by the supercurrent flowing into the rings for an applied external magnetic field corresponding to about half a flux quanta in each ring. The black/white contrast corresponds to induced magnetic moments pointing along or opposite to the direction of the applied magnetic field. The microscope's resolution is sufficient to extract, after a proper deconvolution, the pseudo-spin state of each individual ring (c). (d) Bond order parameter as a function of the fraction of up pseudo-spins in the lattice. The bond order parameter, measuring the departure from a random orientation of pseudo-spin pairs, is zero when this orientation is random, and negative in presence of antiferromagnetic spin-spin correlations, as in the case here. Data sets measured for different geometries of the ring lattices are presented. The bond-order parameter is always negative, but does not reach the low values expected for lattices that would develop full frustration (dotted line), due to the relatively small magnitude of the magnetic moments. Reproduced with permission from Ref. 21.

This allows the community to grasp the physics of artificial systems in all their complexity, even beyond questions of frustration. An emblematic example today is the use of frustrated arrays of magnetic nanostructures for spin-wave computing [40, 41], which takes advantage of the magnetization dynamics in frustrated assemblies of interacting nanomagnets.

Finally, there is another reason why arrays of magnetic nanostructures are employed in frustrated magnetism. At the turn of the 2000s, a number of concepts were developed that are now widespread. The seminal term "spin ice," at the origin of "artificial spin ice," was coined in 1997 [9]. The key notions of Coulomb phase and magnetic monopoles in spin ice appeared between 2005 and 2010 [16, 19]. In this chronology, the conceptual leap that saw the appearance of artificial superconducting lattices as a tool for studying frustration phenomena at the end of the 1990s seems, back then, (too) ahead of its time. Much more than an experimental tour de force, these studies were already moving towards a better understanding of geometrically frustrated systems. Ten years later, in the mid-2000s, the concepts associated with spin ice physics had infused the community and become sufficiently mature so that frustrated arrays of magnetic nanostructures, which appeared around 2005, found a certain resonance.

In this respect, it is worth noting that the community quickly fabricated more complex nanomagnetic arrays than those emulating spin ice physics using the square or kagome geometries (see Section 4.4 and Ref. 42 to have an idea of the variety of geometries explored to date). In the vast majority of cases, the nanomagnets have an elongated shape that forces the anisotropy axis of the artificial Ising spins to be fixed (for energetic reasons, the Ising direction is defined by the long axis of the nanostructure). It is then the lattice geometry that governs the presence or absence of frustration and induces specific magnetic behaviors. Artificial XY spins, i.e., behaving like compass needles, have also been fabricated [43]. For this purpose, thin disks with a typical diameter of a hundred of nanometers can be used to maintain a single domain magnetization state without fixing the anisotropy axis (see Fig. 4.7). Conversely, the use of epitaxial materials allowed to obtain magnetic nanostructures in which the magneto-crystalline anisotropy favors several specific directions,

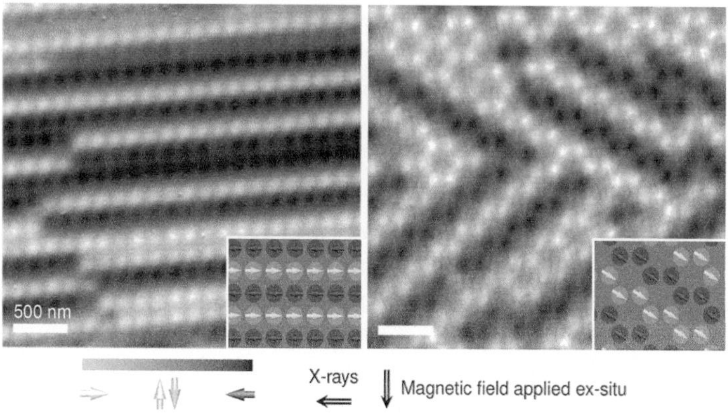

Figure 4.7 Photoemission microscopy images, obtained by subtraction of two images measured with opposite circular polarizations (x-ray magnetic circular dichroism) at a Ni absorption edge, of two lattices of disk-shaped permalloy nanomagnets. The observed contrast relates to the projection of the magnetization along the direction of the x-ray beam (indicated below the images); the lattices have been first saturated with an external magnetic field (direction indicated below the images). The inset cartoons represent the magnetic configurations in the two cases. Reproduced with permission from Ref. 43.

and thus the realization of magnetic elements with more than two states, paving the way of the so-called artificial Potts systems known in statistical physics [44]. Last but not least, even in the case of artificial Ising spins, the use of multilayers or magnetic alloys with perpendicular magnetic anisotropy [45, 46], as well as the use of nanostructures with in-plane magnetization [47–49], enabled the emulation of dipolar versions of the canonical short-range models introduced theoretically in the 1950s [11–13]. In particular, a set of predictions were revisited experimentally and revealed surprising, sometimes counter-intuitive, collective behaviors, simply due to the existence of long-range interactions in nanomagnetic arrays (some of these properties will be discussed in Section 4.4).

To summarize, the degree of maturity of fabrication, characterization and simulation techniques in nanomagnetism have most likely promoted a rapid dissemination of nanomagnetic arrays for the study of spin ice physics. The current level of understanding of the phenomena associated with spin liquid physics is certainly another reason explaining the community's interest in artificial systems. Many of the results obtained with nanomagnetic arrays are now being transposed to other artificial systems having very different sizes, ranging from the atomic or molecular scale [50, 51] to macroscopic dimensions [52, 53]. It is thus interesting to note that the craze for arrays of magnetic nanostructures has opened up a vast field of investigation in frustrated magnetism, extending well beyond magnetic systems.

4.4 Artificial Spin Liquids

We now turn to two emblematic cases for which direct imaging of spin ice properties in arrays of magnetic nanostructures was demonstrated: the artificial square ice and the artificial kagome ice. Both geometries are directly inspired from the three-dimensional structure of ice, illustrated in Fig. 4.1. Projected in the plane, a lattice of corner-sharing tetrahedra becomes a square lattice (actually a checkerboard lattice), and viewed along a $<111>$ crystallographic direction, the same lattice consists of alternating triangular and kagome planes. It is therefore no coincidence that square and kagome lattices were the first to be fabricated and still are the most studied artificial spin ices (reviews on the topic can be found in Refs. 54 and 55). As mentioned above, these two geometries were introduced independently and almost simultaneously by two research teams in the mid-2000s [23–25]. In both cases, while magnetic images revealed ice-rule-obeying configurations, detailed analysis showed quantitative and qualitative differences with the expected behavior, as we shall see.

4.4.1 The Artificial Square Ice

One of the main motivations for fabricating a square array of magnetostatically coupled nanomagnets was to mimic the properties of spin ice compounds in two dimensions and to visualize how Ising spins correlate in a frustrated system [25]. There is however an important subtlety when reducing dimensionality: in the 3D geometry, the interactions between the nearest-neighbor spins (located at the corners of the tetrahedra) are equivalent, but this is no longer the case in 2D. Each node of the 2D square lattice consists of four spins, with pairs of collinear spins and pairs of perpendicular spins (Fig. 4.5). These two kinds of pairs correspond to two different magnetostatic couplings, involving different distances and orientations. This has a strong implication: in 3D, several possible arrangements of the spins respect the ice rule (they are degenerate), but this degeneracy is lifted in 2D. As a result, no spin liquid state (whose realization motivated the fabrication of the 2D square ice) forms; rather, an ordered state is observed, with conventional properties (the ground state is antiferromagnetically ordered, see Fig. 4.5). It was only several years later that the flexibility offered by nanofabrication techniques allowed to circumvent this inherent difficulty of the square geometry and to access an ice physics in a 2D setting [56–58].

The square ice features non-intuitive but relatively-simple-to-grasp physics. Its phase diagram only comprises two phases, but no phase transition. At very high temperatures, the system is a true paramagnet: all spins in the square lattice fluctuate independently. The system can be viewed as a sum of N independent Ising spins, i.e., the magnetic equivalent of an ideal gas. Each spin having two possible states (Ising variable), there are 2^N accessible micro-states, all having the same probability ($1/2^N$) to be selected. The system is uncorrelated. As the temperature is reduced, the system correlates, but remains paramagnetic. The main difference with the high-temperature limit is that the accessible micro-states are only those satisfying the ice rule (the 2-in/2-out constraint). There is an extensively large number of such states: as a consequence, the system never orders, even at 0 K, and keeps fluctuating greatly down to the lowest accessible temperatures. Contrary to the high-temperature paramagnet, pairwise spin correlations are nonzero at low temperatures as the local ice-rule constraint must be satisfied everywhere and at any time.

For nonzero temperatures, the probability of violating the ice-rule constraint is nonzero. As we saw in Section 4.2.2, spin liquids that fulfill two conditions, a divergence-free constraint and a paramagnetic-like behavior, can be described by an emergent electrostatics. Ground-state configurations can then be seen as a charge-neutral vacuum whose excitations (associated with local violation of the ice rule) behave as magnetically charged quasiparticles, the so-called magnetic monopoles [19]. These quasiparticles diffuse freely in the lattice and interact via a Coulomb potential. They develop just above the ground state and can be described as topological, point-like charged defects. Beyond the elegance of describing excitations in terms of quasiparticles, this analogy with electrostatics is insightful in predicting the behavior of these quasiparticle excitations. These predictions can now be confronted to actual experimental images of the quasiparticles on top of their charge-neutral ground state, and certain theoretical predictions can be tested for the first time. To date, however, very few studies have been carried out in this direction.

Let us emphasize that these magnetic monopoles are created in pairs, and that this pair of topological charges is nothing else than the result of a spin reversal in the ground-state manifold (Fig. 4.8). The deconfinement of the two thus-generated (oppositely charged) monopoles corresponds to a succession of individual spin reversals. What is special about these (Coulomb) phases is that, once the monopole pair has been created, this succession of spin reversals takes place at constant energy. At finite temperature, the monopoles can then move throughout the lattice at no additional energy cost, allowing the system to fluctuate. Monopole dynamics is thus directly associated to the spin dynamics of the system. Lowering the temperature of a square ice system from the high-temperature regime, the density of magnetic monopoles decreases and ultimately vanishes when the temperature is sufficiently low. The system dynamics then freezes and the configuration that was reached remains in an arrested state. In other words, when the last monopole/antimonopole pair has annihilated, the lattice is trapped, within a well-defined magnetic micro-state (one among an extensively large number). As the monopole dynamics is the only way for the system to fluctuate in experiments, exploring the highly degenerate ground-state manifold is thus a lost battle at low enough temperature. However, theoreticians have identified other kinds of spin dynamics that allow to recover ergodicity and to navigate within the ground-state manifold of a spin ice system. This dynamics involves collective and cooperative flipping of spins, for instance, the

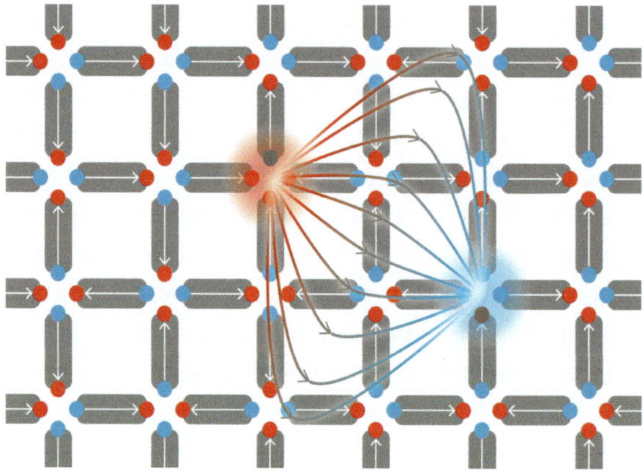

Figure 4.8 Cartoon showing a pair of magnetic monopoles (large red and blue clouds) hosted by a square ice Coulomb phase. Each spin (white arrow in elongated magnets delimited in grey) is described by two opposite magnetic charges (red and blue dots), which sum up to zero at vertices where the ice rule (i.e., the 2-in/2-out constraint) is satisfied. In this configuration, a spin has been flipped to produce of pair of opposite magnetic charges (3-in /1-out and 1-in/3-out, i.e., a pair of ± 2 charges). After this energetically costly spin flip, other spin flips can be performed at no extra energy cost, leading to a spatial separation of the two charges. In a Coulomb phase, these two charges interact via an effective Coulomb-like law at long distances, and may be seen as point-like magnetically charged quasiparticle excitations evolving on a charge neutral vacuum.

reversal of an entire spin loop rather than single spin flip events. Doing so, they demonstrated that new properties emerge associated with unconventional phase transitions (existence of topological sectors, infinite order phase transition, etc.). These loop excitations also involve spins at long distances and are responsible for the slow (algebraic) decay of the magnetic correlations specific to Coulomb spin liquids [59].

The loop dynamics has no experimental counterpart (at least in classical systems) and has been introduced theoretically to access the low-temperature thermodynamics of highly frustrated magnets. In some cases, however, the physics observed experimentally [60] turns out to be well-approximated by loop models. The reason is that at sufficiently low temperatures, when the density of magnetic monopoles vanishes but remains nonzero, the creation and annihilation process of a monopole/antimonopole pair cannot be distinguished from a partial loop dynamics. In other words, a (collective) loop flip can be effectively generated experimentally, provided that the thermal energy is enough to (i) overcome the energy barrier associated with the creation of a monopole/antimonopole pair, (ii) deconfine the two quasiparticles, and (iii) ultimately recombine them. Interestingly, if the spin ice dynamics is ensured by point-like (0D) magnetically charged quasiparticles at finite temperatures, spatially extended 1D loops carrying no net magnetization and no magnetic charge become more relevant at very low temperatures. The fundamental nature of the dynamics thus drastically changes as a spin ice material reaches its ground state, giving rise to exotic properties that can be probed experimentally, at least to some extent [60].

4.4.2 The Artificial Kagome Ice

Like the square ice, the phase diagram of the kagome ice comprises two phases, but no phase transition. At very high temperatures, the system is a true paramagnet: all spins in the lattice fluctuate independently and magnetic correlations average to zero. As the temperature decreases, the system correlates, but remains paramagnetic. Like in the square ice, from one regime to the next, an ice-rule constraint sets in that must be strictly obeyed in the ground-state manifold. The system retains an extensive entropy: it never orders and keeps fluctuating down to the lowest accessible temperatures. Contrary to the high-temperature paramagnet, pair-wise spin correlations are nonzero at low temperatures. However, the odd number of spins meeting at each vertex implies a nonzero local divergence. Hence, the vacuum is here not divergence-free and is effectively charged: the ground-state manifold of the kagome ice is not a Coulomb phase (the spin-spin correlations decay exponentially with the distance and excitations do not behave as magnetic monopoles).

Unlike in the square geometry, nearest-neighbor interactions are naturally equivalent in the kagome lattice as the corresponding spins make the same 120° angle with each other (Fig. 4.9). This is the reason why a disordered physics was first observed experimentally

13. Kanô, K. and Naya, S. (1953). Antiferromagnetism. The kagome Ising net, *Progr. Theor. Phys.*, **10**, pp. 158–172.
14. Baxter, R. J. (1982) *Exactly Solved Models in Statistical Mechanics* (Academic Press Limited, London).
15. Lieb, E. H. (2004) *Statistical Mechanics* (Springer-Verlag, Berlin).
16. Henley, C. L. (2010). The 'Coulomb phase' in frustrated systems. *Annu. Rev. Condens. Matter Phys.*, **1**, pp. 179–210.
17. Rehn, J. and Moessner, R. (2016). Maxwell electromagnetism as an emergent phenomenon in condensed matter, *Phil. Trans. R. Soc. A*, **374**, 20160093.
18. Ryzhkin, I. A. (2005). Magnetic relaxation in rare-earth oxide pyrochlores, *J. Exp. Theor. Phys.*, **101**, pp. 481–486.
19. Castelnovo, C., Moessner, R. and Sondhi, S. L. (2008). Magnetic monopoles in spin ice, *Nature* **451**, pp. 42–45.
20. Davidović, D., Kumar, S., Reich, D. H., Siegel, J., Field, S. B., Tiberio, R. C., Hey, R. and Ploog, K. (1996). Correlations and disorder in arrays of magnetically coupled superconducting rings, *Phys. Rev. Lett.*, **76**, pp. 815–818.
21. Davidović, D., Kumar, S., Reich, D. H., Siegel, J., Field, S. B., Tiberio, R. C., Hey, R. and Ploog, K. (1997). Magnetic correlations, geometrical frustration, and tunable disorder in arrays of superconducting rings, *Phys. Rev. B*, **55**, pp. 6518–6540.
22. Serret, E., Butaud, P. and Pannetier, B. (2002). Vortex correlations in a fully frustrated two-dimensional superconducting network, *EuroPhys. Lett.*, **59**, pp. 225–231.
23. Tanaka, M., Saitoh, E., Miyajima, H., Yamaoka, T. and Iye, Y. (2005). Domain structures and magnetic ice-order in NiFe nano-network with honeycomb structure, *J. Appl. Phys.*, **97**, 10J710.
24. Tanaka, M., Saitoh, E., Miyajima, H., Yamaoka, T. and Iye, Y. (2006). Magnetic interactions in a ferromagnetic honeycomb nanoscale network, *Phys. Rev. B*, **73**, 052411.
25. Wang, R., Nisoli, C., Freitas, R., Li, J., McConville, W., Cooley, B. J., Lund, M. S., Samarth, N., Leighton, C., Crespi, V. H. and Schiffer, P. (2006). Artificial 'spin ice' in a geometrically frustrated lattice of nanoscale ferromagnetic islands, *Nature*, **439**, pp. 303–306.
26. Gilbert, I., Chern, G.-W., Zhang, S., O'Brien, L., Fore, B., Nisoli, C. and Schiffer, P. (2014). Emergent ice rule and magnetic charge screening from vertex frustration in artificial spin ice, *Nat. Phys.*, **10**, pp. 670–675.

27. Morrison, M. J., Nelson, T. R. and Nisoli, C. (2013). Unhappy vertices in artificial spin ice: New degeneracies from vertex frustration. *New J. Phys.,* **15**, 045009.
28. Chern, G.-W., Morrison, M. J. and Nisoli, C. (2013). Degeneracy and criticality from emergent frustration in artificial spin ice, *Phys. Rev. Lett.*, **111**, 177201.
29. Saccone, M., Van den Berg, A., Harding, E., Singh, S., Giblin, S. R., Flicker, F. and Ladak, S. (2023). Exploring the phase diagram of 3D artificial spin-ice, *Commun. Phys.*, **6**, 217.
30. Donnelly, C., Guizar-Sicairos, M., Scagnoli, V., Gliga, S., Holler, M., Raabe, J. and Heyderman, L. J. (2017). Three-dimensional magnetization structures revealed with X-ray vector nanotomography, *Nature*, **547**, pp. 328–331.
31. Marrows, C.H. (2021), Experimental studies of artificial spin ice, in *Spin Ice*, Udagawa, M., Jaubert, L. (Eds) pp 455–478, Springer Series in Solid-State Sciences, vol 197. Springer.
32. Morgan, J. P., Stein, A., Langridge, S. and Marrows, C. H. (2011). Thermal ground-state ordering and elementary excitations in artificial magnetic square ice, *Nat. Phys.,* **7**, pp. 75–79.
33. Kapaklis, V., Arnalds, U. B., Harman-Clarke, A., Papaioannou, E. Th., Karimipour, M., Korelis, P., Taroni, A., Holdsworth, P. C. W., Bramwell, S. T. and B. Hjörvarsson (2012). Melting artificial spin ice, *New J. Phys.,* **14**, 035009.
34. Farhan, A., Derlet, P. M., Kleibert, A., Balan, A., Chopdekar, R. V., Wyss, M., Perron, J., Scholl, A., Nolting, F. and Heyderman, L. J. (2013). Direct observation of thermal relaxation in artificial spin ice, *Phys. Rev. Lett.*, **111**, 057204.
35. Kapaklis, V., Arnalds, U. B., Farhan, A., Chopdekar, R. V., Balan, A., Scholl, A., Heyderman, L. J. and Hjörvarsson, B. (2014). Thermal fluctuations in artificial spin ice, *Nat. Nanotech.*, **9**, pp. 514–519.
36. Montaigne, F., Lacour, D., Chioar, I.-A., Rougemaille, N., Louis, D., Mc Murtry, S., Riahi, H., Santos-Burgos, B., Menteş, T. O., Locatelli, A., Canals, B. and Hehn, M. (2015). Size distribution of magnetic charge domains in thermally activated but out-of-equilibrium artificial spin ice, *Sci. Rep.*, **4**, 5702.
37. Drisko, J., Daunheimer, S. and Cumings, J. (2015). $FePd_3$ as a material for studying thermally active artificial spin ice systems. *Phys. Rev. B*, 91, 224406.

38. Vansteenkiste, A. and Van de Wiele, B. (2011). MuMax: A new high-performance micromagnetic simulation tool, *J. Magn. Magn. Mater.*, **323**, pp. 2585–2591.

39. Jensen, J. H., Strømberg, A., Lykkebø, O. R., Penty, A., Leliaert, J., Själander, M., Folven, E. and Tufte, G. (2022). Flatspin: A large-scale artificial spin ice simulator, *Phys. Rev. B*, **106**, 064408.

40. Gartside, J. C., Stenning, K. D., Vanstone, A., Holder, H. H., Arroo, D. M., Dion, T., Caravelli, F., Kurebayashi, H. and Branford, W. R. (2022). Reconfigurable training and reservoir computing in an artificial spin-vortex ice via spin-wave fingerprinting, *Nat. Nanotech.*, **17**, pp. 460–469.

41. Chumak, A. V. et al. (2022). Advances in magnetics. Roadmap on spin-wave computing, *IEEE Trans. Magn.*, **58**, pp. 1–72.

42. Skjœrvø, S. H., Marrows, C. H., Stamps, R. L., Heyderman, L. J. (2020). Advances in artificial spin ice. *Nat. Rev. Phys.*, **2**, pp. 13–28.

43. Streubel, R., Kent, N., Dhuey, S., Scholl, A., Kevan, S. and Fischer, P. (2018). Spatial and temporal correlations of XY macro spins, *Nano Lett.*, **18**, pp. 7428–7434.

44. Louis, D., Lacour, D., Hehn, M., Lomakin, V., Hauet, T. and Montaigne, F. (2018). A tunable magnetic metamaterial based on the dipolar four-state Potts model, *Nat. Mater.*, **17**, pp. 1076–1080.

45. Zhang, S., Li, J., Gilbert, I., Bartell, J., Erickson, M. J., Pan, Y., Lammert, P. E., Nisoli, C., Kohli, K. K., Misra, R., Crespi, V. H., Samarth, N., Leighton, C. and Schiffer, P. (2012). Perpendicular magnetization and generic realization of the Ising model in artificial spin ice, *Phys. Rev. Lett.*, **109**, 087201.

46. Chioar, I.-A., Rougemaille, N., Grimm, A., Fruchart, O., Wagner, E., Hehn, M., Lacour, D., Montaigne, F. and Canals, B. (2014). Nonuniversality of artificial frustrated spin systems, *Phys. Rev. B*, **90**, 064411.

47. Qi, Y., Brintlinger, T. and Cumings, J. (2008). Direct observation of the ice rule in an artificial kagome spin ice, *Phys. Rev. B*, **77**, 094418.

48. Rougemaille, N., Montaigne, F., Canals, B., Duluard, A., Lacour, D., Hehn, M., Belkhou, R., Fruchart, O., El Moussaoui, S., Bendounan, A. and Maccherozzi, F. (2011). Artificial kagome arrays of nanomagnets: A frozen dipolar spin ice, *Phys. Rev. Lett.*, **106**, 057209.

49. Brunn, O. Perrin, Y., Canals, B. Rougemaille, N. (2021). Signatures of farther neighbor couplings in artificial square ice, *Phys. Rev. B*, **103**, 094405.

50. Azizi, A., Dogan, M., Cain, J. D., Eskandari, R., Yu, X., Glazer, E. C., Cohen, M. L. and Zettl, A. (2020). Frustration and atomic ordering in a monolayer semiconductor alloy, *Phys. Rev. Lett.*, **124**, 096101.

51. Alfonso-Moro, M., Guisset, V., David, P., Canals, B., Coraux, J. and Rougemaille, N. (2023). Geometrical frustration, correlated disorder, and emerging order in a corrugated C_{60} monolayer, *Phys. Rev. Lett.*, **131**, 186201.

52. Mellado, P., Concha, A. and Mahadevan, L. (2012). Macroscopic magnetic frustration, *Phys. Rev. Lett.*, **109**, 257203.

53. Teixeira, H. A., Bernardo, M. F., Nascimento, F. S., Saccone, M. D., Caravelli, F., Nisoli, C. and de Araujo, C. I. L. (2024). Macroscopic magnetic monopoles in a 3D-printed mechano-magnet, *J. Magn. Magn. Mater.*, **596**, 171929.

54. Nisoli, C., Moessner, R. and Schiffer, P. (2013). Colloquium: Artificial spin ice: Designing and imaging magnetic frustration, *Rev. Mod. Phys.*, **85**, pp. 1473–1490.

55. Rougemaille, N. and Canals, B. (2019). Cooperative magnetic phenomena in artificial spin systems: spin liquids, Coulomb phase and fragmentation of magnetism – a colloquium, *Eur. Phys. J. B*, **92**, 62.

56. Perrin, Y., Canals, B. and Rougemaille, N. (2016). Extensive degeneracy, Coulomb phase and magnetic monopoles in artificial square ice, *Nature*, **540**, pp. 410–413.

57. Östman, E., Stopfel, H., Chioar, I.-A., Arnalds, U. B., Stein, A., Kapaklis, V. and Hjörvarsson, B. (2018). Interaction modifiers in artificial spin ices. *Nat. Phys.*, **14**, pp. 375–379.

58. Farhan, A., Saccone, M., Petersen, C. F., Dhuey, S., Chopdekar, R. V., Huang, Y.-L., Kent, N., Chen, Z., Alava, M. J., Lippert, T., Scholl, A. and van Dijken, S. (2019). Emergent magnetic monopole dynamics in macroscopically degenerate artificial spin ice, *Sci. Adv.*, **5**, eaav6380.

59. Balents, L. (2010). Spin liquids in frustrated magnets, *Nature*, **464**, pp. 199–208.

60. Schánilec, V., Brunn, O., Horáček, M., Krátký, S., Meluzín, P., Šikola, T., Canals, B. and Rougemaille, N. (2022). Approaching the topological low-energy physics of the F model in a two- dimensional magnetic lattice, *Phys. Rev. Lett.*, **129**, 027202.

61. Wills, A. S., Ballou, R. and Lacroix, C. (2002). Model of localized highly frustrated ferromagnetism: The kagomé spin ice, *Phys. Rev. B*, **66**, 144407.

62. Möller, G. and Moessner, R. (2009). Magnetic multipole analysis of kagome and artificial spin-ice dipolar arrays, *Phys. Rev. B*, **80**, 140409.

63. Chern, G.-W., Mellado, P. and Tchernyshyov, O. (2011). Two-stage ordering of spins in dipolar spin ice on the kagome lattice, *Phys. Rev. Lett.*, **106**, 207202.

64. Brooks-Bartlett, M. E., Banks, S. T., Jaubert, L. D. C., Harman-Clarke, A. and Holdsworth, P. C. W. (2014). Magnetic-Moment Fragmentation and Monopole Crystallization, *Phys. Rev. X*, **4**, 011007.
65. Jaubert, L. D. C. (2015). Monopole holes in a partially ordered spin liquid, *SPIN*, **05**, 1540005.
66. Zhang, S., Gilbert, I., Nisoli, C., Chern, G.-W., Erickson, M. J., O'Brien, L., Leighton, C., Lammert, P. E., Crespi, V. H. and Schiffer, P. (2013). Crystallites of magnetic charges in artificial spin ice, *Nature*, **500**, pp. 553–557.
67. Canals, B., Chioar, I.-A., Nguyen, V.-D., Hehn, M., Lacour, D., Montaigne, F., Locatelli, A., Menteş, T. O., Santos-Burgos, B. and Rougemaille, N. (2016). Fragmentation of magnetism in artificial kagome dipolar spin ice, *Nat. Commun.*, **7**, 11446.
68. Chioar, I.-A., Canals, B., Lacour, D., Hehn, M., Santos-Burgos, B., Menteş, T. O., Locatelli, A., Montaigne, F. and Rougemaille, N. (2014). Kinetic pathways to the magnetic charge crystal in artificial dipolar spin ice, *Phys. Rev. B*, **90**, 220407(R).
69. Schánilec, V., Canals, B., Uhlír, V., Flajšman, L., Sadílek, J., Šikola, T. and Rougemaille, N. (2020). Bypassing dynamical freezing in artificial kagome ice, *Phys. Rev. Lett.*, **125**, 057203.
70. Farhan, A., Derlet, P. M., Anghinolfi, L., Kleibert, A. and Heyderman, L. J. (2017). Magnetic charge and moment dynamics in artificial kagomé spin ice, *Phys. Rev. B*, **96**, 064409.
71. Brunn, O. (2024), *Field demagnetised artificial square ice magnets: long-range interactions, origin of stochasticity and effective thermodynamics*, PhD thesis (Université Grenoble Alpes and Brno University of Technology).
72. Narten, A. H. (1972). Liquid water: Atom pair correlation functions from neutron and x-ray diffraction, *J. Chem. Phys.*, **56**, pp. 5681–5687.
73. Harris, M. J., Zinkin, M. P., Tun, Z., Wanklyn, B. M. and Swainson, I. P. (1994). Magnetic structure of the spin-liquid state in a frustrated pyrochlore. *Phys. Rev. Lett.*, **73**, pp. 189–192.

Chapter 5

Magnetization Dynamics of Barium Hexaferrite Hollow Spherical Nanoparticles

Maria José Vazquez Bernardez,[a] Daniel Stoeffler,[a] Christophe Lefevre,[a] and Nicolas Vukadinovic[b]

[a] *University of Strasbourg, CNRS, Institute of Materials Physics and Chemistry UMR 7504, 67034 Strasbourg France*
[b] *Dassault Aviation, 92552 St-Cloud, France*
nicolas.vukadinovic@dassault-aviation.com

The hollow magnetic particles are versatile nanostructures promising for future spin-based technologies. The case of submicrometer-sized hollow spherical particles with a large uniaxial magnetic anisotropy is considered in this Chapter. By means of micromagnetic simulations, the equilibrium magnetization configuration and the associated linear response in the microwave range are investigated. Depending on the shell thickness and the external particle diameter, multiple resonant excitations are revealed and identified as spatially nonuniform magnetic modes. The changes induced by the presence of a medial planar domain wall for larger hollow spheres are analyzed with a focus on the magnetic excitations localized within the domain walls. The main experimental routes for fabricating

2D and 3D Nanostructures: Structure, Properties, and Applications
Edited by Jean-Claude Serge Levy

ISBN 978-981-5129-97-7 (Hardcover), 978-1-003-74871-7 (eBook)
www.jennystanford.com

DOI: 10.1201/97810037487-5

hollow magnetic spheres are then presented and the emergent applications are highlighted.

5.1 Introduction

Nowadays, the hollow nano-objects constitute a vast and rich class of nanomaterials. A well-defined morphology, a large specific surface and a low density are valuable features for a large variety of applications: catalysis, gas and pollutant sensors, water treatment, energy storage, nanophotonics, and drug delivery (Wang et al., 2016). Among these hollow nano-objects, the structured magnetic particles are the focus of ever-growing interest. This structuration at small scale offers additional degrees of freedom to induce new static and dynamic magnetic properties. These ones could be exploited in various fields including ultra-high density data storage, spintronics, microwave devices and biomedicine. As an illustration of this concept, this Chapter addresses the case of submicrometer-sized hollow spherical particles with a large uniaxial magnetic anisotropy. This type of core–shell particles is defined by two geometrical parameters, namely, the external diameter and the thickness of the shell. In such systems, the high-frequency (microwave range) linear response characterized by the dynamic susceptibility spectrum exhibits multiple magnetic resonance lines whose features in terms of spectral positions and amplitudes are controlled by the abovementioned parameters. An appropriate choice of the material composition of the shell and geometrical parameters allows shaping the microwave absorption spectra of the nanoparticles, which is attractive for microwave applications.

From the theoretical viewpoint, the microwave magnetization dynamics of solid spherical particles has been first investigated using various approaches and assumptions. Using a macroscopic approach, Aharoni (Aharoni, 1991) demonstrated that for small particle sizes where the exchange energy exceeds the magnetostatic energy (exchange regime), new resonance lines with size dependence emerge, termed exchange modes. This model with appropriate boundary conditions (Aharoni, 1997) allowed interpreting the experimental multipeak dynamic susceptibility spectra measured on assemblies of quasi monodisperse ferromagnetic nanoparticles (Viau et al., 1995; Toneguzzo et al., 1997; Mercier et al., 2000). The case of predominant magnetostatic interaction for large particle sizes (magnetostatic or dipolar regime) has been treated by Walker for elliptical geometries (Walker, 1957) and then by Fletcher and Bell for spherical objects (Fletcher and Bell, 1959). Arias et al. (Arias et al., 2005) dealt with the linear frequency response of ferromagnetic spheres by taking into account both the exchange and dipolar interactions (dipole-exchange regime). These various works assume a spatially uniform static magnetization configuration within the sphere. On the other hand, Nguyen and Cottam (Nguyen and Cottam, 2008) analyzed the spin-wave spectra in spherical nanostructures (solid, truncated and hollow spheres) in the dipole-exchange regime based on a microscopic approach including nonuniform static and dynamic magnetization configurations. They pointed out the existence of discrete spin-wave modes with possible mode mixing between them.

The spherical nanoshell counterpart was theoretically and numerically studied as well. Aharoni (Aharoni, 2002) calculated the resonance frequencies of the exchange modes within the nanoshell. This model was recently extended to treat multilayered nanoparticles with interlayer exchange couplings (McKeever and Aziz, 2022). Sladanhia-Bautista et al. (Sladanhia-Bautista, 2022) computed the resonance frequencies of the existing magnetostatic modes by solving the Walker's equations in the case of hollow spheres. Moreover, micromagnetic simulations of the dynamic susceptibility spectra including both the exchange and magnetostatic interactions have been conducted for ferromagnetic nanoshells with nonuniform magnetization configurations (McKeever et al., 2019).

In this Chapter, we adopt the micromagnetic simulations approach for modelling the microwave magnetization dynamics of highly anisotropic hollow spherical particles. The used magnetic parameters are representative of M-type hexagonal ferrites (Pullar, 2012), materials of growing interest for microwave applications. After a brief recall on the micromagnetism theory, section 5.2 reports some numerical results regarding the zero-field susceptibility spectra of such structured particles and the identification of the revealed size-dependent resonance modes.

From the experimental side, the fabrication of magnetic hollow spherical nanoparticles with well-controlled geometrical parameters using robust methods suitable for large-scale production remains a critical point (Soares et al., 2019). Section 5.3 focuses on recent progress in this field and highlights the most promising routes to reach this goal. Special attention is paid to magnetic oxides and among them the M-type hexagonal ferrites.

Finally, section 5.4 displays some relevant applications of the magnetic spherical hollow nanoparticles and the related future challenges. In particular, the major impact of such nanostructured particles expected in the field of microwave devices is showcased.

5.2 Micromagnetic Simulations of the Static and Dynamic Properties of Hollow Spheres with a Large Uniaxial Magnetic Anisotropy

5.2.1 Micromagnetic Framework (MMF)

In the micromagnetic approach, the magnetic configuration of a ferromagnetic system is described by the magnetization vector field $\overrightarrow{M}(\vec{r})$ having only rotational degrees of freedom ($\vec{r}$ being the position vector). For a homogeneous material of saturation magnetization M_S, $\left|\overrightarrow{M}(\vec{r})\right| = M_S$ so that the reduced $\overrightarrow{m}(\vec{r}) = \overrightarrow{M}(\vec{r})/M_S$ magnetization vector field has a modulus equal to one. The minimization of the total energy expression $E_{\text{tot}}[\overrightarrow{M}(\vec{r})]$, introduced by Brown (Brown, 1963) as a functional of $\overrightarrow{M}(\vec{r})$, allows to determine the equilibrium states. From a variational point of view, the total energy change dE_{tot} due to a small magnetization change $d\vec{m}(\vec{r})$ can be expressed as

$$dE_{\text{tot}} = -M_S \iiint d\overrightarrow{m}(\vec{r}).\overrightarrow{B}_{\text{eff}}(\vec{r})d^3r \qquad (5.1)$$

introducing the effective field $\overrightarrow{B}_{\text{eff}}(\vec{r})$ as the local magnetic field acting on the magnetization. It turns that $\overrightarrow{B}_{\text{eff}}(\vec{r})$ can be expressed as the functional derivative of the magnetic energy density dE_{tot}/dV with respect to the magnetization orientation (V being the volume of the system over which the integration is performed).

$$\vec{B}_{\text{eff}}(\vec{r}) = -\frac{1}{M_S}\frac{d}{d\vec{m}}\frac{dE_{\text{tot}}}{dV}. \quad (5.2)$$

The temporal evolution of the magnetization is obtained solving the Landau–Lifshitz (Landau and Lifshitz, 1935) equation with the corresponding effective magnetic field.

During this work, we use the common total energy expression taking into account the ferromagnetic exchange, the magnetic anisotropy, the applied field and the dipolar field terms.

Ferromagnetic exchange energy term which favors parallel alignment between magnetic moments. In the MMF, its phenomenological expression is

$$\begin{aligned} E_{\text{exch}} &= \iiint A_{\text{ex}}|\vec{\nabla}\vec{m}|^2 d^3r \\ &= \iiint A_{\text{ex}}((\vec{\nabla}m_x(\vec{r}))^2 + (\vec{\nabla}m_y(\vec{r}))^2 + (\vec{\nabla}m_z(\vec{r}))^2)d^3r \end{aligned} \quad (5.3)$$

where A_{ex} is the exchange-stiffness constant (in J/m) which is positive and can be experimentally measured. This term tends to favor solutions where the magnetization varies slowly and is minimized when the magnetization is uniform.

Magnetocrystalline anisotropy energy term which takes into account the occurrence inside the material of so-called easy axes corresponding to preferred directions for the magnetization. Its most general form writes

$$E_{\text{anis}} = \iiint KG(\vec{m}(\vec{r}))d^3r \quad (5.4)$$

where K is the anisotropy constant (in J/m^3) and $G(\vec{m})$ is a dimensionless function which vanishes along the easy axes. For an uniaxial anisotropy, there is only one easy axis of unit vector $\hat{u}$ and G is given by $G(\vec{m}) = 1 - (\vec{m}.\vec{u})^2$ with $K > 0$.

Zeeman energy term which expresses the fact that, when a magnetic field $\vec{B}_A$ is applied on the sample, its magnetization tends to align in the direction of the applied field. Its expression is

$$E_Z = -\iiint \vec{M}(\vec{r}).\vec{B}_A d^3r = -M_S\iiint \vec{m}(\vec{r}).\vec{B}_A d^3r. \quad (5.5)$$

It is the lowest when the magnetization and the applied field are aligned.

None of the energy terms listed above can explain the existence of magnetic domains which is an experimental fact and requires to take into account the dipolar field resulting from the magnetization.

Magnetostatic energy term: the magnetization distribution inside the material introduces an additional dipolar field $\overrightarrow{B}_D(\vec{r})$—the so-called demagnetizing field—which adds to the external applied field. This field is calculated by introducing pseudo magnetic charges densities $\rho_m(\vec{r}) = -\overrightarrow{\nabla}.\overrightarrow{M}(\vec{r})$ in the volume and $\sigma_m(\vec{r}) = \overrightarrow{M}(\vec{r}).\hat{n}(\vec{r})$ on the surface of the sample ($\hat{n}(\vec{r})$ being the unit vector normal to the surface at $\vec{r}$). Maxwell's equations allow us to write the demagnetizing field as

$$\overrightarrow{B}_D(\vec{r}) = \frac{\mu_0}{4\pi}\iiint \rho_m(\overrightarrow{r}')\frac{(\overrightarrow{r} - \overrightarrow{r}')}{|\overrightarrow{r} - \overrightarrow{r}'|^3}d^3r' + \frac{\mu_0}{4\pi}\iint \sigma_m(\overrightarrow{r}')\frac{(\overrightarrow{r} - \overrightarrow{r}')}{|\overrightarrow{r} - \overrightarrow{r}'|^3}d^2r'. \quad (5.6)$$

The magnetostatic energy expression is consequently

$$E_D = -\frac{1}{2}\iiint \overrightarrow{M}(\vec{r}).\overrightarrow{B}_D(\vec{r})d^3r = -\frac{1}{2}M_S\iiint \overrightarrow{m}(\vec{r}).\overrightarrow{B}_D(\vec{r})d^3r. \quad (5.7)$$

The static equilibrium states $\overrightarrow{m}_{\mathrm{equ}}(\vec{r})$ minimize the total energy obtained by summing the four contributions previously introduced

$$E_{\mathrm{tot}}[\overrightarrow{m}(\vec{r})] = \iiint \left[A_{\mathrm{ex}}|\nabla\overrightarrow{m}(\vec{r})|^2 + KG(\overrightarrow{m}(\vec{r}))\right] d^3r - M_S\iiint \overrightarrow{m}(\vec{r}).\left[\overrightarrow{B}_A + \frac{1}{2}\overrightarrow{B}_D[\overrightarrow{m}(\vec{r})]\right] d^3r. \quad (5.8)$$

This minimization is done from a functional point of view where the infinitesimal total energy variation $\delta E_{\mathrm{tot}}[\overrightarrow{m}(\vec{r})] = E_{\mathrm{tot}}[\overrightarrow{m}(\vec{r}) + \delta\overrightarrow{m}(\vec{r})] - E_{\mathrm{tot}}[\overrightarrow{m}(\vec{r})]$ has to remain equal to zero for any infinitesimal variation $\delta\overrightarrow{m}(\vec{r})$ of the magnetization vector field. This leads to the two Brown equations

$$\overrightarrow{m}(\vec{r}) \times \overrightarrow{B}_{\mathrm{eff}}(\vec{r}) = \vec{0},\ \forall\vec{r} \in V \quad (5.9)$$

and

$$\frac{\partial \vec{m}}{\partial n}(\vec{r}) = \vec{0}, \ \forall \vec{r} \in S \tag{5.10}$$

corresponding to the zero torque condition in the volume V and to the magnetization stationary with respect to normal variations condition at the surface S. Using the expression for $E_{\text{tot}}[\vec{m}(\vec{r})]$, the effective field $\vec{B}_{\text{eff}}(\vec{r})$ can be explicitly obtained

$$\vec{B}_{\text{eff}}(\vec{r}) = \frac{2A_{\text{ex}}}{M_S}\Delta\vec{m}(\vec{r}) - \frac{K}{M_S}\vec{\nabla}_{\vec{m}}G(\vec{m}(\vec{r})) + \vec{B}_A + \vec{B}_D(\vec{r}). \tag{5.11}$$

This expression containing the demagnetizing field $\vec{B}_D(\vec{r})$ shows that the calculation of the effective field at each point $\vec{r}$ requires to take into account the magnetization in the whole sample.

To determine the temporal evolution of the magnetization between equilibrium states or when a time varying excitation is applied, the Landau–Lifshitz (LL) equation is used.

$$\begin{aligned}(1+\alpha^2)\frac{\partial}{\partial t}\vec{m}(\vec{r}) = &-\gamma\vec{m}(\vec{r}) \times \vec{B}_{\text{eff}}(\vec{r}) \\ &-\alpha\gamma\vec{m}(\vec{r}) \times (\vec{m}(\vec{r}) \times \vec{B}_{\text{eff}}(\vec{r}))\end{aligned} \tag{5.12}$$

where α is the dimensionless damping constant which represents the energy dissipation and γ is the gyromagnetic ratio (in rad/(T.s)) which determines the time scale of the precession and is usually obtained from Ferromagnetic Resonance (FMR) measurements. In this equation, $-\gamma(\vec{m} \times \vec{B}_{\text{eff}})$ describes the precession of the magnetization around the effective field $\vec{B}_{\text{eff}}$ and the term $-\alpha\gamma(\vec{m} \times (\vec{m} \times \vec{B}_{\text{eff}}))$ describes the relaxation motion of the magnetization towards $\vec{B}_{\text{eff}}$. The interplay between both terms leads to damped oscillations of the magnetization around its local effective field. The damping constant α is a critical parameter for the determination of the magnetization temporal evolution: it can be obtained from the measured FMR linewidth.

The LL equation can also be used to determine the equilibrium states $\vec{m}_{\text{equ}}(\vec{r})$ as an alternative of minimizing the total energy. Indeed, when the zero torque condition $\vec{m}(\vec{r}) \times \vec{B}_{\text{eff}}(\vec{r}) = \vec{0}$ is fulfilled, the right hand side term is zero and $\partial\vec{m}(\vec{r})/\partial t = 0$.

magnetization of the system in its equilibrium state $\vec{m}_{\text{equ}}(\vec{r})$. This perturbation was applied during a sufficiently long time so that all transient configurations were damped. This resulted in the magnetization evolving towards a new equilibrium state (referred to as the excited state $\widetilde{\vec{m}}$). Upon cancellation of the excitation, the magnetization underwent damped precessional motion around its equilibrium position until it returned to the initial equilibrium state $\vec{m}_{\text{equ}}(\vec{r})$. The temporal evolution of the dynamic magnetization $\delta\vec{m}(t) = \langle\vec{m}(\vec{r}, t)\rangle - \langle\vec{m}_{\text{equ}}\rangle$ was calculated by solving the LL equation. The dimensionless magnetic susceptibility $\chi_{xx}(f)$ expresses the relationship between the excitation $\tilde{b}_x$ and the magnetic response $\langle m_x(\vec{r}, t)\rangle$ of the system in the linear Fourier space for the x coordinate.

$$\chi_{xx}(f) = \frac{\mu_0 M_S}{\tilde{b}_x}\left(\tilde{m}_x - 2i\pi f\int_0^{+\infty}\delta\bar{m}_x(t)e^{-2i\pi ft}\,dt\right) \quad (5.13)$$

The imaginary component of the magnetic susceptibility $\chi_{xx}(f)$ was used to determine the system's precessional resonance frequencies:

$$-\text{Im}(\chi_{xx}(f)) = \mu_0 M_S\left(\frac{2\pi f}{\tilde{b}_x}\right)\int_0^{+\infty}\delta\bar{m}_x(t)\cos(2\pi ft)\,dt \quad (5.14)$$

where $\widetilde{b}_x$ is the amplitude of the uniform excitation in the x direction, and $\delta\bar{m}_x(t)$ is the average dynamic magnetization. In this Chapter, the diagonal component χ_{xx} of the susceptibility tensor was only considered for the case of the single-domain hollow spheres because the average static magnetization is aligned with the uniaxial anisotropy K_u along the z direction and the isotropic response remains in the plane perpendicular to the z-axis. For the case of bidomain hollow spheres, both the χ_{xx} and χ_{yy} components were computed for in-plane pumping fields applied perpendicular (x-axis) or parallel (y-axis) to the domain-wall plane, respectively.

5.2.3 Static Magnetic Properties

The equilibrium magnetization configurations in thin spherical shells of soft ferromagnets have been investigated in detail (Sloika et al., 2017; McKeever et al., 2019). As a result, two micromagnetic ground states were evidenced: the onion state and the double-vortex state. In addition, a phase diagram delimiting these ground

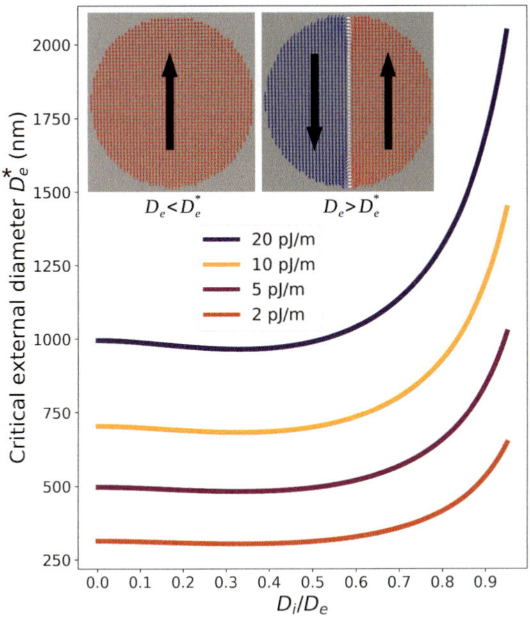

Figure 5.2 Critical value of the external diameter D_e, termed D_e^*, as a function of the ratio D_i/D_e for different values of the exchange parameter A_{ex} following the model developed in (Goll et al., 2004). The magnetization vector field images of the single-domain and bidomain states shown for $D_i = 0$ have been obtained using the Muview2 tool (Rowlands, 2018).

states was built as functions of the geometrical parameters of the nanoshell (Sloika et al., 2017). The situation is quite different for larger hollow particles with a high uniaxial anisotropy. With the parameters reported in Sec. 5.2.2, static micromagnetic simulations indicated that all magnetic nanoshells, with external diameters ranging from 50 nm to 500 nm and thicknesses varying from $D_i/D_e = 0$ to $D_i/D_e = 0.95$ (that we will consider in the main part of this section), were in a single-domain state at equilibrium in accordance with Goll D. et al.'s model (see Fig. 5.2) (Goll et al., 2004). Following the strong uniaxial magnetocrystalline anisotropy of the system, the uniform magnetization is aligned along the z-axis.

For even larger diameters, multidomain solutions occur where domain walls (DW) separating magnetic domains of opposite magnetizations lower the total energy with respect to the uniform solution. Indeed, when increasing the diameter, the dipolar energy gain (approximately proportional to the magnetized volume of the sphere) resulting from the magnetization structuration in domains of opposite magnetizations overcomes the energy loss of the formation of the DW (approximately proportional to the surface of the DW). The critical diameter (D_e^*) where the single-domain to bidomain transition occurs is displayed in Fig. 5.2 as a function of D_i/D_e for several values of the exchange parameter A_{ex} according to the model developed by Goll D. et al. (Goll et al., 2004). As expected, for a given D_i/D_e, the critical diameter is reduced for smaller A_{ex} as a consequence of the dependence of the DW energy as $\sqrt{A_{ex}}$. More important, this model shows that increasing D_i/D_e favors uniform magnetized solutions and that the critical diameter increases rapidly for $D_i/D_e > 0.5$.

For our system with $A_{ex} = 20$ pJ/m, metastable bidomain solutions are obtained for D_e values larger than D_e^* as illustrated in Fig. 5.3 for D_e =1000 nm and $D_i/D_e = 0.8$. In this bidomain magnetic configuration, the planar DW - having a thickness of about 50 nm - separates the magnetized sphere in two half-spheres of opposite nearly uniform magnetizations aligned along the uniaxial

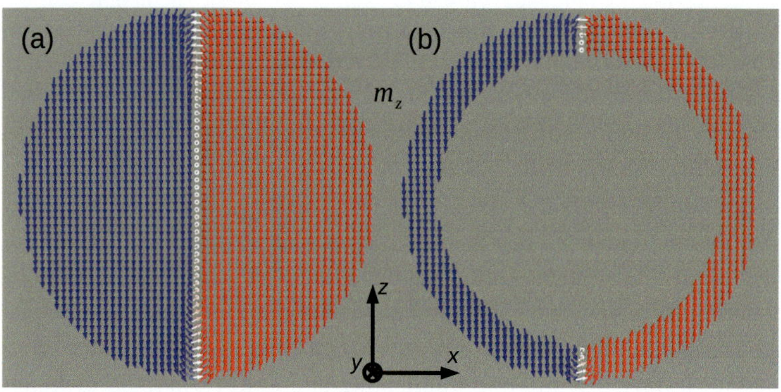

Figure 5.3 Magnetization vector field in the (xOz) plane perpendicular to the domain wall for a bidomain solid sphere with $D_e = 1000$ nm (a) and a bidomain hollow sphere with $D_e = 1000$ nm and $D_i = 800$ nm (b). The usual blue/white/red color scale corresponds to the variations from -1 to $+1$ of the m_z component. The magnetization vector field images have been obtained using the Muview2 tool (Rowlands, 2018).

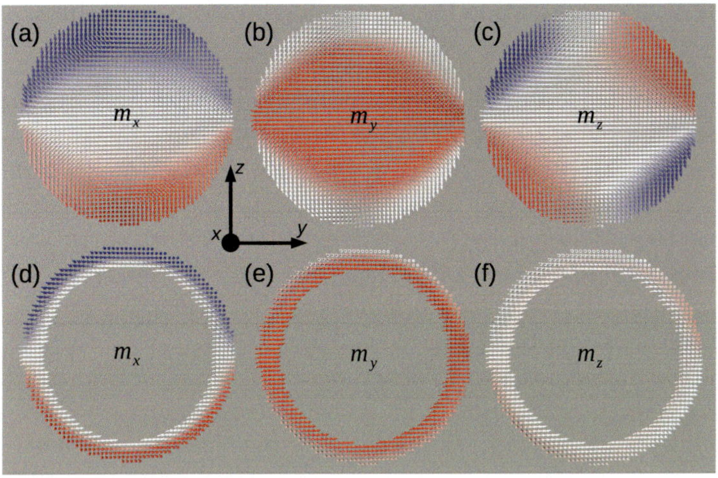

Figure 5.4 Magnetization vector field in the (yOz) domain-wall plane for a bidomain solid sphere with $D_e = 1000$ nm (a–c) and a bidomain hollow sphere with $D_e = 1000$ nm and $D_i = 800$ nm (d–f). The usual blue/white/red color scale corresponds to the variations from -1 to $+1$ of the m_α $(\alpha = x, y, z)$ component as indicated for each map. The magnetization vector field images have been obtained using the Muview2 tool (Rowlands, 2018).

anisotropy z-axis. Building an initial solution with a DW in the (yOz) allows to get a final solution with the DW in the same plane. The comparison of the equilibrium magnetization vector fields obtained for solid (Figs. 5.3(a) and 5.4(a–c)) and hollow spheres with $D_i/D_e = 0.8$ (Figs. 5.3(b) and 5.4(d–f)) shows that both states are structured similarly but more rapid angular variations of the magnetization vector are obtained for hollow spheres due to the thickness confinement in the shell. These more rapid angular variations can be clearly seen by comparing Figs. 5.4(a),(b) and Figs. 5.4(d),(e): for the solid sphere, the vector $\vec{m}$ varies from (0, 1, 0) at the center to (±1, 0, 0) at the surface over a distance of half a diameter (500 nm) whereas, for the hollow sphere, a similar variation occurs over the thickness of the shell (100 nm). This can partially explain the increase of the critical diameter D_e^* with the reduction of the thickness of the shell because, usually, strongly confined domain walls have a higher energy density.

5.2.4 Magnetization Dynamics: Single-Domain Hollow Spheres

The computation of microwave magnetization dynamics revealed that the magnetic hollow structure within the sphere gives rise to multiple resonance modes. The zero-field dynamic susceptibility spectrum of a shell with an external diameter of 500 nm and a shell thickness of 50 nm, $D_i/D_e = 0.8$ exhibits five distinct resonance peaks, with four of them displaying amplitudes comparable to the ferromagnetic resonance mode found in a solid spherical particle of identical composition (refer to Fig. 5.5).

The eigenmode maps illustrating the x component of the dynamic magnetization δm_x at the instant corresponding to $t = n.T + T/4$ with n a large integer and $T = 1/f$ the temporal period of the resonance eigenfrequency f are shown in Fig. 5.6 ($\langle\delta m_x\rangle$ reaches a maximum at one-fourth of the mode's period). All resonance modes, including the lowest frequency mode, are spatially non uniform, whereas the system is in a single-domain state at static equilibrium.

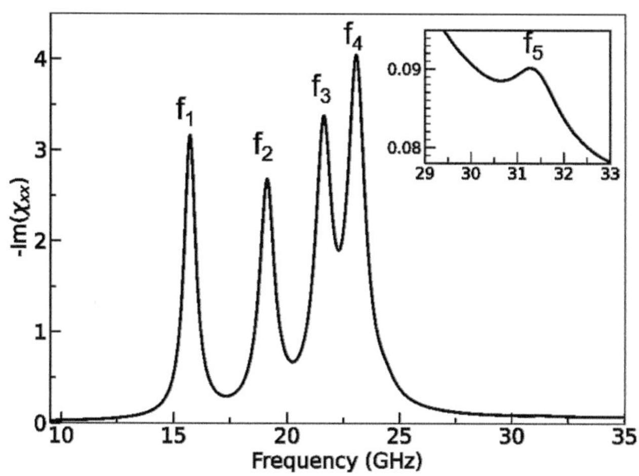

Figure 5.5 Zero-field dynamic susceptibility spectrum (imaginary part of χ_{xx}) for a shell of doped barium hexaferrite with $D_e = 500$ nm and $D_i = 400$ nm ($D_i/D_e = 0.8$). The resonance frequencies of the computed lines are noted f_1, f_2, f_3, f_4, and f_5, respectively. The inset is focused around f_5. Adapted with permission from Vazquez Bernardez et al. (2024), copyright 2024 ACS Publications.

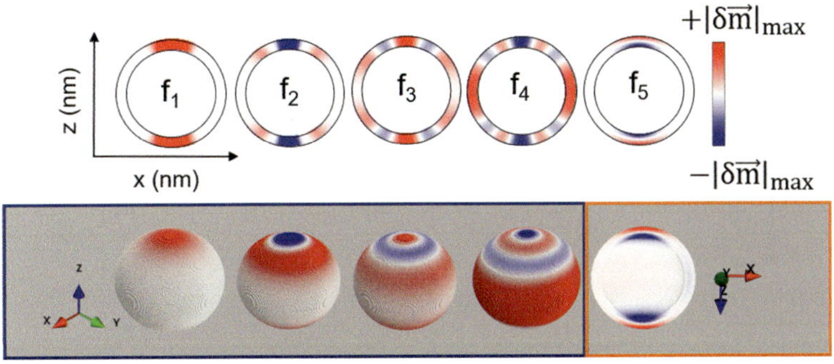

Figure 5.6 Zero-field eigenmode maps of the component δm_x for a shell of doped barium hexaferrite, $D_e = 500$ nm, $D_i = 400$ nm ($D_i/D_e = 0.8$) taken at the instant corresponding to $t = n.T + T/4$ with n a large integer and $T = 1/f$ the temporal period of a resonance. The resonance frequencies for the first four modes are $f_1 = 15.70$ GHz, $f_2 = 19.11$ GHz, $f_3 = 21.61$ GHz, and $f_4 = 23.06$ GHz displaying one, two, three, and four polar dynamical domains, respectively. The fifth resonance frequency is $f_5 = 31.27$ GHz and corresponds to a mode with a radial quantization. The figure on the top shows the cross-section from pole to pole (xOz) plane, and the line on the bottom displays the three-dimensional images of the same eigenmodes at the surface of the nanoshell. For clarity, the last mode (f_5) has been sectioned in the (xOz) plane. Adapted with permission from Vazquez Bernardez et al. (2024), copyright 2024 ACS Publications.

Simulations disclosed that two different types of mode structuration are present at resonance, polar and radial. The peaks corresponding to the frequencies f_1, f_2, f_3, and f_4 present one, two, three, and four polar dynamical domains in one hemisphere, respectively. The fifth resonance mode, of frequency f_5, of very low amplitude, displays only one radial node.

Next, micromagnetic simulations were conducted to investigate the external size dependence of the resonance frequencies for a ratio $D_i/D_e = 0.6$ (Fig. 5.7(a)). As displayed in this figure, the lowest-frequency resonance mode has a singular behavior. Its resonance frequency is almost size-independent for external diameters comprised between 50 and 100 nm with a value of 20.24 GHz close to the ferromagnetic resonance (FMR) frequency of a solid sphere of the same composition. For greater external diameters, its resonance frequency displays a $1/D_e^2$ dependence

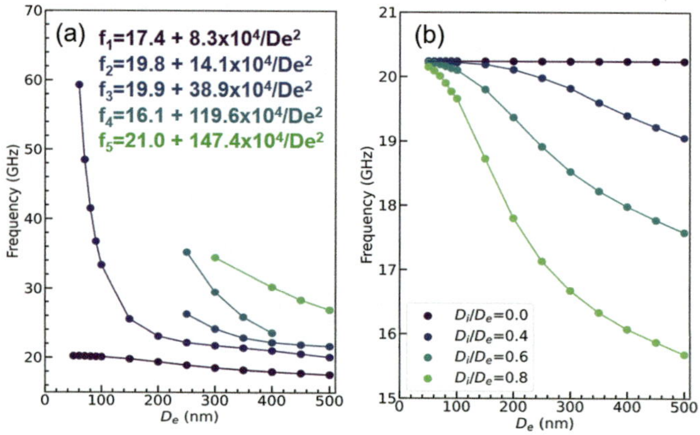

Figure 5.7 External size dependence of the resonance modes. (a) Frequency of the simulated resonance modes as a function of the external diameter D_e for a $BaFe_{12}O_{19}$ core–shell particle with $D_i/D_e = 0.6$, (b) Frequency of the lowest-frequency mode as a function of D_e for $D_i/D_e = 0.0, 0.4, 0.6$ and, 0.8. Adapted with permission from Vazquez Bernardez et al. (2024), copyright 2024 ACS Publications.

upon the emergence of an additional peak in the spectrum. It should be remarked that the evolution of the resonance frequency of the lowest-frequency mode with D_e strongly depends on D_i/D_e (Fig. 5.7(b)). At higher frequencies, several non-evenly spaced resonance modes emerge when the external diameter of the spherical shell is increased (Fig. 5.7(a)). They all exhibit a $1/D_e^2$ dependence consistent with Aharoni's model, demonstrating the predominant exchange regime.

Additionally, as displayed in Fig. 5.7, certain resonance lines are only detectable within a limited range of D_e.

The effect of D_i/D_e was then analyzed for a fixed value of $D_e = 500$ nm. Figure 5.8 shows a cartography of the absolute value of the imaginary part of the dynamic susceptibility χ_{xx} as a function of D_i/D_e and frequency. Decreasing the thickness while maintaining the external diameter constant produces the emergence of new size-dependent resonance modes following a $1/D_e^2$ behavior. It should be noted that some of them exhibit a negative slope for the plot of the resonance frequency versus D_i/D_e for large enough D_i/D_e values. This behavior cannot be reproduced by Aharoni's model valid for

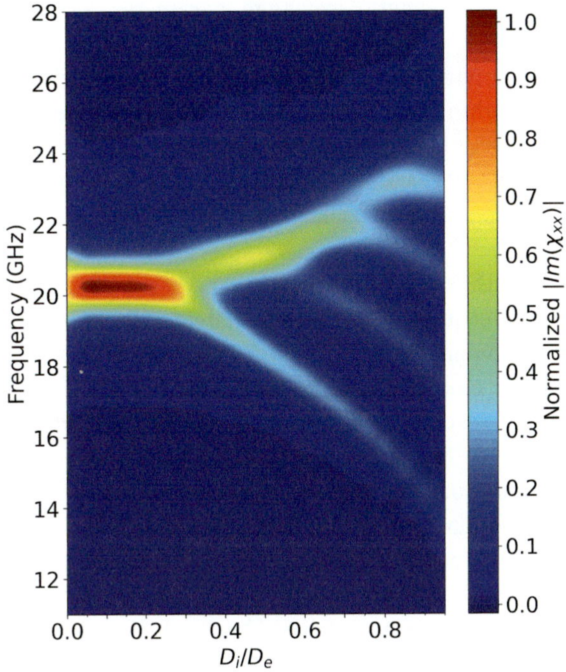

Figure 5.8 Simulated cartography of the normalized absolute value of the imaginary component of the dynamic susceptibility χ_{xx} as functions of D_i/D_e and frequency for a spherical nanoshell of external diameter D_e = 500 nm, normalized by its maximum value for a solid sphere ($D_i/D_e = 0$). Adapted with permission from Vazquez Bernardez et al. (2024), copyright 2024 ACS Publications.

the exchange regime as discussed in section 5.2.6. Similar evolutions have been reported by Nguyen and Cottam (Nguyen and Cottam, 2008) from their microscopic model in the dipole-exchange regime.

These micromagnetic simulations demonstrate the possibility of adjusting the number of resonance peaks and hence, the shape and the spectral bandwidth of the dynamic susceptibility spectra for an M-type ferrite core–shell particle by exclusively manipulating the size and the thickness of its magnetic shell (Fig. 5.8).

Another aspect to consider is the system's ability to withstand real experimental conditions. For instance, the presence of defects and dislocations within the crystal lattice may lead to variations in the system's magnetic exchange parameter. The number of studies that have carried out a complete characterization of the exchange parameter of barium hexaferrite nanoparticles is limited. For $BaFe_{12}O_{19}$ thin films, such a parameter has been established at 5 pJ/m. However, this might discern for three-dimensional systems (Carosella et al., 1992). Under such conditions, we should therefore consider the possibility of having an exchange parameter A_{ex} between 20 and 2 pJ/m. Our micromagnetic simulations demonstrated that within this range of exchange parameter, the spectrum conserves its frequency bandwidth. As A_{ex} decreases, organizing the dynamic magnetization becomes increasingly advantageous, facilitating the emergence of new resonance modes between the ferromagnetic resonance and the highest-order resonance mode (Fig. 5.9).

The existence of size-dependent resonance lines whose number, resonance frequency, and amplitude are controlled by the geometrical parameters of the core–shell system allows shaping the dynamic susceptibility spectra. As illustrative examples, Fig. 5.10 shows two types of spectra governed by the shell parameters: D_e and D_i/D_e. For a damping value α of 0.02, the different resonance lines appear well separated. Increasing the value of the damping constant enlarges each resonance peak causing some of them to merge. The overall imaginary susceptibility spectrum results either in an intensive asymmetric profile (Fig. 5.10(a)) or in a broad absorption (Fig. 5.10(b)).

5.2.5 Magnetization Dynamics: Bidomain Hollow Spheres

The occurrence of a DW in the (yOz) plane splitting the magnetization in two halves of opposite magnetization along z modifies the zero-field dynamic susceptibility spectrum with respect to the single-domain case. These changes concern the size-dependent resonant lines (high-frequency resonances) and the appearance of low-frequency resonances. In addition, the zero-field dynamic susceptibility spectrum of bidomain hollow spheres depend on the in-plane orientation of the pumping field with respect to the DW plane (Fig. 5.11).

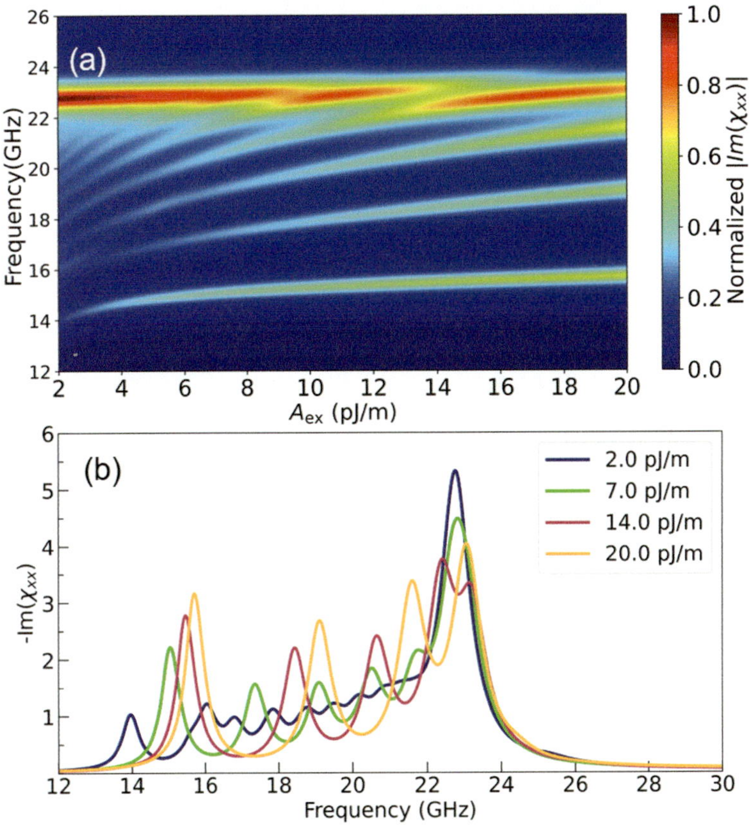

Figure 5.9 Dependence of the frequency of the resonance modes with the exchange parameter A_{ex} for a spherical shell of $D_e = 500$ nm and $D_i/D_e = 0.8$. (a) Cartography of the normalized absolute value of the imaginary component of the dynamic susceptibility as functions of A_{ex} and frequency. (b) Zero-field imaginary component of the dynamic susceptibility spectra for various values of A_{ex} displaying the as-mentioned 'emergent' resonance peaks. Adapted with permission from Vazquez Bernardez et al. (2024), copyright 2024 ACS Publications.

First, we focus on the low-frequency resonances. To better identify the various resonances, a solid sphere of external diameter $D_e = 1000$ nm is considered in a first step as a reference case. Multiple resonances are revealed in the dynamic susceptibility spectra within the frequency range 0.1–4 GHz (Fig. 5.12).

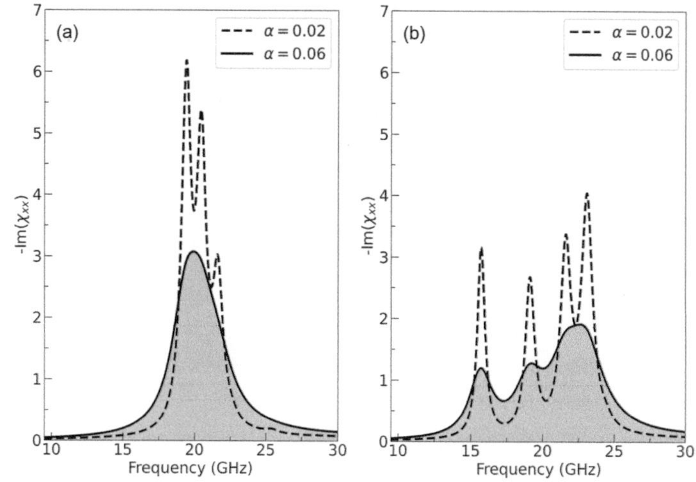

Figure 5.10 Zero-field dynamic susceptibility spectra (imaginary part of χ_{xx}) for a spherical shell with different values of magnetic damping for: (a) a shell of external diameter $D_e = 400$ nm and $D_i / D_e = 0.4$, and (b) a shell of external diameter $D_e = 500$ nm and $D_i / D_e = 0.8$. Adapted with permission from Vazquez Bernardez et al. (2024), copyright 2024 ACS Publications.

The local susceptibility being proportional to the dynamic magnetization, we assimilate $|\delta\vec{m}|$ to the modulus of the local susceptibility and we display its average over one temporal period to exhibit the nodal lines. The cartography of the spatial distribution of the dynamic susceptibility in two vertical planes ((xOz) perpendicular to the DW plane and (yOz) parallel to the DW plane) shows that all the resonances are associated with magnetic excitations localized within the DW (DW modes). For a pumping field applied along the x-axis, (perpendicular to the DW plane), the DW modes are split into two groups. The former one consists of three resonances located below 1 GHz (lines A, B and C). The maps of local dynamic susceptibility indicate that these modes are uniform along the z-axis (vertical axis in the DW plane). They are distinguished by a quantization along the y-axis in the (yOz) plane with one (lines A and B) or two (line C) vertical nodal lines. The second group gathers two resonance lines (lines F and G). The map of local dynamic susceptibility shows that these modes are nonuniform along the z-axis with two nodes symmetrical

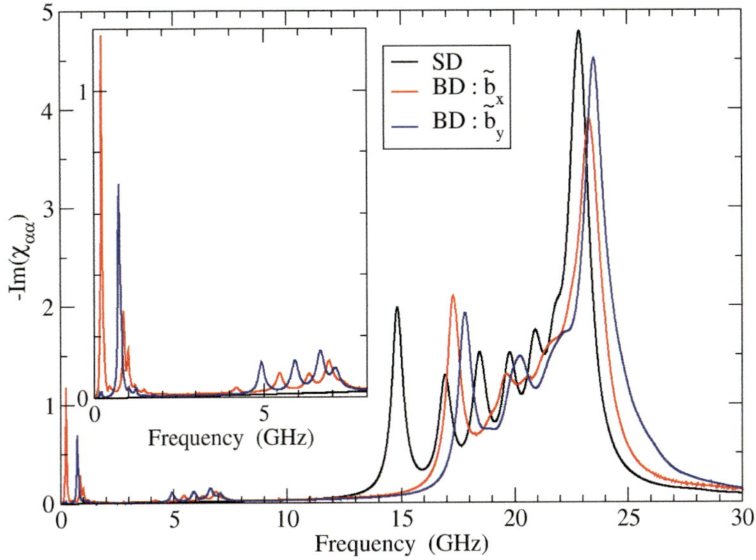

Figure 5.11 Zero-field dynamic susceptibility spectra (imaginary part of $\chi_{\alpha\alpha}$, $\alpha = x, y$) for a shell of doped barium hexaferrite with $D_e = 1000$ nm and $D_i = 800$ nm; single-domain SD solution (black curve) and bidomain BD solutions for an in-plane pumping field along the x-axis (red curve) and along the y-axis (blue curve). The inset is focused on the low-frequency resonances.

with respect to the DW center. However, the distribution in the z coordinate differs between the two modes. The nodes lie near the sphere surface for the line G. These modes are also quantized along the y-axis in the (yOz) plane. For the pumping field applied along the y-axis (parallel to the DW plane), two resonances lines (D and E) emerge. These modes are nonuniform along the z-axis with one node at the DW center and they differ by the quantization along the y-axis. Two additional remarks can be made. (i) Low-amplitude resonances exist also in this frequency range and correspond to high-order DW modes. (ii) The cross-sections of the local dynamic susceptibility in two vertical planes, parallel and perpendicular to the DW plane, give partial information on the 3D modes.

These results bear some similarities with those reported for thick cylindrical nanodots with a perpendicular anisotropy (Vukadinovic and Boust, 2007). In this last case, two main groups of DW

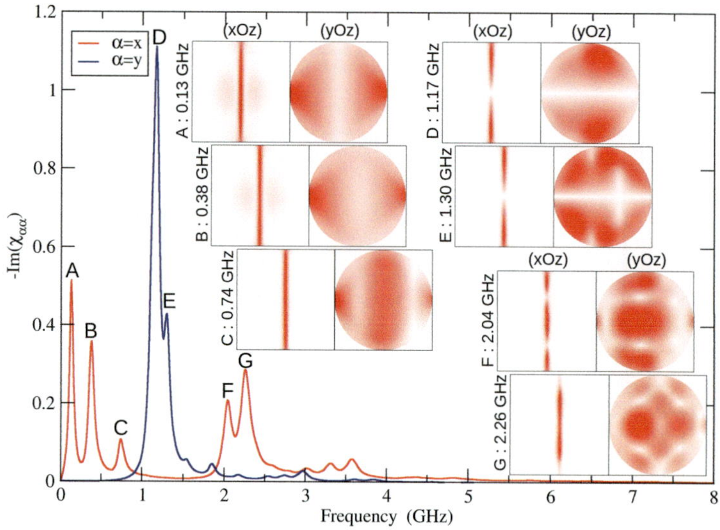

Figure 5.12 Zero-field low frequencies dynamic susceptibility spectra (imaginary part of $\chi_{\alpha\alpha}$, $\alpha = x, y$) for a solid sphere of doped barium hexaferrite of diameter $D = 1000$ nm structured in two domains of opposite magnetization separated by a domain wall (DW). The color insets correspond to eigenmode maps of the modulus of the dynamic magnetization $|\delta\vec{m}|$ averaged over one period for the most intense resonance peaks (cross-section from pole to pole (in the (xOz) plane) or in the (yOz) plane).

resonances exist corresponding to DW modes quantized along the dot thickness and the dot diameter.

Let us now move on to the case of bidomain hollow spheres with $D_e = 1000$ nm and $D_i = 800$ nm (see Fig. 5.13). The confinement of the DW within the shell induces some changes for the DW resonances. For the pumping field applied along the x-axis, the two groups of DW resonances subsist but the amplitude of the lowest-frequency resonance (line A) is strongly increased. In addition, the spectral position of the second group of DW resonances is shifted towards the high frequencies (beyond 4 GHz). The first group of DW resonances is always associated with uniform DW modes along the z-axis and these resonances are distinguished by the polar dependence of the local dynamic susceptibility in the (yOz) plane. The second group of DW resonances (lines C, D and E) is now

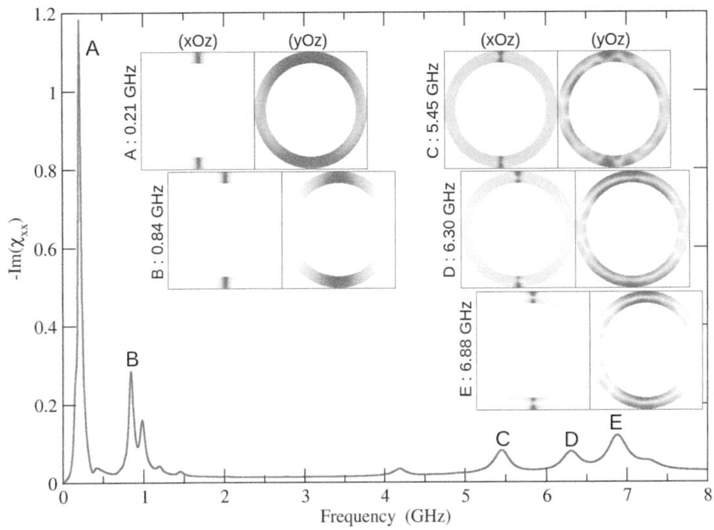

Figure 5.13 Zero-field low frequencies dynamic susceptibility spectrum (imaginary part of χ_{xx}) for a shell of doped barium hexaferrite of diameter $D_e = 1000$ nm and $D_i = 800$ nm structured in two domains of opposite magnetization separated by a domain wall (DW). The color insets correspond to eigenmode maps of the modulus of the dynamic magnetization $|\delta\vec{m}|$ averaged over one period for the most intense resonance peaks (cross-section from pole to pole (in the (xOz) plane) or in the (yOz) plane).

characterized by maps of local susceptibility with one node along the *z*-axis. The resonances at $f = 6.30$ GHz (line C) and at $f = 6.88$ GHz (line D) can be viewed as surface DW modes. The maximal values of the local dynamic susceptibility are located at the inner and outer surfaces of the shell. Similarly, two groups of DW resonances are also observed for the pumping applied along the *y*-axis classified according to the number of nodes along the *z*-axis (Fig. 5.14). Within each group, the DW resonances differ by the polar dependence of the local dynamic susceptibility. The DW resonances at $f = 6.65$ GHz (line D) and $f = 7.07$ GHz (line E) appear as surface DW modes.

The attention is now paid to the high-frequency resonances (size-dependent modes of the hollow spheres). Figure 5.11 shows the comparison between the dynamic susceptibility spectra of a single-domain hollow sphere and a bidomain hollow sphere. Interestingly, the spectral extent of the high-frequency resonances is significantly reduced (approximately by 2 GHz) by the presence of the DW as compared to the single-domain solution: this is mainly due to the shift of +2.5 ($\widetilde{b}_x$) and +3.0 GHz ($\widetilde{b}_y$) of the lowest-frequency resonance, the highest-frequency one being shifted by less than +1.0 GHz. Moreover, when 6 main resonance peaks are obtained for the single-domain solution, only 3 well defined peaks remain for the bidomain solution. This can be understood by analyzing the modes in the bidomain solution displayed in Fig. 5.15. The DW introduces a supplementary node in the dynamic magnetization located in the DW (yOz) plane (see lower set of modes in Fig. 5.15) which splits the polar domain of the dynamic magnetization in two subdomains. This changes the boundary condition that affects directly the frequency of the bidomain modes (by increasing it) and reduces the signal of some modes as compared to the single-domain ones. We conclude that the major improvement of using hollow spheres relative to solid ones, i.e. introducing new resonance peaks and enlarging the absorption band, remains for large spheres even if their magnetization presents domains and domain walls.

5.2.6 Discussion

Micromagnetic simulations have demonstrated that the two principal geometrical characteristics of the hollow sphere intervene in the modulation of the features in the susceptibility spectrum of this system: the magnetic shell size, measured in terms of the external diameter (D_e), and the magnetic shell thickness, given in terms of the D_i/D_e ratio. It is instructive to compare these results with those predicted by Aharoni's model established in the limit of predominant exchange interactions. For a single-domain spherical shell with a uniaxial magnetic anisotropy and in the absence of applied magnetic field, the resonance frequencies of the exchange modes are given by (Aharoni, 1991):

$$f_{kn} = \gamma \left[\frac{2A_{\text{ex}}\mu_{kn}^2}{M_S(D_e/2)^2} + \frac{2K_u}{M_S} \right] \tag{5.15}$$

where γ is the gyromagnetic ratio ($\gamma/2\pi = 29$ GHz/T for the case of $BaFe_{12}O_{19}$), A_{ex} is the magnetic exchange constant, M_S is the magnetization at saturation, and K_u is the uniaxial magnetocrystalline

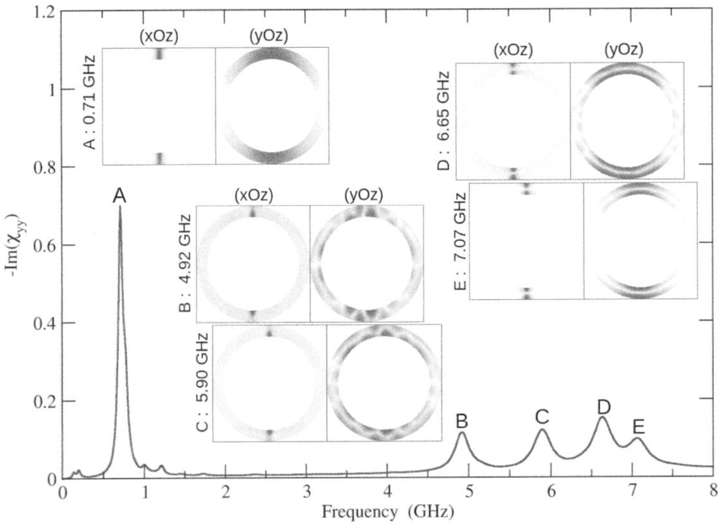

Figure 5.14 Zero-field low frequencies dynamic susceptibility spectrum (imaginary part of χ_{yy}) for a shell of doped barium hexaferrite of diameter $D_e = 1000$ nm and $D_i = 800$ nm structured in two domains of opposite magnetization separated by a domain wall (DW). The color insets correspond to eigenmode maps of the modulus of the dynamic magnetization $|\delta\vec{m}|$ averaged over one period for the most intense resonance peaks (cross-section from pole to pole (in the (xOz) plane) or in the (yOz) plane).

anisotropy constant. The term μ_{kn} are the roots of the derivatives of the Bessel functions intervening in the transcendental equation associated with the appropriate boundary conditions at the shell surfaces (Aharoni, 2002). The index 'n' is the order of the Bessel and describes the spatial discretization in the polar direction of the sphere, counted in terms of dynamical domains, whereas the index 'k' corresponds to different solutions for a given 'n' value.

The first four modes computed by our micromagnetic simulations display $n = 1$, 2, 3, and 4 polar dynamical domains in one hemisphere, respectively, in agreement with Aharoni's model. In contrast, the fifth resonance mode cannot be linked to any value coming from Aharoni's model. Moreover, Aharoni's model fails to reproduce the dependence of the resonance frequencies with D_i/D_e: the origin of this discrepancy can be explained by the inhomogeneous demagnetizing field existing within the shell

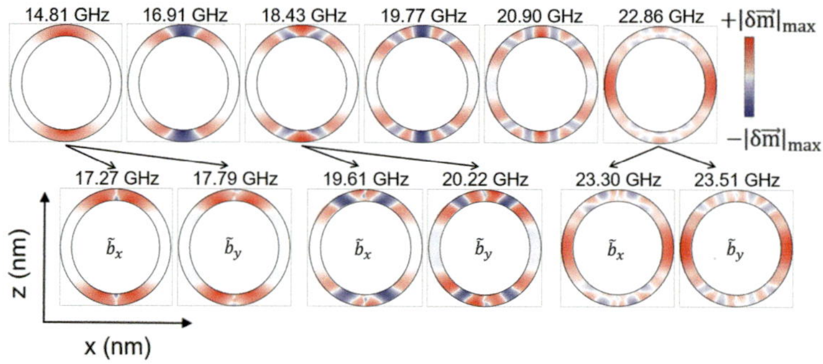

Figure 5.15 Zero-field high frequencies eigenmode maps (cross-section from pole to pole (xOz) plane) of the component δm_x (or δm_y when the excitation field $\tilde{b}$ is applied along the y-axis) for a shell of doped barium hexaferrite, $D_e = 1000$ nm, $D_i = 800$ nm. The rule for imaging the resonance modes is the same as in Fig. 5.6. The upper set of pictures displays the 6 main resonance modes obtained for the single-domain solution whereas the lower set displays the 3 pairs of modes obtained for the bidomain solution according to the pumping field orientation, $\tilde{b}_x$ and $\tilde{b}_y$. The arrows connect modes in both sets that have similar spatial distributions.

and neglected in Eq. 5.15. A corrective term has been recently reported (Vazquez Bernardez et al., 2024). Nevertheless, these results underscore the necessity of a more sophisticated model for hollow magnetic spheres in which the demagnetizing field is treated as a spatially dependent function.

5.3 Elaboration and Examples of Magnetic Oxides Hollow Spheres

Hollow spheres have attracted significant attention from the scientific community for their potential applications since pioneer works which were made by Jerzy Kowalski and colleagues at Rohm and Haas company in the 1980s for the synthesizing of hollow latex spheres (Kowalski et al., 1984). These unique structures have indeed a hollow core surrounded by a thin shell, which confers them several advantages over their solid counterparts such as a high surface area-to-volume ratio, leading to a burgeoning body of research focused on their synthesis and applications. Among these synthesis methods,

Figure 5.16 Scheme of the hard template strategy.

the template approach emerges as the most used strategy. This generally comprises two key stages: firstly, modifying the template to enable the deposition of the precursors onto its surface, and secondly, eliminating the template which can be accomplished either through thermal means or chemical processes, resulting in the formation of a hollow shell (Hu et al., 2011) (Zhu et al., 2021). These templates are broadly classified into hard and soft categories which are summarized hereafter:

(a) Hard template approach refers to a sacrificial structures that define the shape and size of the desired nanoparticles. The template can be a variety of materials, including polystyrene composition, silica spheres, or metal nanoparticles. When utilizing hard templates, the resulting hollow product mirrors the structure of the chosen template, exhibiting a distinct and uniform morphology (see Fig. 5.16).

As an illustrating example, submicron-size hollow Co_3O_4 spheres with a uniform size distribution have been obtained following this method using polystyrene (PS)-bead templates (Ye et al., 2010). Cobalt salt was homogenously precipitated on the surfaces of the PS beads. Aging at 85°C for 24 h allowed inorganic nanocrystals to grow on the surfaces. After calcining the precursors to remove the PS-bead cores, homogenous spheres were obtained, as shown in Fig. 5.17.

(b) The soft template method is an another approach for fabricating hollow materials and offers versatility in generating hollow structures with controlled size. Many soft templates such as emulsions, micelles (see Fig. 5.18) or bubbles can be used for the elaboration of hollow spheres. However, those templates are sensitive to external factors such as pH, temperature, solvent, or by other synthesis conditions. In contrast to hard templates, soft templates are readily detachable, minimizing structural damage and

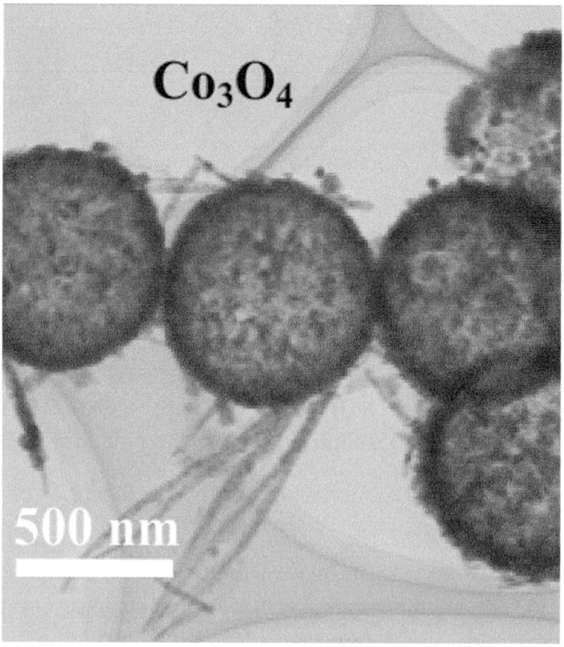

Figure 5.17 Co_3O_4 hollow particles obtained with the hard template strategy. Adapted with permission from Ye et al. (2010), copyright 2010 Multidisciplinary Digital Publishing Institute.

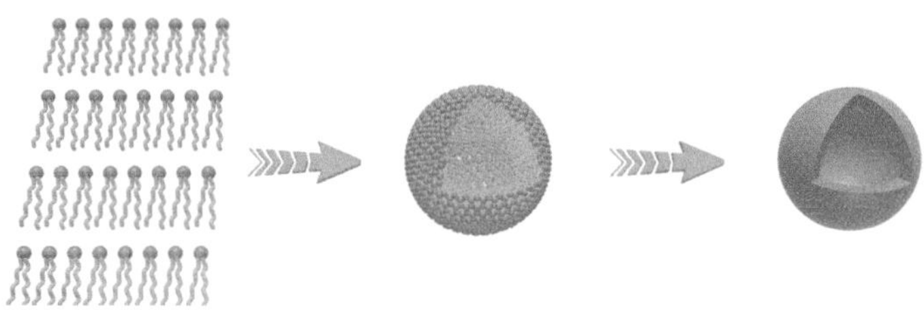

Figure 5.18 Scheme of the soft template strategy using micelles.

eliminating the need for harsh etching agents, making this method relatively safe.

Hollow $BaFe_{12}O_{19}$ (BaM) micrometric particles were successfully synthesized by Choi et al. using a soft template approach (Choi et al., 2014). Firstly and for forming hollow micelle consisted with hydrophobic groups inside and hydrophilic groups outside, a strong

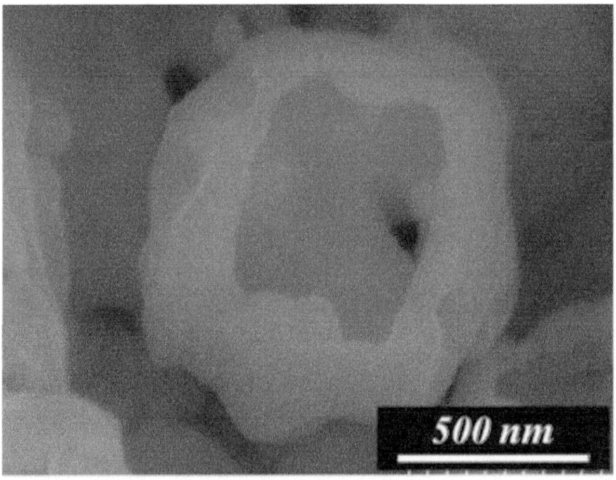

Figure 5.19 Hollow submicrometric particles of BaM. Adapted with permission from Choi et al. (2014), copyright 2014 ScienceDirect Elsevier.

basic surfactant of diethylene-amine ($C_4H_{11}N$) was dispersed into deionized water. Then barium precursor is dispersed in the previous solution allowing to the barium ions to be adsorbed onto the surface of the spherical shape formed by the micelles. Further, ferrite precursor is added, they combine with the barium ions that have adhered to the surface and form a spherical template composed of BaM. Finally, the aqueous solution is stirred and heated to 800–1000°C under air conditions. Such elaboration allows to obtain hollow submicronic spheres shown in Fig. 5.19.

(c) In addition to the previous introduced methods, a trendy approach nowadays is the solvothermal (hydrothermal if solvent is water) synthesis. This method involves using autoclaves to heat and pressurize reaction mixtures containing metal salts and organic solvents at temperatures up to 1000°C and pressures up to 10,000 bars. In the solvothermal methodology, the reaction precursors are mixed in a specific stoichiometric ratio and stirred continuously for a predetermined period. The resultant mixture is then transferred to a Teflon-lined vessel within an autoclave. As the pressure within the autoclave increases, the temperature can surpass the boiling point of the solvent, which promotes the dissolution and reactivity of the nanoparticles within the functional framework (Demazeau, 2010).

Figure 5.20 SEM images of $BiFeO_3$ samples with 12 h of hydrothermal treatment. Adapted with permission from Cadenbach et al. (2023), copyright 2023 Multidisciplinary Digital Publishing Institute.

Many functional oxide materials have been successfully obtained thanks to this method such as $BiFeO_3$, Fe_3O_4 or $TiFe_2O_5$ (Figs. 5.20, 5.21 and 5.22).

5.4 Applications of Magnetic Hollow Spheres

5.4.1 General Applications

Hollow sphere nanoparticles are a versatile class of nanomaterials with a wide range of potential applications. Their unique structure, consisting of a hollow core surrounded by a thin shell, provides them with several advantages over solid nanoparticles. Hereafter, the main applications are briefly reviewed considering first nonmagnetic hollow spheres and then magnetic hollow spheres.

5.4.1.1 Nonmagnetic hollow spheres

The leading applications include :

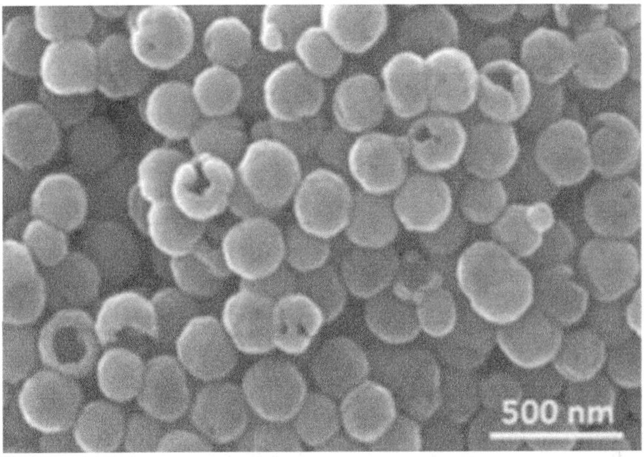

Figure 5.21 Fe_3O_4 hollow sphere. Adapted with permission from Sarkar et al. (2015), copyright 2015 ScienceDirect Elsevier.

Catalysis Hollow sphere nanoparticles are being explored as catalysts for a variety of reactions. Their high surface area and enhanced light scattering properties make them more efficient catalysts than solid nanoparticles allowing them to be used more optimally in a variety of applications such as: (1) catalyze the oxidation of almost 100% carbon monoxide by hematite hollow spheres (Hu et al., 2006); (2) high-yield synthesis of compounds using the Susuki coupling reaction for pharmacological applications using palladium hollow spheres (Sang-Wook et al., 2002); (3) improving the degradation of many types of organic pollutants in water, enabling the production of hydrogen by photocatalysis using TiO_2 hollow spheres (Ren et al., 2003).

Photonics Electrically responsive photonic crystals possess the remarkable ability to alter crystal structures and modify intrinsic structural colors when exposed to external electrical stimuli. This characteristic makes them promising candidates for integration into various technological applications as optically active components. Innovatively inspired by the adaptive properties found in cephalopod skins, Zhong et al. synthesized nanometer-sized hollow silica sphere (3,4-ethylenedioxythiophene)-polystyrene sulfonate photonic crystals which showcases a distinct structural color that

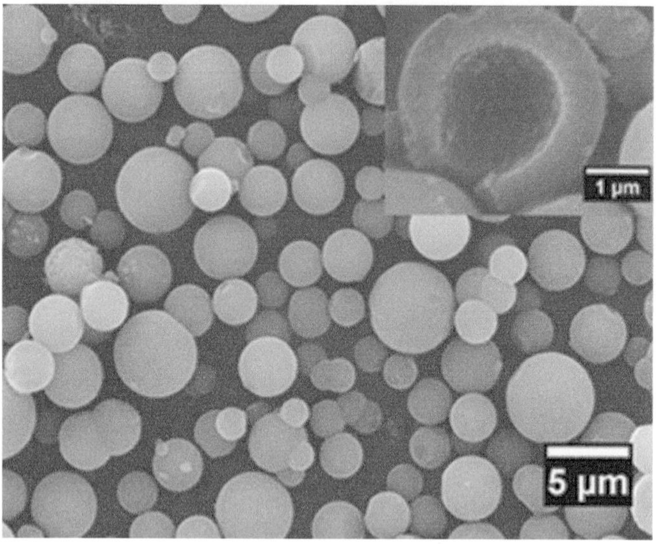

Figure 5.22 Ferric pseudobrookite hollow sphere. Adapted with permission from Zhang et al. (2019), copyright 2019 IOP Publishing.

can be dynamically tuned by applying various voltages (Zhong et al., 2021).

Solar energy harvesting Hollow sphere nanoparticles are being investigated as materials for solar cells. Their high surface area and ability to absorb light make them efficient in capturing solar energy (Ma et al., 2022).

Anode based materials Efforts have been directed towards enhancing the cycle performance of nanosized tin (Sn) particles used as anode material in lithium secondary batteries. This improvement is achieved through the encapsulation of Sn particles with spherical hollow carbon, which can additionally function independently as an active material for additional lithium (Li) ion storage (Lee et al., 2003).

Transducers The potential uses of hollow-sphere transducers in medical ultrasound is quite promising. In that way, hollow spheres of PZT have been elaborated by Meyer et al. They evidenced a good dielectric and piezoelectric properties of the elaborated hollow spheres, making them suitable for medical ultrasound applications. Indeed, their small size, high-frequency thickness modes, and omnidirectional characteristics allow them to be used as sensors independent of receiving direction making them ideal for use in medical ultrasonic guidance systems and small invasive devices used in medical ultrasonic guidance systems (Meyer Jr. et al., 1994).

Drug delivery Number of publication dedicated to Hollow sphere nanoparticles in the biomedical field considerably increase during this last decade focussing on therapy and diagnostic application. For example, hollow-sphere nanoparticles can be used to deliver drugs to specific sites in the body. The hollow core of the nanoparticles can be filled with the drug, and the thin shell can be functionalized with molecules that target specific cells or tissues. This targeted delivery can improve the efficacy of drugs and reduce side effects (Kortesuo et al., 2000; Ziarani et al., 2019).

5.4.1.2 Magnetic hollow spheres

The major applications deal with:

Magnetic Resonance Imaging (MRI) contrast agent Researchers have recently investigated hollow manganese oxide (MnO) nanoparticles as a novel T1 MRI contrast agent. These nanoparticles offer a high surface area-to-volume ratio and allow water access through their pores. In a study by Bae et al., hollow manganese oxide nanoparticles (HMON) were utilized for both cancer diagnosis via MRI imaging and therapy by targeted siRNA delivery (Bae et al., 2011). The HMONs were modified with Herceptin, a therapeutic monoclonal antibody, with the assistance of polyethylenimine-DOPA (3,4-dihydroxy-l-phenylalanine), enabling specific cancer detection and therapy using T1-weighted MRI and intracellular delivery of siRNA, respectively.

Sensors Hollow sphere nanoparticles can be used to create sensors that can detect a variety of analytes, such as gases, chemicals, and biological molecules. For example, the presence of elevated levels of ethanol gas poses a significant safety risk, and individuals may encounter considerable discomfort upon inhalation. Thus, the development of effective ethanol sensors holds considerable importance for ensuring safety and well-being. In that way, Li et al. reported the use of NiO hollow spheres based sensors for ethanol sensing (Li et al., 2022).

5.4.2 Microwave Devices

As mentioned in the introduction, the capacity to fine-tune the frequency dependence of the magnetic susceptibility could deeply impact the design of future microwave devices including miniaturization of antennas using magneto-dielectric composite substrates (Gogoi et al., 2016; Lekshmi et al., 2022; Thakur et al., 2024), the components for spintronic devices (Kumar et al., 2013; Liu et al., 2015), the electromagnetic interference shielding materials and microwave absorbers.

For the latter case, the growing electromagnetic pollution caused by the deployment of telecommunication networks requires the search for new lightweight materials that can absorb the unwanted electromagnetic radiation within a broad frequency range (typically from 100 MHz up to 50 GHz) (Zeng et al., 2020). In this context, composite polymers with core–shell particles appear as a promising solution (Bhattacharjee and Bose, 2021). The core–shell structure offers the opportunity to fit the levels of dielectric and magnetic losses and to improve the impedance matching. The hollow spherical particles can be viewed as a limit case of the core–shell particles. Composite materials including hollow spheres of magnetite (Li and Yang, 2015; Shunenkov et al., 2017) and of spinel ferrite (Mandal and Mandal, 2021; Rafique et al., 2023) have been reported as efficient broadband absorbers below 20 GHz. The case of hollow spheres of M-type hexagonal ferrite selected to exemplify this Chapter is a well-suited solution for designing high-frequency absorbers within the 20-50 GHz frequency range at zero-applied magnetic field. The targeted frequency range is selected by an appropriate choice of the substituted ions for the hexaferrite shell. The presence of multiple size-dependent resonance modes makes possible the shaping of the microwave absorption spectra in terms of bandwidth and absorption level through the geometrical parameters of the shell.

5.5 Conclusion and Perspectives

This Chapter highlights the rising interest in the hollow spherical magnetic nanoparticles. In particular, these numerical findings show how to tailor the magnetic susceptibility spectra in the microwave range using size-dependent resonance modes. The features of the associated resonance lines in terms of spectral positions and amplitudes can be controlled through the geometrical parameters of the magnetic shell. These properties can be exploited for creating or optimizing microwave devices operating within the Ku, K and Ka bands (from 12 GHz up to 40 GHz) when the nanoshell possesses a large uniaxial magnetic anisotropy. In this context, composite materials including a polymer matrix and hollow spherical nanoparticles of doped M-type hexaferrites appear as a promising solution for designing new lightweight microwave absorbers or magneto-dielectric substrates for the miniaturization of antennas. Such applications require to characterize in parallel the dielectric susceptibility spectra. This can be done by means of waveguide measurements in the Ku, K and Ka bands.

To go further, a critical experimental point is the large-scale production of such hollow spherical nanoparticles with a precise control of the thickness and porosity of the magnetic shell. The various synthesis routes presented in this chapter show the importance of the selected method depending on the composition of magnetic oxides. It should be noted that facetted hollow nanoparticles can be generally obtained instead of perfect spherical ones for hexaferrite particles (Baaziz et al., 2014; Zou et al., 2019). Control of the particles' morphology is still challenging. However, numerical results (Vazquez Bernardez et al., 2024) demonstrated that the size-dependent modes subsist for the investigated faceted hollow spherical particles. Prospectively, other routes that use mesoporous materials and in particular zeolites (Peng et al., 2022; Xiao et al., 2022) have been suggested in literature. One must consider that these methods lead to porous materials of weak thickness which could disturb their magnetic properties. Further research on this topic is needed to conclude whether these structures would be suitable for magnetic applications or not.

Two perspectives deserve to be mentioned. First, the multidomain states existing in spherical-like or faceted micrometer-sized particles with a large uniaxial anisotropy could be deeply studied. More complex domain structures (for instance maze-like domains with non rectilinear domain walls) have been theoretically predicted (Lévy and Ghazali, 2002) and experimentally observed in M-type hexaferrite particles (Goto et al., 1980) but little is known about the microwave response of such domain structures. In this context, the preliminary results displayed in this Chapter regarding bidomain solid and hollow spherical particles can be viewed as a first step in this direction. These seemingly simple situations already reveal multiple domain wall resonances in the low-frequency part of the spectra that impact the high-frequency resonances. Second, the hollow spherical nanoparticles are the limit case of core–shell nanoparticles. Recent results (McKeever and Aziz, 2022) on multilayered core–shell structures with interlayer couplings between the constitutive shells suggest additional degrees of freedom to manage the microwave response of nanoparticles. Exploration of this large area has to be pursued.

Acknowledgments

The authors acknowledge the joint research laboratory MOLIERE for supporting this work. The authors would like to acknowledge the High-Performance Computing Center of the University of Strasbourg for supporting this work by providing access to computing resources. Part of the computing resources were funded by the Equipex Equip@Meso project (Programme Investissements d'Avenir) and the CPER Alsacalcul/Big Data.

References

Aharoni, A. (1991). Exchange resonance modes in a ferromagnetic sphere, *J. Appl. Phys.* **69**, 11, pp. 7762–7764.

Aharoni, A. (1997). Effect of surface anisotropy on the exchange resonance modes, *J. Appl. Phys.* **81**, p. 830.

Aharoni, A. (2002). Exchange resonance modes in a hollow sphere, *Phys. status solidi (b)* **231**, 2, pp. 547–553.

Arias, R., Chu, P. and Mills, D. L. (2005). Dipole exchange spin waves and microwave response of ferromagnetic spheres, *Phys. Rev. B* **71**, p. 224410.

Baaziz, W., Pichon, B. P., Liu, Y., Grenéche, J.-M., Ulhaq-Bouillet, C., Terrier, E., Bergeard, N., Halté, V., Boeglin, C., Choueikani, F., Toumi, M., Mhiri, T. and Bégin-Colin, S. (2014). Tuning of synthesis conditions by thermal decomposition toward core–shell $Co_xFe_{1-x}O@Co_yFe_{3-y}O_4$ and $CoFe_2O_4$ nanoparticles with spherical and cubic shapes, *Chem. Mater.* **26**, 17, pp. 5063–5073.

Bae, K. H., Lee, K., Kim, C. and Park, T. G. (2011). Surface functionalized hollow manganese oxide nanoparticles for cancer targeted siRNA delivery and magnetic resonance imaging, *Biomaterials* **32**, pp. 176–184.

Bhattacharjee, Y. and Bose, S. (2021). Core-shell nanomaterials for microwave absorption and electromagnetic interference shieldings: A review, *ACS Appl. Nano. Mater.* **4**, p. 949.

Brown, W. F. J. (1963). *Micromagnetics*, New York: Wiley. ISBN 978-0-88275-665-3.

Cadenbach, T. et al. (2023). Hydrothermal synthesis of bismuth ferrite hollow spheres with enhanced visible-light photocatalytic activity, *Molecules* **28(13)**, p. 5079.

Carosella, C. A., Chrisey, D. B., Lubitz, P., Horwitz, J. S., Dorsey, P., Seed, R. and Vittoria, C. (1992). Pulsed laser deposition of epitaxial $BaFe_{12}O_{19}$ thin films, *J. Appl. Phys.* **71**, 10, pp. 5107–5110.

Choi, M. et al. (2014). Synthesis and characterization of hollow $BaFe_{12}O_{19}$ submicron spheres for advance functional magnetic materials, *Curr. Appl. Phys.* **14(9)**, pp. 1208–1211.

Demazeau, G. (2010). Review. solvothermal processes: Definition, key factors governing the involved chemical reactions and new trends, *Zeitschrift für Naturforschung B* **65(8)**, pp. 999–1006.

Djuhana, D., Oktri, D. C. C. and Kim, D. H. (2014). Micromagnetic Simulation on Ground State Domain Structures of Barium Hexaferrite (BaFe12O19), *Adv. Mater. Res.* **896**, pp. 414–417.

Fletcher, P. C. and Bell, R. O. (1959). Ferrimagnetic resonance modes in spheres, *J. Appl. Phys.* **30**, p. 687.

Gogoi, P., Rabha, M. M., Bhattacharyva, S. and Bhattacharyva, S. H. (2016). Miniaturization of body worn antenna using nano magneto-dielectric composite as substrate in c-band, *J. Magn. Magn. Mater.* **414**, p. 209.

Goll, D., Berkowitz, A. E. and Bertram, H. N. (2004). Critical sizes for ferromagnetic spherical hollow nanoparticles, *Phys. Rev. B* **70**, 18, p. 184432.

Goto, K., Ito, M. and Sakurai, T. (1980). Studies on magnetic domains of small particles of barium ferrite by colloid-SEM method, *Jpn. J. Appl. Phys.* **19**, p. 1339.

Hu, C., Gao, Z. and Yang, X. (2006). Hematite hollow spheres with excellent catalytic performance for removal of carbon monoxide, *Chemistry Letters* **35(11)**, pp. 1288–1289.

Hu, J. et al. (2011). Fabrication and application of inorganic hollow spheres, *Chemical Society Reviews* **40(11)**, pp. 5472–5491.

Kortesuo, P. et al. (2000). Silica xerogel as an implantable carrier for controlled drug delivery—evaluation of drug distribution and tissue effects after implantation, *Biomaterials* **21(2)**, pp. 193–198.

Kowalski, A., Vogel, M. and Blankenship, R. M. (1984). Sequential heteropolymer dispersion and a particulate material obtainable therefrom, useful in coating compositions as a thickening and/or opacifying agent, URL https://patents.google.com/patent/US4427836A/en.

Kumar, P. A., Ray, S., Chakraverty, S. and Sarma, D. D. (2013). Engineered spin-valve type magnetoresistance in Fe_3O_4 core shell nanoparticles, *Appl. Phys. Lett.* **103**, p. 102406.

Landau, L. D. and Lifshitz, E. M. (1935). On the theory of the dispersion of magnetic permeability in ferromagnetic bodies, *Physikalische Zeitschrift Der Sowjetunion* **8**, pp. 153–169.

Lee, K., Jung, Y. and Oh, S. (2003). Synthesis of tin-encapsulated spherical hollow carbon for anode material in lithium secondary batteries, *J. Am. Chem. Soc.* **125(19**, pp. 5652–5653.

Lekshmi, D. R., Adarsh, S. P., Bayal, M., Nair, S. S. and Surendran, K. P. (2022). Functionally graded magnetodielectric composite substrates for massive miniaturization of microstrip antennas, *Mater. Adv.* **3**, p. 2380.

Li, Q. et al. (2022). Highly sensitive ethanol sensing using NiO hollow spheres synthesized via hydrothermal method, *Chemosensors* **10(8)**, p. 341.

Li, Z. W. and Yang, Z. H. (2015). Microwave absorption properties and mechanism for hollow Fe3O4 nanosphere composites, *J. Magn. Magn. Mater.* **387**, p. 131.

Li, Z. W., Yang, Z. H., Kong, L. B. and Zhang, Y. J. (2013). High-frequency magnetic properties at K and Ka bands for barium-ferrite/silicone composites, *J. Magn. Magn. Mater.* **325**, pp. 82–86.

Liu, E., Yuan, H., Kou, Z., Wu, X., Xu, Q., Zhai, Y., Sui, Y., You, B., Du, J. and Zhai, H. (2015). Investigation on spin dependent transport properties of core-shell structural Fe_3O_4/ZnS nanocomposites for spintronic application, *Sci. Rep.* **5**, p. 11164.

Lévy, J.-C. S. and Ghazali, A. (2002). Dipolar effects in magnetic nanostructures, *Phys. Rev. B* Nanostructured Magnetic Materials and their Applications, Shi et al. (Eds.) LNP 593, Springer-Verlag Berlin Heidelberg, pp. 183–202.

Ma, Y. et al. (2022). Highly efficient solar photothermal conversion of graphene-coated conjugated microporous polymers hollow spheres, *J. Colloid Interface Sci.* **623**, pp. 856–869.

Mandal, G. and Mandal, K. (2021). Enhancement of electromagnetic wave absorption in $MnFe_2O_4$ nano-hollow sphere, *J. Appl. Phys* **129**, p. 071902.

McKeever, C. and Aziz, M. M. (2022). Effects of multilayered structure on the static and dynamic properties of magnetic nanospheres, *ACS Appl. Mater. Interfaces* **14**, p. 35177.

McKeever, C., Ogrin, F. Y. and Aziz, M. M. (2019). Microwave magnetization dynamics in ferromagnetic spherical nanoshells, *Phys. Rev. B* **100**, 5, p. 054425.

Mercier, D., Lévy, J.-C. S., Viau, G., Fiévet-Vincent, F., Fiévet, F., Toneguzzo, P. and Acher, O. (2000). Magnetic resonance in spherical Co-Ni and Fe-Co-Ni particles, *Phys. Rev. B* **62**, 1, pp. 532–544.

Meyer Jr., R. et al. (1994). Lead zirconate titanate hollow-sphere transducers, *J. Am. Ceramic Soc.* **77(6)**, pp. 1669–1672.

Nguyen, H. T. and Cottam, M. G. (2008). Dipole-exchange spin waves in ferromagnetic nanostructures with spherical geometries, *Surf. Rev. Lett.* **15**, 06, pp. 727–744.

Peng, Y., Song, S., Liu, F., Yin, Z., Zhong, Y., Yi, X., Zheng, A., Schüth, F. and Gu, D. (2022). General surface-casting synthesis of mesoporous metal oxides with hollow structures and ultrahigh surface areas, *Chem. Mater.* **34**, pp. 7042–7057.

Pullar, R. C. (2012). Hexagonal ferrites: A review of the synthesis, properties and applications of hexaferrite ceramics, *Prog. Mater. Sci.* **57**, 7, pp. 1191–1334.

Rafique, T., Atif, M., Cuenca, J., Mandal, S., Meydan, T., Ali, Z., Khalid, W. and Naddem, M. (2023). Fabrication of hollow $Ni_{1-x}Zn_xFe_2O_4$ spheres (x=0.0, 0.5, 1.0) a higher performance electromagnetic absorber material, *J. Magn. Magn. Mater.* **572**, p. 170647.

Ren, T.-Z., Yuan, Z.-Y. and Su, B.-L. (2003). Surfactant-assisted preparation of hollow microspheres of mesoporous TiO_2, *Chem. Phys. Lett.* **374(1)**, pp. 170–175.

Rowlands, G. (2018). Muview2 3d visualization of micromagnetic data from oommf and mumax, URL https://grahamrow.github.io/Muview2/.

Sang-Wook, K. et al. (2002). Fabrication of hollow palladium spheres and their successful application to the recyclable heterogeneous catalyst for Suzuki coupling reactions, *J. Am. Chem. Soc.* **124(26)**, pp. 7642–7643.

Sarkar, D. et al. (2015). Magnetic properties of Fe_3O_4 nano-hollow spheres, *J. Magn. Magn. Mater.* **393**, pp. 192–198.

Shunenkov, I., Sivkov, A., Ivashutenko, A., Zhuravlev, V., Guo, Q., Li, L., Li, G., Wei, G. and Han, W. (2017). Magnetite hollow microspheres with a broad absorption bandwidth of 11.9 GHz toward promising lightweight electromagnetic microwave absorption, *Phys. Chem. Chem. Phys.* **19**, p. 19975.

Sladanhia-Bautista, D. E. (2022). Magnetostatic modes in a hollow ferromagnetic sphere, *Phys. Lett. A* **451**, p. 128494.

Sloika, M. I., Sheka, D. D., Kravchuk, V. P., Pylypovskyi, O. V. and Gaididei, Y. (2017). Geometry induced phase transitions in magnetic spherical shell, *J. Magn. Magn. Mater.* **443**, pp. 404–412.

Soares, S. F., Fernandes, T., Daniel-da Silva, A. L. and T., T. (2019). The controlled synthesis of complex hollow nanostructures and prospective applications, *Proc. Math. Phys. Eng. Sci.* **475**, 2224, p. 20180677.

Thakur, A., Verma, R., Chauhan, A., Wan, F. and Thakur, P. (2024). Sustained electromagnetic parameters of barium ferrite and epoxy nanocomposites for patch antenna miniaturization over GHz frequency range, *Mater. Lett.* **355**, p. 135516.

Toneguzzo, P., Acher, O., Viau, G., Fiévet-Vincent, F. and Fiévet, F. (1997). Observations of exchange resonance modes on submicrometer sized ferromagnetic particles, *J. Appl. Phys.* **81**, 8, pp. 5546–5548.

Vansteenkiste, A., Leliaert, J., Dvornik, M., Helsen, M., Garcia-Sanchez, F. and Van Waeyenberge, B. (2014). The design and verification of MuMax3, *AIP Adv.* **4**, 10, p. 107133.

Vazquez Bernardez, M. J., Vukadinovic, N., Lefevre, C. and Stoeffler, D. (2024). Tunable dynamic susceptibility in barium hexaferrite core-shell nanoparticles through size-dependent resonance modes, Accepted in *ACS Appl. Electron. Mater.* (2024).

Viau, G., Ravel, F., Acher, O., Fiévet-Vincent, F. and Fiévet, F. (1995). Preparation and microwave characterization of spherical and monodisperse Co@Ni particles, *J. Magn. Magn. Mater.* **140–144**, pp. 377–378.

Vukadinovic, N. and Boust, F. (2007). Three-dimensional micromagnetic simulations of magnetic excitations in cylindrical nanodots with perpendicular anisotropy, *Phys. Rev. B* **75**, p. 014420.

Walker, L. R. (1957). Magnetostatic modes in ferromagnetic resonance, *Phys. Rev.* **105**, p. 390.

Wang, X., Feng, J., Bai, Y. and Yin, Y. (2016). Synthesis, properties and applications of hollow micro-/nanostructures, *Chem. Rev.* **116(18)**, pp. 10983–11060.

Xiao, J., Cheng, K., Xie, X., Wang, M., Xing, S., Liu, Y., Hartman, T., Fu, D., Bossers, K., van Huis, M., van Blaaderen, A., Wang, Y. and Weckhuysen, B. M. (2022). Tandem catalysis with double-shelled hollow spheres, *Nat. Mater.* **21**, pp. 572–579.

Ye, Q.-L., Yoshikawa, H. and Awaga, K. (2010). Magnetic and optical properties of submicron-size hollow spheres, *Materials* **3(2)**, pp. 1244–1268.

Zeng, X., Cheng, X., Yu, R. and Stucky, G. D. (2020). Electromagnetic microwave absorption theory and recent achievements in microwave absorbers, *Carbon* **168**, pp. 606–623.

Zhang, S. et al. (2019). Fabricating Fe_2TiO_5 hollow microspheres with enhanced visible light photocatalytic activity, *Mater. Res. Express* **6(9)**, p. 095505.

Zhong, K. et al. (2021). Dual photonic bandgap hollow sphere colloidal photonic crystals for real-time fluorescence enhancement in living cells, *Biosens. Bioelectron.* **194**, p. 113577.

Zhu, M. et al. (2021). A review of synthetic approaches to hollow nanostructures, *Mater. Chem. Front.* **5(6)**, pp. 2552–2587.

Ziarani, G. et al. (2019). The role of hollow magnetic nanoparticles in drug delivery, *RSC Adv.* **9(43)**, pp. 25094–25106.

Zou, J., Wu, M., Ning, S., Huang, L., Kang, X. and Chen, S. (2019). Ru@Pt Core–Shell Nanoparticles: Impact of the Atomic Ordering of the Ru Metal Core on the Electrocatalytic Activity of the Pt Shell, *ACS Sustainable Chem. Eng.* **7**, 9, pp. 9007–9016.

Chapter 6

Gyroid Ferromagnetic Nanostructures in 3D Magnonics

Mateusz Gołębiewski and Maciej Krawczyk

Institute of Spintronics and Quantum Information,
Faculty of Physics and Astronomy, Adam Mickiewicz University,
Uniwersytetu Poznanskiego 2, 61-614 Poznan, Poland
mateusz.golebiewski@amu.edu.pl

This chapter provides a comprehensive analysis of the magnetic properties of ferromagnetic gyroidal nanostructures, based on both original research and a review of the state-of-the-art literature in the field of 3D magnonics. Through static and dynamic magnetization studies, we highlight how non-trivial shape anisotropy, chirality, and inhomogeneous demagnetization fields within gyroidal structures contribute to the formation of low-energy magnetization textures, spin-wave mode localization, and controllable spin-wave propagation. In addition, the chapter reviews key studies on the magnetic behavior of individual gyroidal components, further advancing our understanding of the gyroid's interactions with external fields and its effective magnetic properties. The combination of original research with a review of recent advances contributes to the ongoing exploration of complex gyroidal systems and their potential in 3D magnonics and spintronics.

2D and 3D Nanostructures: Structure, Properties, and Applications
Edited by Jean-Claude Serge Levy

ISBN 978-981-5129-97-7 (Hardcover), 978-1-003-74871-7 (eBook)
www.jennystanford.com

DOI: 10.1201/97810037487-6

6.1 Introduction

The exploration of spin waves (SWs) and their interactions within magnetic materials covers the interdisciplinary fields of magnonics and spintronics (Kruglyak et al., 2010; Chumak, 2019; Dieny et al., 2020). This convergence has opened avenues for advanced research, particularly in harnessing SWs for signal transport avoiding Joule-Lenz heat emission (Serga et al., 2010; Yan and Bauer, 2013; Barker and Bauer, 2016). The spectrum of SW wavelengths, ranging from micrometers to tens of nanometers, and corresponding to frequencies from a few to hundreds of GHz (Schneider et al., 2008; Maendl et al., 2017), also promises advances in microwave technologies with a versatile platform for manipulating their dispersion and group velocities (Garcia-Sanchez et al., 2015; Wagner et al., 2016; Duerr et al., 2012; Lan et al., 2015; Krawczyk and Grundler, 2014). This versatility, combined with high energy efficiency without sacrificing processing speed (Chumak et al., 2015; Kruglyak et al., 2010; Serga et al., 2010; Lenk et al., 2011; Mahmoud et al., 2020) positions SWs as a compelling alternative to traditional electric currents, especially in computing and microwave systems.

The design of advanced magnonic systems integrates multiple factors, such as geometry, topology, and material properties. Together with the magnitude and direction of the external magnetic field, these factors influence key properties of SWs, including supported SW frequencies, their anisotropy, and ferromagnetic response. This integrated approach enables precise control of magnon dynamics for targeted applications. Magnonic crystals (MCs) represent an innovative intersection of magnetic film technology and periodic patterning, serving as a platform for exploring wave dynamics in both two-dimensional (2D) and three-dimensional (3D) domains (Krawczyk and Grundler, 2014; Chumak et al., 2017). In the case of 2D MCs, the characteristic in-plane dimensions usually span several hundred nanometers, while the magnetic films maintain thicknesses on the order of tens of nanometers (Rychły et al., 2015; Mamica et al., 2019). In these ferromagnetic materials, SWs are primarily governed by isotropic exchange and anisotropic magnetostatic interactions (Krawczyk et al., 2013), making the physics of the system more complex than that of their photonic (light-based) (Butt et al., 2021) and phononic (sound or elastic wave-based) (Vasileiadis et al., 2021) counterparts. Similar to other periodic composites, MCs allow for customization of the dispersion relation through structural and material composition adjustments, allowing for precise control over the velocity, direction, and magnonic gap width of propagating SWs (Krawczyk and Grundler, 2014). A notable aspect of MCs is the way the SWs are affected by the geometry of the system through the internal magnetic fields. This is called the demagnetization field and can be tuned by the orientation of an external magnetic field, often resulting in a reduction of symmetry compared to the original structural design (Neusser et al., 2011; Tacchi et al., 2015; Gross et al., 2021; Ji et al., 2022; Mamica, 2023). The other possibility, unique to MC, is the modulation of the SW properties by the magnetization texture, with or without nanostructuralization. For example, the magnetization domain structure in the form of periodic stripes can have a periodicity up to 100 nm, forming a periodic potential for the propagation of SWs, which can be considered as a fully reprogrammable MC (Banerjee et al., 2017; Gruszecki et al., 2019; Yu et al., 2021; Petti et al., 2022; Szulc et al., 2022). MCs can also be formed from chains (Ma et al., 2015; Mruczkiewicz et al., 2016; Szulc et al., 2024) or arrays (Wang et al., 2020; Tang et al., 2023) of skyrmions. Owing to their small size, skyrmions can be stabilized with periods deep in the nanoscale range.

The MCs discussed above are predominantly planar structures that are uniform along the thickness, and due to its small value, the magnetization is also largely homogeneous. If the geometry or the magnetization texture becomes inhomogeneous along the thickness, we can call the system 3D (Beginin et al., 2019; Gubbiotti et al., 2021; Sadovnikov et al., 2022; Girardi et al., 2024). Strictly speaking, 3D MCs are characterized by a three-dimensional periodic distribution of two materials consisting of inclusion and matrix. Enriched by the additional dimension, the properties of these systems were the focus of theoretical studies, including the exploration of magnonic bandgap phenomena (Krawczyk and Puszkarski, 2006; Mamica et al., 2012; Romero Vivas et al., 2012). They have demonstrated critical aspects such as the threshold of magnetic contrast required between the constituent materials to open a band gap and its dependence on the crystallographic structure of the crystal (Krawczyk et al., 2010). In addition, it has been shown that the selection of an appropriate MC structure and filling fraction can lead to the design of anisotropic and strongly wavevector-dependent effective damping (Romero Vivas et al., 2012), demonstrating the potential of 3D MCs to advance the field of magnonics and spintronics through nuanced control of wave propagation mechanisms in 3D space (Gubbiotti, 2019).

In addition to propagating waves that create bands in an MC and occupy the whole magnetic volume, an important aspect of the nanoscale magnonic systems is also the SW localization (Serha et al., 2022; Liu et al., 2020). A well-described type of SW localization occurs on the surface of thin films in the Damon–Eshbach (DE) configuration (Damon and Eshbach, 1961; Hurben and Patton, 1995), i.e., when the propagation of SWs is perpendicular to an applied external magnetic field that lies in the plane of the film and saturate the sample. The interplay between the dynamic stray fields from the surface and volume magnetic charges formed by oscillating magnetization in the film geometry leads to an asymmetry in the internal magnetic field distribution, creating a gradient in the effective magnetic field across the thickness (the non-zero wavenumber k is required for this effect). As a result, the SWs have a higher intensity near one of the surfaces of the film compared to the inner part and opposite surface. The intensity of these surface-localized modes decays exponentially from the surface of the film toward its center, with the decay length depending on the SW wavelength, i.e., being proportional to k. The side of localization changes with the reversal of the direction of the wavevector (or magnetic field orientation). This type of SW localization offers several advantages for magnonic applications, e.g., increasing the efficiency of SW excitation with inherent nonreciprocity and detection at the film surfaces (Schneider et al., 2008; Jamali et al., 2013), which is particularly beneficial for devices that rely on surface-based SW manipulation (Inoue et al., 2011; Bessonov et al., 2015). In addition, confining the SWs to the surfaces reduces volumetric scattering and damping, potentially leading to lower energy loss and improved propagation characteristics (Yamamoto et al., 2019). Finally, the

DE configuration enables the design of devices with directional propagation properties, i.e., nonreciprocal, which can be exploited in the development of directional magnonic waveguides and logic elements.

SW localization can also be caused by static demagnetization fields. A notable manifestation of this is the emergence of edge modes (Jorzick et al., 2002; Park et al., 2002; Bayer et al., 2006; Bailleul et al., 2003; Kruglyak et al., 2006), where SWs are confined or propagated (Gruszecki et al., 2021, 2022) along the edges of a system, to which static magnetization is perpendicular. The reduction of the internal magnetic field at the edges of the ferromagnetic element can be modified by the magnetization orientation and the shape of the edge, allowing to increase or decrease the localization of low-frequency SWs (Gołębiewski et al., 2023). This makes the edge SWs sensitive to the properties at the very edge of ferromagnets, which allows us to use them for sensing the magnetic properties (Maranville et al., 2006, 2007).

Another interesting type of localization is the topologically protected propagation of SWs along customized paths, facilitated by the MC edges or interfaces between two MCs with different topologies (Shindou et al., 2013; Li et al., 2018; McClarty, 2022; Zhuo et al., 2023). These edges or interfaces act as ideal waveguides, guiding the SWs in one direction and being immune to backscattering, resulting in high coherence even along sharply curved paths (Wang et al., 2018; Feilhauer et al., 2023). Such topological phases are characterized by so-called topological invariants (Kane and Mele, 2005; König et al., 2007), such as Chern numbers, which describe the topological connection between bulk and boundary dynamics (Bansil et al., 2016). Recently, an extension to higher-order topological states has become cutting-edge research (Schindler et al., 2018), realized especially with the use of the artificial crystal in photonics (El Hassan et al., 2019) or acoustics (Ni et al., 2019), and only initiated in magnonics showing corner SWs in 2D antiskyrmion crystal (Hirosawa et al., 2020), and remains completely unexplored in 3D magnonic systems.

The study of 3D MCs holds great promise not only for SW localization phenomena. By tuning the geometry and lattice periods of their unit cells, these structures introduce an additional dimension for novel interactions, including various topological and geometric effects, and emergent material properties (Gubbiotti, 2019; Fischer et al., 2020; Makarov et al., 2022; Cheenikundil et al., 2022; Fernández-Pacheco et al., 2017). Recent advancements in techniques such as X-ray vector nanotomography (Donnelly et al., 2017), magnetic laminography (Donnelly et al., 2022), two-photon lithography (Hunt et al., 2020; van den Berg et al., 2023), focused electron beam deposition (Skoric et al., 2020), and block copolymer templating (Llandro et al., 2020) have enabled the fabrication of complex 3D magnetic systems at the nanoscale, providing the potential for rapid development of the field. Intriguing effects have already been observed in nanorods arranged in diamond-bond networks (May et al., 2019; Stenning et al., 2023), facilitating the analysis of near-degenerate states and laying the groundwork for reconfigurable magnonic devices. Recently, gyroids (Luzzati and Spegt, 1967; Schoen, 1970; Han and Che, 2018), characterized by their chiral triple junctions and fully interconnected 3D network, have been fabricated from ferromagnetic metals (Llandro et al., 2020). Due to the interplay of geometric, chiral and potential topological properties in the nanoscale, the ferromagnetic gyroids can be regarded as 3D MCs possessing all the above-mentioned properties, making them an ideal structure for studying SW dynamics (Gołębiewski et al., 2024). In this chapter, we summarize the current state of research on ferromagnetic gyroids and present our preliminary results on SW dynamics within these structures.

6.2 Geometric Properties of Gyroidal Networks

Since their discovery in 1970 (Schoen, 1970), gyroids have been of interest to research in mathematics, material engineering, and photonics. These structures, belonging to the $I4_132$ space group (No. 214) (Lambert et al., 1996), owe their unique morphology to the cooperative formation at inorganic–organic interfaces, as evidenced in surfactant–silicate mesostructures (Monnier et al., 1993). This architecture has inspired a variety of studies in photonics, where gyroids have been employed as chiral beamsplitters, nonlinear

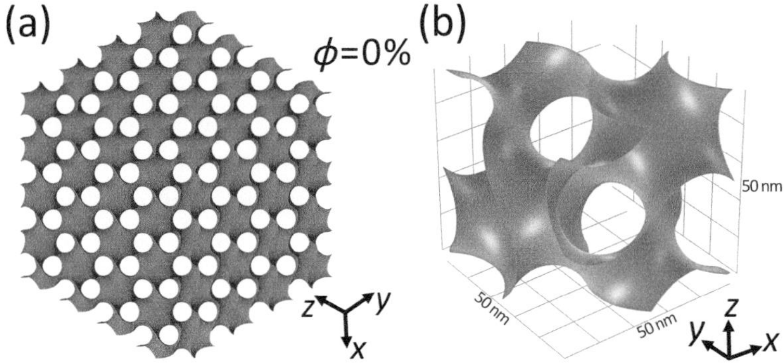

Figure 6.1 The gyroid surface model, illustrating the arrangement along the [111] crystallographic direction through an orthographic projection (a), revealing the hexagonal patterning inherent to the interconnected channels of the system. Panel (b) shows the unit cell in a perspective projection, providing a dynamic view of its geometric configuration.

optical metamaterials, and photonic crystals (Turner et al., 2013; Vignolini et al., 2012; Dolan et al., 2015; Michielsen and Stavenga, 2008). Recent advancements in fabrication techniques (see Introduction 6.1) have also facilitated the creation of artificial systems inspired by gyroid geometry (Yan et al., 2012; Yánez et al., 2016; Turner et al., 2013).

As detailed in Refs. Schoen, (1970, 2012) and Rosi et al. (2020), the gyroid is characterized as a unique triply periodic minimal surface (see Fig. 6.1) possessing zero mean curvature, meaning that each point on the surface functions as a saddle point with equal and opposite principal curvatures (Dacorogna, 2014). This surface extends periodically along three orthogonal vectors and exhibits inherent chirality, lacking any symmetry plane or center, yet retaining rotational symmetry elements (Wohlgemuth et al., 2001). Due to its periodic nature, the gyroid is conceptualized as a crystalline structure (Chen et al., 2013), adhering to the body-centered cubic (bcc) Bravais lattice and associated with the point group O, or described as $I4_132$ in Hermann–Maugin notation. This cubic group, being purely rotational, emphasizes the chiral nature of the gyroid, and the prefix I indicates a body-centered arrangement.

Historically, the gyroidal surface was first described using the conjugate surface construction method (Karcher, 1989), with its embedding later confirmed theoretically in (Große-Brauckmann and Wohlgemuth, 1996). Subsequent studies dealt with the volume fractions of gyroids, especially those with constant mean curvature (Große-Brauckmann, 1997). In a broader scientific context, the gyroid is recognized as Laves' graph of girth ten (Sunada, 2008) and the K_4 crystal (Hyde et al., 2008; Mizuno et al., 2019), known for their efficient space-filling properties and high symmetry. The connection to Laves' graph comes from the gyroid's intricate network of vertices and edges, which mimics the uniformity and connectivity of the graph. Similarly, the local arrangement of atoms in materials that form gyroid structures may be analogous to the tetrahedral coordination of the K_4 crystal.

Its complex design features cubic unit cells interconnected by elliptical cross-section nanorods, as detailed in (Dolan et al., 2015). With a volume filling fraction of $\phi = 0\%$, shown in Fig. 6.1, the gyroid surface divides space into two distinct labyrinths, intersecting at 70.5°. Their mathematical representation is given by the trigonometric equation:

$$\begin{aligned} &\sin(2\pi x/L)\cos(2\pi y/L) \\ &+\sin(2\pi y/L)\cos(2\pi z/L) \\ &+\sin(2\pi z/L)\cos(2\pi x/L) \leq (101.5 - 2\phi)/68.1, \end{aligned} \tag{6.1}$$

where L represents the unit cell dimension of the gyroid. For a more detailed derivation see Tselikas et al., 1996.

6.2.1 Block Copolymer Self-Assembly and Fabrication Techniques

Gyroidal ferromagnetic structures can be fabricated through the self-assembly of block copolymers, which naturally form intricate, periodic morphologies at the nanoscale. Block copolymers are composed of two or more different polymer segments or blocks, covalently bonded in a linear sequence, as presented in Fig. 6.2. These blocks are derived from different types of monomers, small repeating molecular units, allowing for a wide range of chemical compositions within a single polymer chain (Feng et al.,

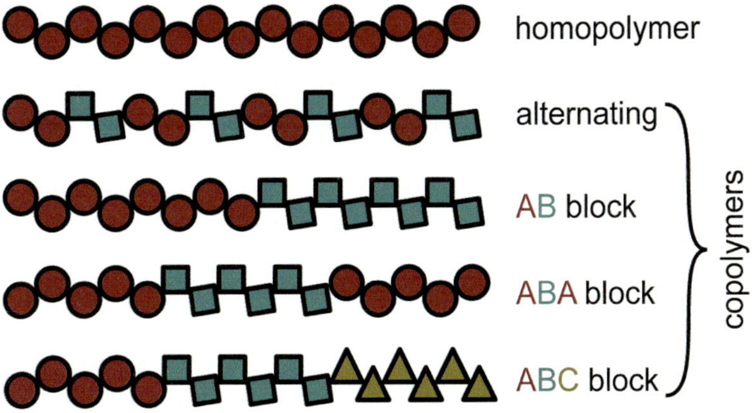

Figure 6.2 Schematic illustration of various polymer types. Block copolymers consist of two or more homopolymer segments (A) connected by covalent bonds. Diblock copolymers contain two distinct segments (AB), while triblock copolymers have three (ABC). A block is technically defined as a segment of a macromolecule, made up of numerous repeating units, with at least one characteristic absent in adjacent segments.

2017). The inherent property of block copolymers is their ability to spontaneously organize into well-defined nano- to microscale structures when subjected to certain conditions, such as changes in temperature. This self-assembly results from the incompatibility between the different blocks, leading to the formation of ordered morphologies such as spheres, cylinders, lamellae and complex 3D structures within a continuous matrix—Fig. 6.3. The architecture of the block copolymer (e.g., diblock, triblock, multiblock) and the volume fraction of each block are critical factors that influence the resulting morphology.

Due to their customizable chemical compositions and the ability to form a wide array of structures, block copolymers have found applications across numerous fields. They can be tailored for use in nanoscale templates for electronic devices, photonic crystals, drug delivery systems, and materials with specific mechanical, optical, or conductive properties. The versatility and tunability of block copolymers, due to the variety of monomers from which they can be made and their unique self-assembly properties, make them a powerful tool in materials science and nanotechnology. For a comprehensive overview, refer to Dolan et al., 2015; Hyde et al., 1997; Segalman, 2005; Cheng et al., 2006.

A simple example of covalently bonded homopolymers are the linear AB diblock copolymers, representing the essential behavior of more complex systems like linear triblock terpolymers. The morphology of diblock copolymers in thermodynamic equilibrium is influenced by three key parameters: N (total degree of polymerization, i.e., the number of monomer units in a polymer chain), f_A or f_B (volume fractions of each homopolymer), and χ (the Flory–Huggins interaction parameter, indicating monomer repulsion). In block copolymers, a higher χ value signifies greater incompatibility between the blocks, driving them to segregate and self-assemble into complex structures to minimize the free energy of the system, e.g., gyroids (Fig. 6.3—G/G'). The interplay between χ, which promotes phase separation, and N, which influences the entropy and enthalpy of the system, guides the self-assembly process.

As the block copolymers self-organize, the gyroid structure emerges as a result of the balance between the repulsive interactions (dictated by χ) and the chain length of the polymers (defined by N). This balance leads to the formation of continuous, triply periodic minimal surfaces, which are characteristic of the gyroid structure. This architecture, with interconnected labyrinths of both materials, is thus a direct consequence of the molecular properties of the block copolymers and their drive to achieve a thermodynamically favorable configuration. Larger gyroid structures, scaling from millimeters to sub-millimeters, are typically fabricated using top-down methods (Yan et al., 2012; Yánez et al., 2016; Turner et al.,

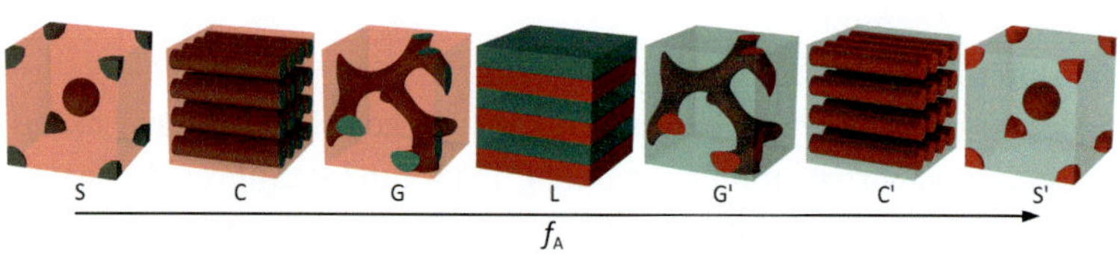

Figure 6.3 Equilibrium morphologies of AB diblock copolymers in a bulk, varying as a function of the volume fraction f_A of the blocks and the interaction parameter χ. The morphologies are categorized as follows: S/S', body-centered cubic spheres; C/C', hexagonally packed cylinders; G/G', bicontinuous gyroids; and L, lamellae.

2013; Gan et al., 2016). Conversely, the creation of nanostructures with unit cells below 100 nm is predominantly achieved through bottom-up techniques (Armatas and Kanatzidis, 2006; Kresge et al., 1992). The process involves microphase separation in copolymers, resulting in gyroid networks of a minority polymer block within a majority block matrix (Ross et al., 2014; She et al., 2013).

The formation of complex structures using block copolymers as templates involves a multi-step process that exploits the self-assembly properties and the technique of electrodeposition to introduce ferromagnetic materials, such as permalloy (Py) or nickel (Ni), into the predetermined patterns. One of the polymer blocks is selectively removed to create a porous template that retains the architecture. This can be achieved by various methods, including chemical etching or UV degradation, depending on the nature of the block copolymer. With the porous template in place, electrodeposition is used to fill the voids with a ferromagnetic material, as illustrated in Fig. 6.4. In this instance, a diblock copolymer of poly(4-fluorostyrene) (PFS) and 38% poly(lactic acid) (PLA) is employed, with PFS and PLA serving as the majority and minority blocks, respectively (Scherer et al., 2014; Llandro et al., 2020). During electrodeposition, the template is immersed in an electrolytic solution containing ions of the ferromagnetic material. When a voltage is applied, the ions are reduced and deposited onto

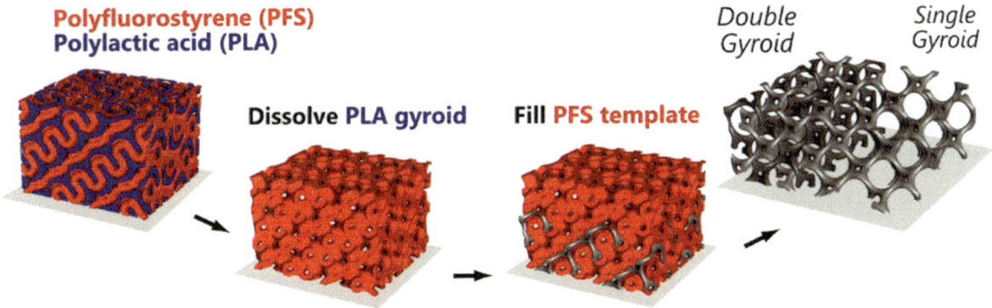

Figure 6.4 Fabrication and structural characterization of gyroid nanostructures. This scheme shows the process for creating free-standing metallic gyroid nanostructures, involving the thermal annealing of a block copolymer template, followed by the selective removal of the minority block and subsequent electrodeposition. Reprinted (adapted) with permission from *Nano Lett.* 2020, 20, 5, 3642–3650. Copyright 2024 American Chemical Society.

the conductive areas of the template, gradually filling the structure. After the electrodeposition process is complete, the remaining block copolymer template is removed, often by solvent washing or thermal decomposition, leaving a freestanding structure composed entirely of the ferromagnetic material. The end product is a 3D structure of ferromagnetic material that replicates the geometry of the original template.

In Fig. 6.4, the fabrication of magnetic gyroid nanostructures from Llandro et al., 2020 featuring unit cell dimensions smaller than 50 nm, is demonstrated. These nanostructures are realized through electrodeposition into block copolymer templates, yielding $Ni_{75}Fe_{25}$ gyroids. The self-assembly of 3D nanostructures, especially when integrated with inorganic components (Simon et al., 2001), allows unprecedented morphological control at the nanoscale, opening new avenues for magnetism and magnonics. The SW interactions within gyroids are complex and involve, among others, shape anisotropy, inhomogeneous demagnetization fields, curved surfaces, and chirality. Our research aims to show the complexity of these interactions, provide an understanding of the magnetic behavior in gyroids, and lay the framework for future explorations in 3D magnonics.

6.3 Gyroid Structures in Magnetism

Short SWs, with their nanoscale wavelengths and microwave frequencies, enable applications of miniaturized magnonic devices (Chumak et al., 2015). The concepts of SWs, linked both to the chirality inherent in magnetization dynamics (as discussed above and reviewed in Kruglyak, 2021) and to chirality introduced by the Dzyaloshinskii–Moriya interaction (DMI) (Tacchi et al., 2023) or by geometric curvature (Otálora et al., 2017; Sheka et al., 2022), may play an important role in the further development of magnonics-based information processing technologies (Chen et al., 2020, 2021). Chirality, which refers to the preference of spin structures to twist in a particular direction, facilitates also directional SW propagation, opening the door to non-reciprocal devices. DMI,

which exists in systems with broken inversion symmetry (Gan et al., 2016), stabilizes chiral magnetic textures, improving control over spin structure and enabling the design of robust, energy-efficient memory devices, the structures which can also be exploited for magnonics (Garcia-Sanchez et al., 2014). Similar properties are expected to exist in curvilinear ferromagnetic systems, since curvature, when introduced on a length comparable to the exchange length, introduces the anisotropic exchange in the same form as DMI (Sheka et al., 2022). Therefore, the complex architecture of gyroidal structures provides a unique, three-dimensional platform for exploiting all of these phenomena. Nonetheless, the study of magnetization dynamics within periodic 3D nanostructures is still in its early stages. While extensive research exists on SWs in uniform 1D and 2D structures, investigations into collective SW dynamics in 3D artificial systems (May et al., 2021; Guo et al., 2023; Gubbiotti, 2019; Fernández-Pacheco et al., 2017), particularly gyroids (Gołębiewski et al., 2024), remain limited. We attempt to summarize existing and explore new research directions on the collective dynamics of SWs in gyroidal nanostructures.

As described above, the inherent chirality and curvature of nanoscale magnetic gyroids offer a promising pathway for controlling non-collinear spin textures. Recent studies have underscored this potential (Lich et al., 2023), and the experimental visualization of magnetic structures in $Ni_{75}Fe_{25}$ gyroid networks has provided further evidence (Llandro et al., 2020). However, the DMI and curvature-induced anisotropy, which add another layer of complexity to research in 3D ferromagnetic structures (Hertel, 2013; Gaididei et al., 2014), have not yet been explored in gyroidal structures. Similarly, novel physical effects observed in curved magnetic wires and films (Sheka, 2021), and the integration of chiral and topological properties (Shindou et al., 2013; McClarty, 2022), have not been tested in gyroids. Furthermore, their potential extends further, with numerous energetically equivalent stable magnetization states, suggesting suitability for the realization of artificial spin-ice systems in 3D. Indeed, gyroidal magnetic nanostructures, particularly those with at least one dimension in the exchange length scale, have emerged as key to the realization of in-volume system phenomena such as monopole-type excitations (Wannier, 1950; Diep, 2013; Lacroix et al., 2011), which were previously observed only in two-dimensional (2D) systems (Skjærvø et al., 2020; Nisoli et al., 2017).

Taking the magnonic perspective, we use micromagnetic simulations to reveal the collective dynamics of SWs in gyroids. This approach is consistent with the findings in Demidov et al., 2010, where the importance of direct observation and mapping of SWs, especially in nano-oscillators, is highlighted. A notable aspect of our investigations is the study of the ferromagnetic resonance and SW dispersion within gyroids. The frequency–wavevector relationship of SWs is of particular interest in complex 3D geometries, where the multidimensional nature can give rise to tunable, magnetic field-directed band gaps, mode crossings/hybridizations, and unconventional group velocities. By analyzing these properties, we aim to unveil the potential of gyroids as magnonic metamaterials and efficient SW carriers.

6.3.1 Static Magnetization Configuration

The study of static magnetization in 3D gyroidal ferromagnetic nanostructures apart from learning the specifics of these structures, also can advance our understanding of complex magnetic systems in general, and open up new technological possibilities, as they have already been recognized for their intricate, frustrated remanent states (Llandro et al., 2020). These states are inherently ferromagnetic, but do not conform to a unique equilibrium configuration. The detailed imaging and mapping of gyroids, shown in Fig. 6.5, provide a deeper insight into their magnetic structure, highlighting the complex interplay between the magnetic fields and the intricate single- and double-gyroid architecture. Figure 6.5(b) shows the magnetic induction map of the double-gyroid nanostructure. The external stray field surrounding the gyroid exhibits dipole-like behavior aligned with the saturating field (**H**). Within the structure, multiple flux-closed loops are evident, some of which enclose only a single unit cell. However, identifying the chirality of these loops is challenging due to the integration of the magnetic induction map across the thickness of several unit cells and the entangled nature of the double-gyroid networks.

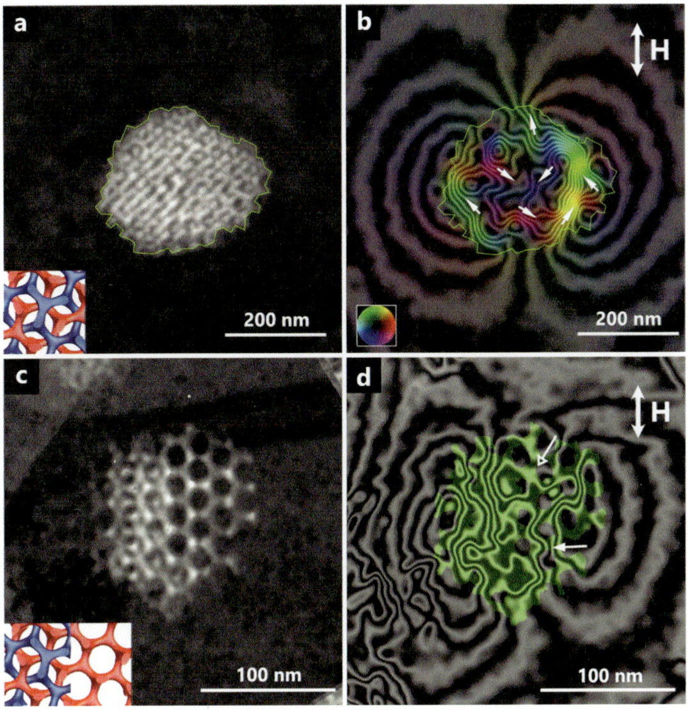

Figure 6.5 Magnetographic analysis of Py double- and single-gyroid structures with unit cell size $L = 42$ nm. Insets illustrate simulated models of double-gyroids and hybrid double/single-gyroid structures, with each gyroid network marked in red and blue. Panels (a) and (b) display the phase contribution of the mean inner potential and the map of magnetic induction for a double-gyroid. Bidirectional arrows indicate the saturation magnetic field (**H**) direction. Stray magnetic fields around the structure are traced by contour lines, which within the struts are non-uniform, as indicated by unidirectional arrows representing varied field directions. The structure's outline, derived from the phase shift's mean inner potential contribution, is shown in (b). In panels (c) and (d), the phase contribution of the mean inner potential and the corresponding magnetic flux contour map are depicted for a sample containing both double-gyroid and single-gyroid regions. In (d), the gyroid particle is masked in green and superimposed with contour lines to illustrate the relationship between field lines and gyroid struts. Points where flux contours intersect gyroid struts or loop around vertices are marked with closed and open unidirectional arrows, respectively. The contour interval for both (b) and (d) is set at $2\pi/64$ radians. Reprinted (adapted) with permission from *Nano Lett.* 2020, 20, 5, 3642–3650. Copyright 2024 American Chemical Society.

In the single-gyroid region [Fig. 6.5(d)], the correlation between the structure and the magnetization of the ferromagnetic gyroid becomes more pronounced. The external stray field remains dipole-like, but the remanent state of the single gyroid is more complex than a simple network of flux-closed loops. The contours are generally aligned with the structure, but in some regions (marked with the arrows) the flux lines cut directly across the struts, implying that the magnetization is transverse to the strut axis. In other regions, the flux contours encircle specific vertices, possibly indicating that the magnetization of the three constituent struts converge or diverge from the vertex center.

In Lich et al., 2023, the authors have presented an insightful analysis through phase-field simulation studies, focusing on the formation and switching behavior of magnetization textures within Py gyroid nanostructures with a periodicity of 50 nm. The study uncovers the coexistence of left-handed (LH) and right-handed (RH) helices in these structures, a feature particularly pronounced in gyroids with small solid volume fractions. Moreover, it has been demonstrated that at small ϕ, the magnetization textures conform to the ice rule. However, as ϕ increases, this rule is disrupted, leading to magnetic field texture frustration, a phenomenon that becomes prominent at a critical volume fraction of about 30%. Furthermore, the findings indicate an increase in the coercive field as the volume fraction rises from 10 to 50%, followed by a reversal of this trend with further increases in ϕ. This behavior is attributed to the balance between LH and RH magnetic helices, with global switching events occurring near $\phi = 50\%$, while local switching is observed in other volume fractions. They summarize the relationship between solid volume fraction and static magnetization distribution in gyroids.

Exploring the realm of static magnetism within complex 3D structures (Hsueh et al., 2011) opens up promising avenues of research, particularly in the creation and control of topological spin textures such as skyrmions, torons (Ji et al., 2022), hopfions (Tai and Smalyukh, 2018), and others (Wu and Smalyukh, 2022). The curvature and topology inherent in gyroids provide a novel platform for studying how these structures can stabilize such textures. Additional research focusing on static magnetization configurations under external fields could deepen our understanding of domain distribution and orientation. Also, the process of magnetization reversal in gyroids, influenced by geometry and thermal fluctuations, could provide valuable insights into material optimization for precise magnetic switching. Broadening the scope to include hybrid structures by integrating materials such as superconductors or topological insulators with gyroidal ferromagnets opens up other research opportunities.

Studying the magnetization texture and dynamics at the level of individual nanostructures, unit cells, vertices and struts, is essential for a comprehensive understanding of their magnetic behavior. Current research suggests the possibility of multiple equivalent magnetization configurations within the 3D gyroid network, which is particularly promising for applications such as reservoir computing (Tanaka et al., 2019), where systems benefit from high interconnectivity and a variety of stable states. The ability to access and manipulate these different states could enable new paradigms in nanoscale computing by exploiting the properties of these complex magnetic 3D structures.

6.3.2 Magnetization Dynamics

As already mentioned in the Introduction, the study of SW dynamics in 3D nanostructures is at a very early stage (Gubbiotti, 2019; Cheenikundil et al., 2022; Li et al., 2023), especially experimentally (Sahoo et al., 2018; Donnelly et al., 2020; Guo et al., 2023; Girardi et al., 2024). In the following subsections, we will summarize our research on Ni-based gyroids, including experimental ferromagnetic resonance (FMR) studies and numerical simulations. Here, the gyroid sample's unit cell is defined with a dimension of 50 nm and $\phi = 10\%$, as shown in Fig. 6.6. This specific dimensionality leads to strut diameters that align with crucial nanoscale measures, e.g., the exchange length (around 9 nm for Ni).

6.3.2.1 Gyroidal crystallography vs. ferromagnetic response

Our recent work (Gołębiewski et al., 2024) presents a synergistic approach combining micromagnetic simulations and FMR measurements, employed to examine the magnetic properties of three-

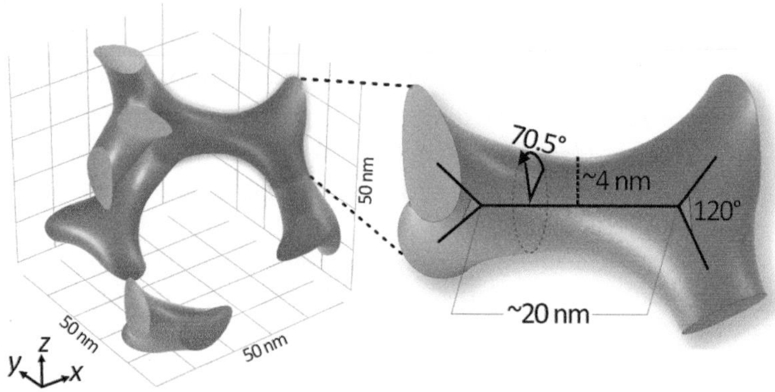

Figure 6.6 Illustration of the cubic gyroid unit cell tailored for micromagnetic simulations, with dimensions $L = 50$ nm and a volume fraction of $\phi = 10\%$. An enlarged inset highlights the chiral linkage between two primary gyroid nodes, emphasizing key geometric features.

dimensional gyroidal Ni nanostructures. This investigation reveals several distinctive characteristics of the gyroid network, particularly the pronounced influence of the static, external magnetic field orientation on the ferromagnetic response. This relationship is closely linked to the crystallographic alignment of the gyroid structure, as presented in Fig. 6.7.

We utilize the finite element method (FEM) micromagnetic simulations in *tetmag*, both in the time domain for determining static magnetization structures (Hertel, 2023), and, using add-on frequency-domain algorithm for analyzing oscillatory, magnonic properties (d'Aquino and Hertel, 2023)—see more details in Appendix A.1. The following Ni material parameters have been used: the saturation magnetization $M_s = 480$ kA/m, the exchange stiffness $A_{ex} = 13$ pJ/m, the Gilbert damping coefficient $\alpha = 0.008$, and the g-factor of 2.14 (Coey, 2010; Singh et al., 1976).

We examine the impact of three specific field directions, aligned with the crystallographic axes [100], [110], and [111], on finite-sized gyroids, see Fig. 6.7. The findings from the resonance spectra simulations attribute the observed shifts in resonance signals in the rotating gyroid sample to crystallographic anisotropy. Moreover, several bulk SW modes, as identified in this work, exhibit a

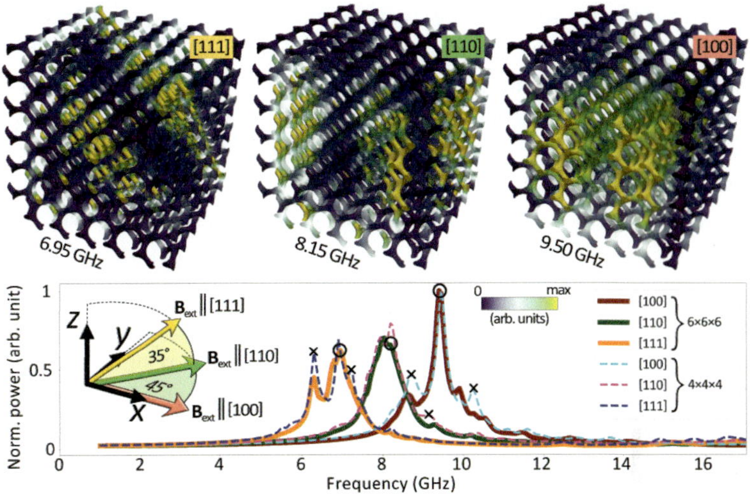

Figure 6.7 Spectral analysis of SW modes in a 6 × 6 × 6 gyroid structure. The lower part of the figure shows the frequency spectra with high-intensity volume modes highlighted by black circles. For comparison, spectra of more compact structures (dotted lines) are superimposed, showing a significant reduction in the intensity of edge modes (indicated by crosses) with increasing structure size. The color gradient in the mode visualization corresponds to the imaginary part of the magnetic susceptibility. Reprinted (adapted) with permission from *ACS Appl. Mater. Interfaces* 2024, 16, 17, 22177–22188, licensed under CC-BY 4.0. Copyright 2024 American Chemical Society.

predominantly localization on the different regions across the perpendicular axis of the gyroid structure, on the facets perpendicular to the [110] and [100] magnetic field directions. This behavior likely stems from a combined effect of crystallographic influence and shape anisotropy impacting the localization of resonant modes across various regions, potentially influenced by the system's chirality. Furthermore, we identified SW modes localized on the outer surfaces of the gyroid structure (see the modes marked with 'x' in the spectra shown in Fig. 6.7). This also indicates the important influence of the shape and outer surfaces of the 3D nanostructure on the SW dynamics. This observation prompts us to investigate the surface effects further. The preliminary results of our numerical studies of surface-localized SW modes are summarized in the next section.

The *tetmag* simulations were complemented by meticulous experimental measurements using broadband FMR, where a multi-domain (multiple, interconnected sub-parts with different crystallographic orientations) gyroid sample was positioned on a coplanar waveguide (CPW) line. This setup facilitated the analysis of distinctions and correlations between a solid, uniform Ni layer and the gyroid-structured segment of the sample. The Ni gyroid nanostructure analyzed was fabricated through the thermal annealing of a block copolymer template [detailed in Section 6.2.1 (Fig. 6.4) and Llandro et al., 2020], followed by selective dissolution of one of the gyroid-forming blocks and filling the resultant right-handed gyroid network with Ni by electrodeposition ($L = 50$ nm, $\phi = 10\%$). Performed measurements and provided numerical analysis allow us to conceptualize gyroid films as homogeneous materials or magnonic metamaterials, where the effective saturation magnetization is reduced in proportion to the gyroid's filling factor—see Fig. 6.8. We can see, that the gyroid signal exhibits a broader frequency full-width at half-maximum (FWHM) compared to that of homogeneous Ni. For instance, at a magnetic field of 450 mT, the FWHM is 3.69 GHz for the gyroid and 3.09 GHz for Ni. This broader signal is likely due to the multidomain nature of the sample. Specifically, the observed signal from the gyroid represents an average from multiple domains with varying crystallographic orientations located above the CPW. From the FMR measurements shown in Fig. 6.8(e), we deduce an average FWHM difference of 1.3 GHz across various external magnetic field values. Micromagnetic simulations for a cube-shaped sample (Fig. 6.7) indicate a maximum peak separation of 2.95 GHz at 300 mT, specifically between the [100] and [111] crystallographic directions. These simulations evaluated the crystallographic orientations that represent the most divergent configurations of the gyroid lattice relative to the applied field, resulting in the largest possible separation of resonant frequencies. In field-swept FMR experiments conducted at a fixed frequency, the absorption line FWHM conforms to the equation $\mu_0 \Delta H \propto 4\pi\alpha f/\gamma$. This relationship holds when the magnetization vector is aligned with the applied magnetic field, either in-plane or perpendicular to it, leading to a linewidth that scales proportionally with frequency. The slope of this scaling

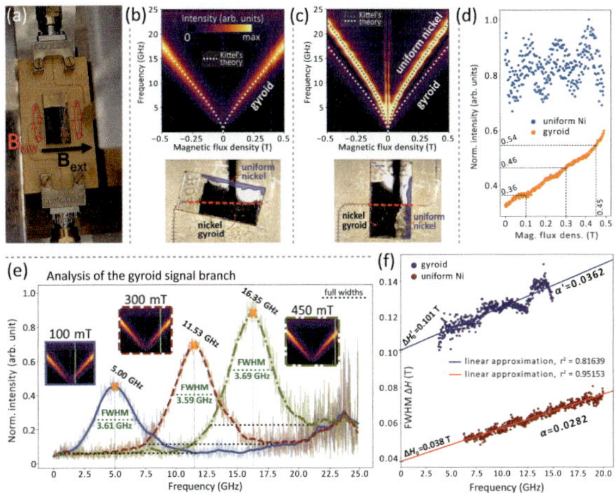

Figure 6.8 A broadband FMR measurement on a Ni gyroid structure. The sample was repositioned relative to the CPW to study the effect of an additional homogeneous Ni layer. Two configurations were used to assess energy absorption from the microwave field B_{MW}, applied perpendicular to the static magnetic field (a). Dynamic magnetization amplitude as a function of static magnetic flux density and frequency for selected configurations are shown in (b) and (c). A strong signal from the gyroid layer appears when the CPW aligns directly below it (b), while a higher-frequency signal from the uniform Ni layer is observed when the CPW intersects its projected position (highlighted in purple) (c). Dotted lines represent theoretical fits using the Kittel formula (Eq. A.6). For the uniform Ni, parameters from micromagnetic simulations were used; for the gyroid, calculated effective parameters ($M_{\mathrm{eff}} = 132$ kA/m, g-factor of 2.2) were applied to the formula. Plot (d) summarizes peak intensities of FMR signals (blue dots for uniform Ni, orange dots for gyroid) as a function of external magnetic field strength, with normalized values for 100 mT, 300 mT, and 45 mT. Plot (e) provides a cross-sectional analysis of FMR signals at different external magnetic fields: solid blue line for $B_{\mathrm{ext}} = 100$ mT, dashed brown line for $B_{\mathrm{ext}} = 300$ mT, and dash-dotted green line for $B_{\mathrm{ext}} = 450$ mT. Horizontal dashed green lines indicate the FWHM for each section. Orange crosses mark peak maxima and corresponding frequencies. Insets show intensity plots from FMR measurements with green vertical lines marking specific section locations. Plot (f) shows the magnetic field FWHMs as a function of frequency for FMR signals of gyroid (purple dots) and uniform Ni (dark red dots). Linear regression, based on experimental data and Eq. A.7, estimates determination coefficient r^2, ΔH_0 (from the abscissa), and damping values α (from the slope). Parameters for the gyroid are marked with a prime (′). Reprinted (adapted) with permission from *ACS Appl. Mater. Interfaces* 2024, 16, 17, 22177–22188, licensed under CC-BY 4.0. Copyright 2024 American Chemical Society.

is defined by the Gilbert damping parameter α. In addition to this intrinsic contribution, empirical data suggest the presence of a frequency-independent term, which contributes to the overall linewidth observed in the experiments. Experimental investigations have also enabled a linear regression analysis of FWHM across a spectrum of B_{MW} frequencies, as depicted in Fig. 6.8(f) for the designated orientations of the sample over CPW. This analytical approach is instrumental in ascertaining the damping values α and the inhomogeneous linewidth contribution ΔH_0 for both the homogeneous Ni layer and the gyroidal structure, respectively. In the first case, the derived values are $\alpha = 0.0282 \pm 2.54\%$ and $\Delta H_0 = 0.038 \pm 1.28\%$ T. Comparatively, the gyroidal structure exhibits $\alpha' = 0.0362 \pm 6.49\%$ and $\Delta H_0' = 0.101 \pm 1.05\%$ T. Larger α' for the gyroid structure is likely attributable to the scattering of SW modes within the nanowires, which are much thinner than the bulk Ni and exhibit complex noncollinear interconnections.

6.3.2.2 Localization properties

To study surface effects in gyroids, we simulate the gyroid in a thin film geometry in dependence on the thickness of the gyroid structure and the orientation of the external magnetic field. To avoid the influence of the lateral edges effects on the SW spectra we employ Bloch–Floquet boundary conditions at the unit cell boundaries. The simulations are performed with COMSOL Multiphysics software. In this implementation, we assume full magnetization saturation and solve the Landau–Lifshitz–Gilbert equation in the linear approximation as an eigenproblem. Further details about this numerical approach can be found in Refs Mruczkiewicz et al., 2013; Rychły et al., 2018 in 2D implementation, and in the Appendix A.1. With that, we model an infinite gyroidal plane with varying thicknesses: 1, 3, and 6 unit cells, corresponding to dimensions of 50, 150, and 300 nm, respectively (Fig. 6.9). Material parameters identical to those in the previous section were used. Throughout our simulations, the external magnetic field, $\mathbf{B}_{\mathrm{ext}}$, was maintained constant at 500 mT and oriented in the plane of the gyroidal plane along the crystallographic direction [100]. This field strength, corroborated by Llandro et al., 2020, supports our assumption

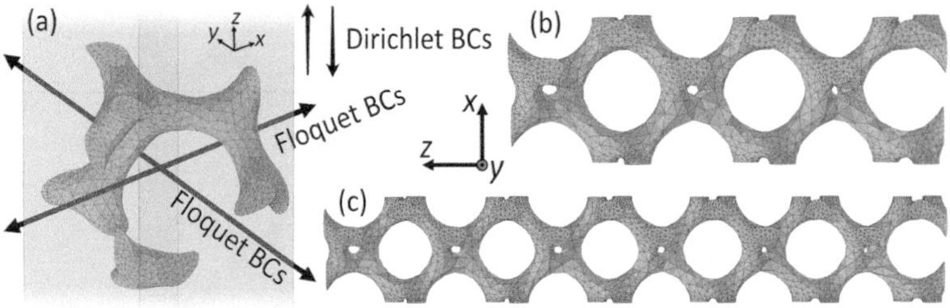

Figure 6.9 Visualization of discretization grids for gyroid structures utilized in COMSOL micromagnetic simulations. Arrows indicate the Bloch–Floquet boundary conditions (BCs) along the x and y axes, defining the structure's plane. In contrast, along the z-axis, perpendicular and at a significant distance from the plane, Dirichlet boundary conditions (BCs) are implemented. This setup is exemplified for a single-unit thick $(1 \times 1 \times 1)$ gyroid layer in (a), and similarly for other models: $1 \times 1 \times 3$ in (b), and $1 \times 1 \times 6$ in (c). Note that the relative sizes of the structures are not to scale in this representation.

of complete saturation of the structure aligned with the external magnetic field's direction.

During the analysis of the micromagnetic simulation results, a surprising pattern emerged in the behavior of the lowest frequency mode, which exhibited the highest intensity, particularly under the influence of the rotating in-plane magnetic field. It is, an intriguing switch in the area of localization of SWs from the bottom [at 30° and 60° in Fig. 6.10(b) and (c)] to the top of the layer (at 120° and 150°). Between these angles, specifically at 0, 90, and 180°, this mode exhibits a bulk character. This shift in mode amplitude concentration regions suggests a complex interplay between magnetic field orientation and intrinsic properties of the gyroid structure, involving dipolar and exchange interactions. The change of the amplitude localization is followed by the change of the resonance frequency [see Fig. 6.10(d)], the lowest (about 13.6 GHz at the field angle of 45–50° and 125–130°) occurs when the mode has a surface character and increases with the transition into the bulk region.

Preliminary analysis indicates that this localization phenomenon is more pronounced in structures with increased thickness, i.e., characterized by a higher number of unit cells per thickness, as can

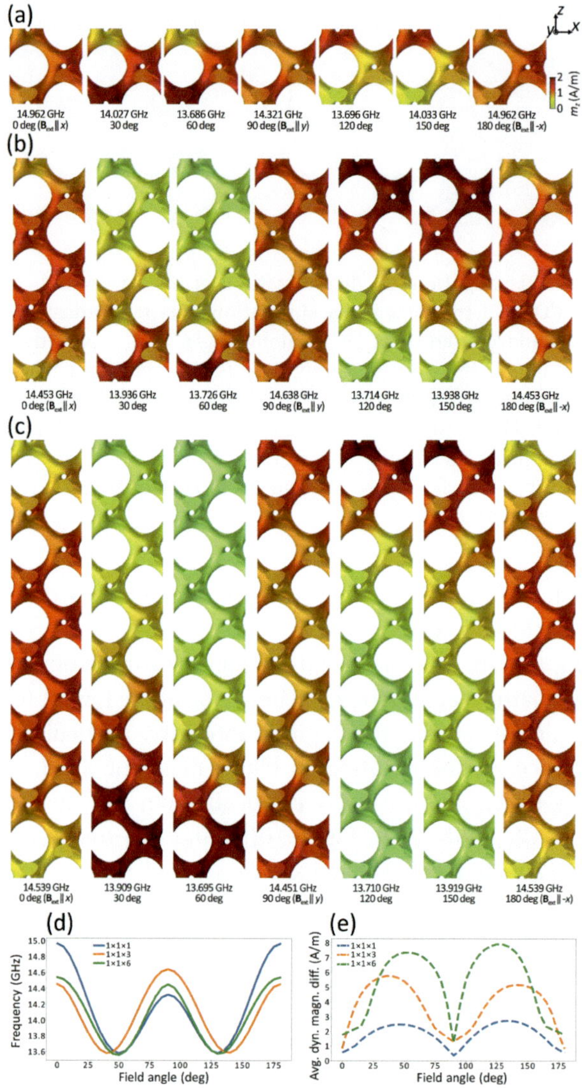

Figure 6.10 Distribution of low frequency, the most intensive mode in (a) $1\times1\times1$, (b) $1\times1\times3$, and $1\times1\times6$ gyroids ($\phi = 10\%$) and their localization as a function of magnetic field rotation ($B_{ext} = 0.5$ T) with respect to the x-axis. The plots below illustrate the impact of the rotating field's angle relative to the structure on the eigenfrequency value (d) and the variation between the minimum and maximum dynamic magnetization components within the unit cell (e).

be seen by comprising SW amplitude distribution in Fig. 6.10(a)–(c). This has been quantified by plotting the difference between the maximum and minimum value of the dynamic magnetization component in the unit cell in Fig. 6.10(d). The largest value of this difference indicates stronger amplitude localization. Interestingly, the angle at which maximum localization occurs also shifts with changes in film thickness. In such cases, the crystallography exerts a greater influence on the SW amplitude distribution and indicates a possible influence of chirality.

Understanding the dynamics controlling the localization of SW modes in these structures has the potential to open up new applications in 3D magnonics, such as sensing or controlling dynamical coupling by rotating the magnetic field.

6.3.2.3 Dispersion relation

The surface character of the SW mode at $k = 0$ has not been found in homogeneous thin ferromagnetic films so far. Thus, it is interesting to see how this and other modes in the gyroid structure evolve with the wavevector. The introduced Bloch–Floquet boundary conditions in COMSOL Multiphysics are well suited for the calculation of the dispersion relation, just by parametrize the solution with the Bloch wavevector, k. In reciprocal lattice space, the range $-\pi/L < k < \pi/L$ defines the 1st Brillouin zone along the main direction of the cubic lattice. The primary focus is on this reciprocal-space region, as the bands from the subsequent zones are folded back to the 1st zone, and the whole magnonic band structure has a periodicity with the period equal to the reciprocal lattice vector $2\pi/L$. Our research delves into the SW dispersion in a finite-height gyroid structure, as defined in the previous section. As evidenced in Fig. 6.11, the direction of the magnetic field relative to SW propagation significantly influences the gyroid's band structure, leading to the opening and closing of band gaps. This phenomenon is exemplified in our simulations of the $1 \times 1 \times 1$ [Fig. 6.11(a)] and $1 \times 1 \times 3$ [Fig. 6.11(b)] gyroid structures for two relative orientations of the in-plane magnetic field and wavevector, i.e., DE configuration in which both directions are orthogonal, and backward volume (BV) configuration in which these vectors are parallel. In the DE configuration, in $1 \times$

1 × 1 structure, we observe a wide band gap from 18.1 to 23.6 GHz, limited by the flat band, and another one just above, from about 23.8 to 27.9 GHz. On the other hand, in BV configuration, there are only two very narrow gaps at low frequencies and a wide band gap between 37.6 GHz and 51.2 GHz. The spectra change significantly when we increase the film thickness to 3 unit cells [Fig. 6.11(b)]. In this case, the band structure is much denser due to the decreasing frequency of the SW modes quantized along the thickness. The band gaps at low frequencies are now closed for DE, but it is opened for BV configuration, although it is rather small, only 1.8 GHz wide (at around 22 GHz). In higher frequencies, the band gap in DE configuration opens (38.6–41.9 GHz). Interestingly, in the frequency range from 23 to 38.6 GHz, the DE bands show a pronounced slope with numerous crossings but no hybridizations, whereas in the BV configuration, extensive hybridizations result in a wavy band structure as a function of *k*, separated by distinct band gaps.

The preliminary analyses reveal the complexities of SW propagation in gyroid structures, including not only the emergence of band gaps and mode crossings, but also the potential to dynamically tune these gaps and modify their widths through external control parameters. The distinctive dispersion properties of gyroids hold great promise for further research aimed at improving the transmission capabilities of magnonic 3D devices.

6.4 Conclusions

To summarize the research described in this chapter, it is evident that ferromagnetic gyroidal nanostructures exhibit interesting properties with significant implications for the field of 3D magnonics. The presented studies, both static and dynamic magnetization based, have shown that factors such as non-trivial shape anisotropy, chirality, and inhomogeneous demagnetization fields within gyroidal structures are determined by the specific crystallography, and can lead to the creation of multiple low-energy state magnetization textures, SW mode localization, and controllable SW propagation. Though preliminary, these investigations have begun to unveil the substantial potential inherent in these structures.

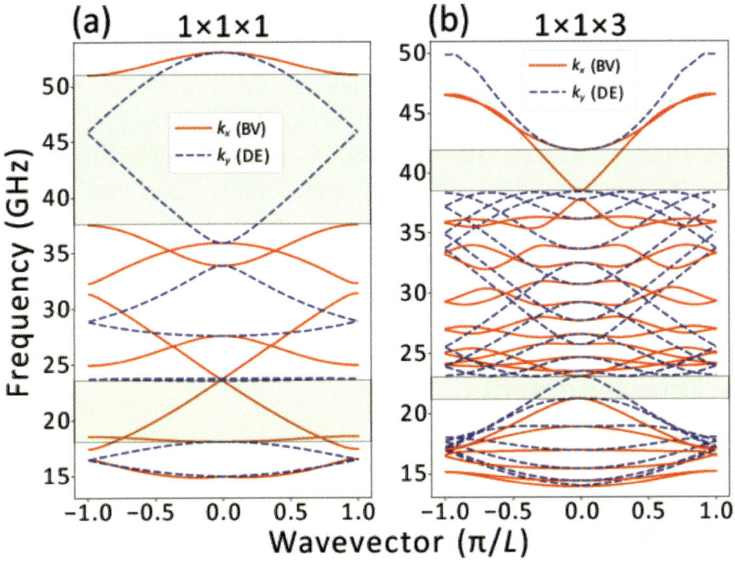

Figure 6.11 The dispersion relations within the 1st Brillouin zone for a gyroid structure along the x-axis ([100] crystallographic direction), characterized by a 10% filling factor and 1 (a), and 3 (b) unit cells per thickness. The data is color-coded to distinguish between different configurations: the BV configuration is represented in red, while the DE configuration is depicted in blue. These plots have been generated in the presence of an in-plane external magnetic field, $B_{\mathrm{ext}} = 0.5$ T, oriented along the x-axis. Wider band gaps specific to one of the configurations are highlighted using green rectangles.

Further studies of individual nanostructures forming gyroids, such as chiral triple junctions, are crucial for a comprehensive understanding of the magnetic behavior of gyroids and their interactions with external fields. Much like how micro/nano texturing in metamaterials influences their macroscopic effective properties, these detailed investigations are crucial for elucidating the intricate relationships between gyroid geometry, chirality, and their effective magnetic properties, particularly at microwave frequencies. This is evidenced by the study of resonance frequencies in gyroid samples under rotational field manipulation, demonstrating the significant influence of geometric anisotropy on the FMR signal strength and the structure's effective properties. Gyroids, with their inherently chiral structure, nanoscale unit cell dimensions, and exchange-length bond sizes, show promise for 3D spin-ice systems, exhibiting intriguing collective SW properties such as band gaps and magnon mode hybridizations. However, further experimental studies are essential to validate these properties and demonstrate their practical applications. Thus, the results presented in this chapter establish our current understanding of ferromagnetic gyroidal nanostructures and pave the way for their further investigation.

Acknowledgments

The research leading to these results was funded by the National Science Centre of Poland, Projects No. UMO-2020/39/I/ST3/02413 and No. UMO-2023/49/N/ST3/03032.

Conflict of Interest

The authors have no conflicts to disclose.

Data Availability Statement

The data underlying this study are openly available in Zenodo at https://doi.org/10.5281/zenodo.13143799.

Appendix

A.1 Micromagnetic Simulations

In Section 6.3.2.1, for the numerical investigation, we used *tetmag*—the GPU-accelerated open-source FEM micromagnetic solver (Hertel, 2023). We exploit its ability to solve magnetostatic open boundary problems in large-scale micromagnetic simulations via a hybrid finite-element/boundary-element formalism (Hertel et al.,

2019). In this analysis, however, we do not assume periodic boundary conditions. All *tetmag*-based micromagnetic simulations were performed in two steps, where the first one was to compute the stable magnetic configuration at a given field value. After the magnetization relaxation, we conducted simulations of the ferromagnetic resonance using a dedicated frequency domain algorithm (d'Aquino and Hertel, 2023) based on a formulation proposed by (d'Aquino et al., 2009).

To investigate the SW modes within the gyroid structure described in Sections. 6.3.2.2 and 6.3.2.3, we utilized the capabilities of COMSOL Multiphysics software. It employs FEM to solve complex systems of coupled partial differential equations. This includes both the Landau–Lifshitz–Gilbert and the Maxwell equations under the magnetostatic approximation. In these simulations, each magnetic moment within the unit cells of the gyroid is represented as a normalized unit vector denoted by $\mathbf{m} = \mathbf{M}/M_s$, where $\mathbf{M}$ is the magnetization distribution function in space and time, and M_s is the saturation magnetization of the ferromagnetic material.

The core of our approach is solving the Landau–Lifshitz–Gilbert equation in its explicit form:

$$\frac{d\mathbf{m}}{dt} = \gamma \frac{1}{1+\alpha^2} \left(\mathbf{m} \times \mathbf{B}_{\text{eff}} + \alpha \left(\mathbf{m} \times \left(\mathbf{m} \times \mathbf{B}_{\text{eff}}\right)\right)\right), \tag{A.1}$$

where $d\mathbf{m}/dt$ is the time evolution of the reduced magnetization, γ is the gyromagnetic ratio, and α represents the dimensionless damping coefficient. The effective magnetic flux density field, $\mathbf{B}_{\text{eff}}$, integrates the externally applied field, $\mathbf{B}_{\text{ext}}$, with the magnetostatic demagnetizing field, $\mathbf{B}_{\text{d}}$, and the Heisenberg exchange field, $\mathbf{B}_{\text{exch}}$:

$$\mathbf{B}_{\text{eff}} = \mathbf{B}_{\text{ext}} + \mathbf{B}_{\text{d}} + \mathbf{B}_{\text{exch}}. \tag{A.2}$$

The demagnetizing field strength, $\mathbf{H}_{\text{d}}$ (equivalent to $\mathbf{B}_{\text{d}}/\mu_0$, where μ_0 is the vacuum permeability), is governed by Ampre's law, and is derived from the magnetic scalar potential gradient, U_{m}:

$$\mathbf{H}_{\text{d}} = -\nabla U_{\text{m}}, \tag{A.3}$$

which further evolves within the magnetic body as:

$$\nabla^2 U_{\text{m}} = \nabla \cdot \mathbf{M}. \tag{A.4}$$

In our COMSOL implementation, we addressed the eigenproblem derived from Eqs. A.1–A.4. Assuming complete magnetization saturation by the magnetic field, we adopted a linear approximation to decompose the total magnetization vector $\mathbf{m}'$ into static equilibrium and dynamic components $\mathbf{m}'(\mathbf{r}, t) = M_s\hat{\mathbf{n}}(\mathbf{r}) + \mathbf{m}(\mathbf{r}, t)\forall\mathbf{m} \perp \hat{\mathbf{n}}$ (local unit vector $\hat{\mathbf{n}}(\mathbf{r})$ defines the equilibrium orientation of the magnetization), disregarding all nonlinear terms in the dynamic magnetization $\mathbf{m}(\mathbf{r}, t)$. This method aligns with foundational research such as Refs. Mruczkiewicz et al., 2013; Rychły et al., 2018, which provides a comprehensive understanding of the underlying physical principles.

The Bloch–Floquet boundary conditions, applied to parallel surfaces on the opposite sides of the unit cell, are defined as:

$$\mathbf{m}_{\mathrm{dst}} = \mathbf{m}_{\mathrm{src}} e^{-ik(r_{\mathrm{dst}} - r_{\mathrm{src}})} \tag{A.5}$$

where k represents the wavenumber, $\mathbf{m}$ is the normalized dynamic component of the magnetization vector at both sides of the unit cell: the destination (dst) and the source (src), r denotes the spatial coordinates of the boundaries where the boundary conditions are applied, and i is the imaginary unit. For a unit cell encompassing periodicity, these boundary conditions are applied to its corresponding parallel faces (as illustrated in Fig. 6.9). By sweeping the wavenumber parametrically, eigenfrequencies are calculated at each interval, resulting in wavenumber versus frequency plots that reveal the dispersion curves for the periodic structure (Hakoda et al., 2018; Collet et al., 2011). These curves characteristically show periodicity in relation to the wavenumber, with a repeating pattern every $2\pi/L$.

For the distant planes parallel to the axes of the examined gyroid layer, Dirichlet boundary conditions are implemented to nullify the scalar magnetic potential, U_{m}. These first-type boundary conditions specify the precise value a variable must take at the boundary during the PDE solution process. Here, we set $U_{\mathrm{m}}|_{\mathrm{src}} = U_{\mathrm{m}}|_{\mathrm{dst}} = 0$. To ensure the simulation's physical fidelity and convergence, these conditions are placed sufficiently far from the specimen. In our COMSOL simulations, the cell's height was configured to be 40 times the thickness of the gyroid layer.

A.2 Broadband FMR Experiment

By adjusting the position of the sample on the CPW, as described in Section 6.3.2.1, apart from gyroidal structures, we observed a higher frequency signal specifically associated with the homogeneous Ni layer at one edge of the sample. It was confirmed by a good match between the detected signal and the theoretical prediction from the Kittel formula for the resonance frequency:

$$f = \frac{\gamma}{2\pi}\sqrt{B_{\text{ext}}(B_{\text{ext}} + \mu_0 M_{\text{s}})}. \tag{A.6}$$

In field-swept broadband FMR experiments at a fixed microwave field frequency, the FWHM of the absorption linewidth follows $\mu_0 \Delta H = 4\pi\alpha f/\gamma$. This relationship holds when the magnetization vector aligns with the applied magnetic field, either in-plane or perpendicular. The linewidth scales linearly with frequency, with the slope determined by the Gilbert damping parameter α. Besides this, empirical data show an additional frequency-independent term:

$$\Delta H = \Delta H_0 + \frac{4\pi\alpha}{\mu_0\gamma} f, \tag{A.7}$$

where ΔH_0 represents inhomogeneous contributions, adding to the overall linewidth observed in the experiments.

References

Armatas, G. S. and Kanatzidis, M. G. (2006). Mesostructured germanium with cubic pore symmetry, *Nature* **441**, 7097, p. 1122–1125.

Bailleul, M., Olligs, D. and Fermon, C. (2003). Micromagnetic phase transitions and spin wave excitations in a ferromagnetic stripe, *Phys. Rev. Lett.* **91**, p. 137204.

Banerjee, C., Gruszecki, P., Klos, J. W., Hellwig, O., Krawczyk, M. and Barman, A. (2017). Magnonic band structure in a Co/Pd stripe domain system investigated by Brillouin light scattering and micromagnetic simulations, *Phys. Rev. B* **96**, p. 024421.

Bansil, A., Lin, H. and Das, T. (2016). Colloquium: Topological band theory, *Rev. Mod. Phys.* **88**, p. 021004.

Barker, J. and Bauer, G. E. (2016). Thermal spin dynamics of yttrium iron garnet, *Phys. Rev. Lett.* **117**, 21, p. 217201.

Bayer, C., Jorzick, J., Demokritov, S. O., Slavin, A. N., Guslienko, K. Y., Berkov, D. V., Gorn, N. L., Kostylev, M. P. and Hillebrands, B. (2006). *Spin-Wave Excitations in Finite Rectangular Elements* (Springer Berlin Heidelberg, Berlin, Heidelberg), ISBN 978-3-540-39842-4, pp. 57–103.

Beginin, E., Sadovnikov, A., Sakharov, V., Stognij, A., Khivintsev, Y. and Nikitov, S. (2019). Collective and localized modes in 3D magnonic crystals, *J. Magn. Magn. Mater.* **492**, p. 165647.

Bessonov, V. D., Mruczkiewicz, M., Gieniusz, R., Guzowska, U., Maziewski, A., Stognij, A. I. and Krawczyk, M. (2015). Magnonic band gaps in YIG-based one-dimensional magnonic crystals: An array of grooves versus an array of metallic stripes, *Phys. Rev. B* **91**, p. 104421.

Butt, M., Khonina, S. and Kazanskiy, N. (2021). Recent advances in photonic crystal optical devices: A review, *Optics and Laser Technology* **142**, p. 107265.

Cheenikundil, R., Bauer, J., Goharyan, M., d'Aquino, M. and Hertel, R. (2022). High-frequency modes in a magnetic buckyball nanoarchitecture, *APL Mater.* **10**, 8, p. 081106.

Chen, J., Hu, J. and Yu, H. (2020). Chiral magnonics: Reprogrammable nanoscale spin wave networks based on chiral domain walls, *iScience* **23**, 6, p. 101153.

Chen, J., Yu, H. and Gubbiotti, G. (2021). Unidirectional spin-wave propagation and devices, *J. Phys. D: Appl. Phys.* **55**, 12, p. 123001.

Chen, Y., Yao, H. and Wang, L. (2013). Acoustic band gaps of three-dimensional periodic polymer cellular solids with cubic symmetry, *J. Appl. Phys.* **114**, 4, p. 043521.

Cheng, J., Ross, C., Smith, H. and Thomas, E. (2006). Templated self-assembly of block copolymers: Top-down helps bottom-up, *Adv. Mater.* **18**, 19, pp. 2505–2521.

Chumak, A. V. (2019). Magnon spintronics, in E. Y. Tsymbal and I. Žutić (eds.), *Spintronics Handbook: Spin Transport and Magnetism, Second Edition*, 2nd edn., chap. 6 (CRC Press, Boca Raton), ISBN 9780429423079, pp. 247–302.

Chumak, A. V., Serga, A. A. and Hillebrands, B. (2017). Magnonic crystals for data processing, *J. Phys. D: Appl. Phys.* **50**, 24, p. 244001.

Chumak, A. V., Vasyuchka, V. I., Serga, A. A. and Hillebrands, B. (2015). Magnon spintronics, *Nat. Phys.* **11**, 6, pp. 453–461.

Coey, J. M. (2010). *Magnetism and Magnetic Materials* (Cambridge University Press), ISBN 9780511845000.

Collet, M., Ouisse, M., Ruzzene, M. and Ichchou, M. (2011). Floquet-Bloch decomposition for the computation of dispersion of two-dimensional periodic, damped mechanical systems, *Int. J. Solids Struct.* **48**, 20, pp. 2837–2848.

Dacorogna, B. (2014). *Introduction to the Calculus of Variations*, 3rd edn. (Imperial College Press).

Damon, R. and Eshbach, J. (1961). Magnetostatic modes of a ferromagnet slab, *J. Phys. Chem. Solids* **19**, 3, pp. 308–320.

d'Aquino, M. and Hertel, R. (2023). Micromagnetic frequency-domain simulation methods for magnonic systems, *J. Appl. Phys.* **133**, 3, p. 033902.

d'Aquino, M., Serpico, C., Miano, G. and Forestiere, C. (2009). A novel formulation for the numerical computation of magnetization modes in complex micromagnetic systems, *J. Comput. Phys.* **228**, 17, pp. 6130–6149.

Demidov, V. E., Urazhdin, S. and Demokritov, S. O. (2010). Direct observation and mapping of spin waves emitted by spin-torque nano-oscillators, *Nat. Mater.* **9**, pp. 984–988.

Dieny, B., Prejbeanu, I. L., Garello, K., Gambardella, P., Freitas, P., Lehndorff, R., Raberg, W., Ebels, U., Demokritov, S. O., Akerman, J., Deac, A., Pirro, P., Adelmann, C., Anane, A., Chumak, A. V., Hirohata, A., Mangin, S., Valenzuela, S. O., Onbasli, M. C., d'Aquino, M., Prenat, G., Finocchio, G., Lopez-Diaz, L., Chantrell, R., Chubykalo-Fesenko, O. and Bortolotti, P. (2020). Opportunities and challenges for spintronics in the microelectronics industry, *Nature Electronics* **3**, 8, pp. 446–459.

Diep, H. (2013). *Frustrated Spin Systems* (World Scientific), ISBN 9789814440745.

Dolan, J. A., Wilts, B. D., Vignolini, S., Baumberg, J. J., Steiner, U. and Wilkinson, T. D. (2015). Optical properties of gyroid structured materials: From photonic crystals to metamaterials, *Adv. Opt. Mater.* **3**, 1, pp. 12–32.

Donnelly, C., Finizio, S., Gliga, S., Holler, M., Hrabec, A., Odstrcil, M., Mayr, S., Scagnoli, V., Heyderman, L. J., Guizar-Sicairos, M. and Raabe, J. (2020). Time-resolved imaging of three-dimensional nanoscale magnetization dynamics, *Nat. Nanotechnol.* **15**, 5, pp. 356–360.

Donnelly, C., Guizar-Sicairos, M., Scagnoli, V., Gliga, S., Holler, M., Raabe, J. and Heyderman, L. (2017). Three-dimensional magnetization structures revealed with X-ray vector nanotomography, *Nature* **547**, pp. 328–331.

Donnelly, C., Hierro-Rodríguez, A., Abert, C., Witte, K., Skoric, L., Sanz-Hernández, D., Finizio, S., Meng, F., McVitie, S., Raabe, J., Suess, D., Cowburn, R. and Fernández-Pacheco, A. (2022). Complex free-space magnetic field textures induced by three-dimensional magnetic nanostructures, *Nat. Nanotechnol.* **17**, 2, pp. 136–142.

Duerr, G., Thurner, K., Topp, J., Huber, R. and Grundler, D. (2012). Enhanced transmission through squeezed modes in a self-cladding magnonic waveguide, *Phys. Rev. Lett.* **108**, 22, p. 227202.

El Hassan, A., Kunst, F. K., Moritz, A., Andler, G., Bergholtz, E. J. and Bourennane, M. (2019). Corner states of light in photonic waveguides, *Nat. Photonics* **13**, 10, pp. 697–700.

Feilhauer, J., Zelent, M., Zhang, Z., Christensen, J. and Mruczkiewicz, M. (2023). Unidirectional spin-wave edge modes in magnonic crystal, *APL Mater.* **11**, 2, p. 021104.

Feng, H., Lu, X., Wang, W., Kang, N.-G. and Mays, J. W. (2017). Block copolymers: Synthesis, self-assembly, and applications, *Polymers* **9**, 10.

Fernández-Pacheco, A., Streubel, R., Fruchart, O., Hertel, R., Fischer, P. and Cowburn, R. P. (2017). Three-dimensional nanomagnetism, *Nat. Commun.* **8**, 1, pp. 1–14.

Fischer, P., Sanz-Hernández, D., Streubel, R. and Fernández-Pacheco, A. (2020). Launching a new dimension with 3D magnetic nanostructures, *APL Mater.* **8**, 1, p. 010701.

Gaididei, Y., Kravchuk, V. P. and Sheka, D. D. (2014). Curvature effects in thin magnetic shells, *Phys. Rev. Lett.* **112**, 25, p. 257203, publisher: American Physical Society.

Gan, Z., Turner, M. D. and Gu, M. (2016). Biomimetic gyroid nanostructures exceeding their natural origins, *Sci. Adv.* **2**, 5, p. e1600084.

Garcia-Sanchez, F., Borys, P., Soucaille, R., Adam, J. P., Stamps, R. L. and Kim, J. V. (2015). Narrow magnonic waveguides based on domain walls, *Phys. Rev. Lett.* **114**, 24, p. 247206.

Garcia-Sanchez, F., Borys, P., Vansteenkiste, A., Kim, J.-V. and Stamps, R. L. (2014). Nonreciprocal spin-wave channeling along textures driven by the Dzyaloshinskii-Moriya interaction, *Phys. Rev. B* **89**, p. 224408.

Girardi, D., Finizio, S., Donnelly, C., Rubini, G., Mayr, S., Levati, V., Cuccurullo, S., Maspero, F., Raabe, J., Petti, D. and Albisetti, E. (2024). Three-dimensional spin-wave dynamics, localization and interference in a synthetic antiferromagnet, *Nat. Commun.* **15**, 1, p. 3057.

Gołębiewski, M., Hertel, R., d'Aquino, M., Vasyuchka, V., Weiler, M., Pirro, P., Krawczyk, M., Fukami, S., Ohno, H. and Llandro, J. (2024). Collective spin-wave dynamics in gyroid ferromagnetic nanostructures, *ACS Appl. Mater. Interfaces* **16**, 17, pp. 22177–22188.

Gołębiewski, M., Reshetniak, H., Makartsou, U., Krawczyk, M., van den Berg, A., Ladak, S. and Barman, A. (2023). Spin-wave spectral analysis in crescent-shaped ferromagnetic nanorods, *Phys. Rev. Appl.* **19**, p. 064045.

Gross, F., Zelent, M., Gangwar, A., Mamica, S., Gruszecki, P., Werner, M., Schutz, G., Weigand, M., Goering, E. J., Back, C. H., Krawczyk, M. and GrAfe, J. (2021). Phase resolved observation of spin wave modes in antidot lattices, *Appl. Phys. Lett.* **118**, 23, p. 232403.

Große-Brauckmann, K. (1997). Gyroids of constant mean curvature, *Exp. Math.* **6**, 1, pp. 33–50.

Große-Brauckmann, K. and Wohlgemuth, M. (1996). The gyroid is embedded and has constant mean curvature companions, *Calc. Var. Partial Differ. Equ.* **4**, 6, pp. 499–523.

Gruszecki, P., Banerjee, C., Mruczkiewicz, M., Hellwig, O., Barman, A. and Krawczyk, M. (2019). Chapter two—the influence of the internal domain wall structure on spin wave band structure in periodic magnetic stripe domain patterns, in R. L. Stamps and H. Schultheiß (eds.), *Recent Advances in Topological Ferroics and their Dynamics, Solid State Physics*, Vol. 70 (Academic Press), pp. 79–132.

Gruszecki, P., Guslienko, K. Y., Lyubchanskii, I. L. and Krawczyk, M. (2022). Inelastic spin-wave beam scattering by edge-localized spin waves in a ferromagnetic thin film, *Phys. Rev. Appl.* **17**, p. 044038.

Gruszecki, P., Lyubchanskii, I. L., Guslienko, K. Y. and Krawczyk, M. (2021). Local non-linear excitation of sub-100 nm bulk-type spin waves by edge-localized spin waves in magnetic films, *Appl. Phys. Lett.* **118**, 6, p. 062408.

Gubbiotti, G. (2019). *Three-Dimensional Magnonics*, 1st edn. (Jenny Stanford Publishing, New York), ISBN 9780429299155.

Gubbiotti, G., Sadovnikov, A., Beginin, E., Nikitov, S., Wan, D., Gupta, A., Kundu, S., Talmelli, G., Carpenter, R., Asselberghs, I., Radu, I. P., Adelmann, C. and Ciubotaru, F. (2021). Magnonic band structure in vertical meander-shaped $Co_{40}Fe_{40}B_{20}$ thin films, *Phys. Rev. Appl.* **15**, p. 014061.

Guo, H., Deenen, A. J. M., Xu, M., Hamdi, M. and Grundler, D. (2023). Realization and control of bulk and surface modes in 3D nanomagnonic networks by additive manufacturing of ferromagnets, *Adv. Mater.*, p. 2303292.

Hakoda, C., Rose, J., Shokouhi, P. and Lissenden, C. (2018). Using Floquet periodicity to easily calculate dispersion curves and wave structures of homogeneous waveguides, *AIP Conf. Proc.* **1949**, 1, p. 020016.

Han, L. and Che, S. (2018). An overview of materials with triply periodic minimal surfaces and related geometry: From biological structures to self-assembled systems, *Adv. Mater.* **30**, 17, p. 1705708.

Hertel, R. (2013). Curvature-induced magnetochirality, *SPIN* **03**, 03, p. 1340009, publisher: World Scientific Publishing Co.

Hertel, R. (2023). tetmag, https://github.com/R-Hertel/tetmag.

Hertel, R., Christophersen, S. and Börm, S. (2019). Large-scale magnetostatic field calculation in finite element micromagnetics with H2-matrices, *J. Magn. Magn. Mater.* **477**, pp. 118–123.

Hirosawa, T., Díaz, S. A., Klinovaja, J. and Loss, D. (2020). Magnonic quadrupole topological insulator in antiskyrmion crystals, *Phys. Rev. Lett.* **125**, p. 207204.

Hsueh, H.-Y., Huang, Y.-C., Ho, R.-M., Lai, C.-H., Makida, T. and Hasegawa, H. (2011). Nanoporous gyroid nickel from block copolymer templates via electroless plating, *Adv. Mater.* **23**, 27, pp. 3041–3046.

Hunt, M., Taverne, M., Askey, J., May, A., Van Den Berg, A., Ho, Y. L. D., Rarity, J. and Ladak, S. (2020). Harnessing multi-photon absorption to produce three-dimensional magnetic structures at the nanoscale, *Materials* **13**, 3, p. 761.

Hurben, M. and Patton, C. (1995). Theory of magnetostatic waves for in-plane magnetized isotropic films, *J. Magn. Magn. Mater.* **139**, 3, pp. 263–291.

Hyde, S., Ninham, B. W., Andersson, S., Larsson, K., Landh, T., Blum, Z. and Lidin, S. (1997a). Chapter 1—the mathematics of curvature, in S. Hyde, B. W. Ninham, S. Andersson, K. Larsson, T. Landh, Z. Blum and S. Lidin (eds.), *The Language of Shape* (Elsevier Science B.V., Amsterdam), ISBN 978-0-444-81538-5, pp. 1–42.

Hyde, S., Ninham, B. W., Andersson, S., Larsson, K., Landh, T., Blum, Z. and Lidin, S. (1997b). Chapter 3—molecular forces and self-assembly, in S. Hyde, B. W. Ninham, S. Andersson, K. Larsson, T. Landh, Z. Blum and S. Lidin (eds.), *The Language of Shape* (Elsevier Science B.V., Amsterdam), ISBN 978-0-444-81538-5, pp. 87–140.

Hyde, S., Ninham, B. W., Andersson, S., Larsson, K., Landh, T., Blum, Z. and Lidin, S. (1997c). Chapter 4—beyond flatland: The geometric forms due to self-assembly, in S. Hyde, B. W. Ninham, S. Andersson, K. Larsson, T. Landh, Z. Blum and S. Lidin (eds.), *The Language of Shape* (Elsevier Science B.V., Amsterdam), ISBN 978-0-444-81538-5, pp. 141–197.

Hyde, S. T., O'Keeffe, M. and Proserpio, D. M. (2008). A short history of an elusive yet ubiquitous structure in chemistry, materials, and mathematics, *Angew. Chem. Int. Ed.* **47**, 42, pp. 7996–8000.

Inoue, M., Baryshev, A., Takagi, H., Lim, P. B., Hatafuku, K., Noda, J. and Togo, K. (2011). Investigating the use of magnonic crystals as extremely sensitive magnetic field sensors at room temperature, *Appl. Phys. Lett.* **98**, 13, p. 132511.

Jamali, M., Kwon, J. H., Seo, S.-M., Lee, K.-J. and Yang, H. (2013). Spin wave nonreciprocity for logic device applications, *Sci. Rep.* **3**, 1, p. 3160.

Ji, L., Zhao, R., Hu, X., Hu, C., Shen, X., Liu, X., Zhao, X., Zhang, J., Chen, W. and Zhang, X. (2022). Reconfigurable ferromagnetic resonances by engineering inhomogeneous magnetic textures in artificial magnonic crystals, *Adv. Funct. Mater.* **32**, 20, p. 2112956.

Jorzick, J., Demokritov, S. O., Hillebrands, B., Bailleul, M., Fermon, C., Guslienko, K. Y., Slavin, A. N., Berkov, D. V. and Gorn, N. L. (2002). Spin wave wells in nonellipsoidal micrometer size magnetic elements, *Phys. Rev. Lett.* **88**, p. 047204.

Kane, C. L. and Mele, E. J. (2005a). Quantum spin Hall effect in graphene, *Phys. Rev. Lett.* **95**, p. 226801.

Kane, C. L. and Mele, E. J. (2005b). Z_2 topological order and the quantum spin Hall effect, *Phys. Rev. Lett.* **95**, p. 146802.

Karcher, H. (1989). The triply periodic minimal surfaces of Alan Schoen and their constant mean curvature companions, *Manuscripta Math.* **64**, 3, pp. 291–357.

König, M., Wiedmann, S., Brüne, C., Roth, A., Buhmann, H., Molenkamp, L. W., Qi, X.-L. and Zhang, S.-C. (2007). Quantum spin Hall insulator state in hgte quantum wells, *Science* **318**, 5851, p. 766–770.

Krawczyk, M. and Grundler, D. (2014). Review and prospects of magnonic crystals and devices with reprogrammable band structure, *J. Phys. Condens. Matter* **26**, 12, p. 123202.

Krawczyk, M., Klos, J., Sokolovskyy, M. L. and Mamica, S. (2010). Materials optimization of the magnonic gap in three-dimensional magnonic crystals with spheres in hexagonal structure, *J. Appl. Phys.* **108**, 9, p. 093909.

Krawczyk, M., Mamica, S., Mruczkiewicz, M., Klos, J. W., Tacchi, S., Madami, M., Gubbiotti, G., Duerr, G. and Grundler, D. (2013). Magnonic band structures in two-dimensional bi-component magnonic crystals with in-plane magnetization, *J. Phys. D: Appl. Phys.* **46**, 49, p. 495003.

Krawczyk, M. and Puszkarski, H. (2006). Magnonic excitations versus three-dimensional structural periodicity in magnetic composites, *Crystal Research and Technology* **41**, 6, pp. 547–552.

Kresge, C. T., Leonowicz, M. E., Roth, W. J., Vartuli, J. C. and Beck, J. S. (1992). Ordered mesoporous molecular sieves synthesized by a liquid-crystal template mechanism, *Nature* **359**, 6397, pp. 710–712.

Kruglyak, V. V. (2021). Chiral magnonic resonators: Rediscovering the basic magnetic chirality in magnonics, *Appl. Phys. Lett.* **119**, 20, p. 200502.

Kruglyak, V. V., Demokritov, S. O. and Grundler, D. (2010). Magnonics, *J. Phys. D: Appl. Phys.* **43**, 26, p. 264001.

Kruglyak, V. V., Keatley, P. S., Hicken, R. J., Childress, J. R. and Katine, J. A. (2006). Time resolved studies of edge modes in magnetic nanoelements, *J. Appl. Phys.* **99**, 8, p. 08F306.

Lacroix, C., Mendels, P. and Mila, F. (2011). *Introduction to Frustrated Magnetism: Materials, Experiments, Theory*, Springer Series in Solid-State Sciences (Springer Berlin Heidelberg), ISBN 9783642105890.

Lambert, C. A., Radzilowski, L. H. and Thomas, E. L. (1996). Triply periodic level surfaces as models for cubic tricontinuous block copolymer morphologies, *Philos. Trans. R. Soc. A* **354**, 1715, pp. 2009–2023.

Lan, J., Yu, W., Wu, R. and Xiao, J. (2015). Spin-wave diode, *Phys. Rev. X* **5**, 4, p. 041049.

Lenk, B., Ulrichs, H., Garbs, F. and Münzenberg, M. (2011). The building blocks of magnonics, *Phys. Rep.* **507**, 4-5, pp. 107–136.

Li, X., Wang, Z., Lei, Z., Ding, W., Shi, X., Yan, J. and Ku, J. (2023). Magnetic characterization techniques and micromagnetic simulations of magnetic nanostructures: from zero to three dimensions, *Nanoscale* **15**, pp. 19448–19468.

Li, Y.-M., Xiao, J. and Chang, K. (2018). Topological magnon modes in patterned ferrimagnetic insulator thin films, *Nano Lett.* **18**.

Lich, L. V., Hue, D. T. H., Giang, D. T. H., Duc, N. H., Shimada, T., Kitamura, T. and Dinh, V. H. (2023). Formation and switching of chiral magnetic field textures in three-dimensional gyroid nanostructures, *Acta Mater.* **249**, p. 118802.

Liu, J., Wang, L. and Shen, K. (2020). Dipolar spin waves in uniaxial easy-axis antiferromagnets: A natural topological nodal-line semimetal, *Phys. Rev. Res.* **2**, p. 023282.

Llandro, J., Love, D. M., Kovács, A., Caron, J., Vyas, K. N., Kákay, A., Salikhov, R., Lenz, K., Fassbender, J., Scherer, M. R., Cimorra, C., Steiner, U.,

Barnes, C. H., Dunin-Borkowski, R. E., Fukami, S. and Ohno, H. (2020). Visualizing magnetic structure in 3D nanoscale Ni-Fe gyroid networks, *Nano Lett.* **20**, 5, pp. 3642–3650.

Luzzati, V. and Spegt, P. A. (1967). Polymorphism of lipids, *Nature* **215**, pp. 701–704.

Ma, F., Zhou, Y., Braun, H. B. and Lew, W. S. (2015). Skyrmion-based dynamic magnonic crystal, *Nano Lett.* **15**, 6, pp. 4029–4036.

Maendl, S., Stasinopoulos, I. and Grundler, D. (2017). Spin waves with large decay length and few 100 nm wavelengths in thin yttrium iron garnet grown at the wafer scale, *Appl. Phys. Lett.* **111**, 1, p. 012403.

Mahmoud, A., Ciubotaru, F., Vanderveken, F., Chumak, A. V., Hamdioui, S., Adelmann, C. and Cotofana, S. (2020). Introduction to spin wave computing, *J. Appl. Phys.* **128**, 16, p. 161101.

Makarov, D., Volkov, O. M., Kákay, A., Pylypovskyi, O. V., Budinská, B. and Dobrovolskiy, O. V. (2022). New dimension in magnetism and superconductivity: 3D and curvilinear nanoarchitectures, *Adv. Mater.* **34**, 3, p. 2101758.

Mamica, S. (2023). The influence of the demagnetizing field on the concentration of spin wave energy in two-dimensional magnonic crystals, *J. Magn. Magn. Mater.* **588**, p. 171395.

Mamica, S., Krawczyk, M. and Grundler, D. (2019). Nonuniform spin-wave softening in two-dimensional magnonic crystals as a tool for opening omnidirectional magnonic band gaps, *Phys. Rev. Appl.* **11**, p. 054011.

Mamica, S., Krawczyk, M., Sokolovskyy, M. L. and Romero-Vivas, J. (2012). Large magnonic band gaps and spectra evolution in three-dimensional magnonic crystals based on magnetoferritin nanoparticles, *Phys. Rev. B* **86**, p. 144402.

Maranville, B. B., McMichael, R. D. and Abraham, D. W. (2007). Variation of thin film edge magnetic properties with patterning process conditions in $Ni_{80}Fe_{20}$ stripes, *Appl. Phys. Lett.* **90**, 23, p. 232504.

Maranville, B. B., McMichael, R. D., Kim, S. A., Johnson, W. L., Ross, C. A. and Cheng, J. Y. (2006). Characterization of magnetic properties at edges by edge-mode dynamics, *J. Appl. Phys.* **99**, 8, p. 08C703.

May, A., Hunt, M., van den Berg, A., Hejazi, A. and Ladak, S. (2019). Realisation of a frustrated 3D magnetic nanowire lattice, *Commun. Phys.* **2**.

May, A., Saccone, M., van den Berg, A., Askey, J., Hunt, M. and Ladak, S. (2021). Magnetic charge propagation upon a 3D artificial spin-ice, *Nat. Commun.* **12**, 1, pp. 1–10.

McClarty, P. A. (2022). Topological magnons: A review, *Annu. Rev. Condens. Matter Phys.* **13**, 1, pp. 171–190.

Michielsen, K. and Stavenga, D. G. (2008). Gyroid cuticular structures in butterfly wing scales: Biological photonic crystals, *J. R. Soc. Interface.* **5**, 18, pp. 85–94.

Mizuno, A., Shuku, Y. and Awaga, K. (2019). Recent developments in molecular spin gyroid research, *Bull. Chem. Soc. Jpn.* **92**, 6, pp. 1068–1093.

Monnier, A., Schüth, F., Huo, Q., Kumar, D., Margolese, D., Maxwell, R. S., Stucky, G. D., Krishnamurty, M., Petroff, P. M., Firouzi, A., Janicke, M. T. and Chmelka, B. F. (1993). Cooperative formation of inorganic-organic interfaces in the synthesis of silicate mesostructures, *Science* **261**, 5126, pp. 1299–1303.

Mruczkiewicz, M., Gruszecki, P., Zelent, M. and Krawczyk, M. (2016). Collective dynamical skyrmion excitations in a magnonic crystal, *Phys. Rev. B* **93**, p. 174429.

Mruczkiewicz, M., Krawczyk, M., Sakharov, V. K., Khivintsev, Y. V., Filimonov, Y. A. and Nikitov, S. A. (2013). Standing spin waves in magnonic crystals, *J. Appl. Phys.* **113**, 9, p. 093908.

Neusser, S., Duerr, G., Tacchi, S., Madami, M., Sokolovskyy, M. L., Gubbiotti, G., Krawczyk, M. and Grundler, D. (2011). Magnonic minibands in antidot lattices with large spin-wave propagation velocities, *Phys. Rev. B* **84**, p. 094454.

Ni, X., Weiner, M., Alú, A. and Khanikaev, A. B. (2019). Observation of higher-order topological acoustic states protected by generalized chiral symmetry, *Nat. Mater.* **18**, 2, pp. 113–120.

Nisoli, C., Kapaklis, V. and Schiffer, P. (2017). Deliberate exotic magnetism via frustration and topology, *Nat. Phys.* **13**, 3, pp. 200–203.

Otálora, J. A., Yan, M., Schultheiss, H., Hertel, R. and Kákay, A. (2017). Asymmetric spin-wave dispersion in ferromagnetic nanotubes induced by surface curvature, *Phys. Rev. B* **95**, p. 184415.

Park, J. P., Eames, P., Engebretson, D. M., Berezovsky, J. and Crowell, P. A. (2002). Spatially resolved dynamics of localized spin-wave modes in ferromagnetic wires, *Phys. Rev. Lett.* **89**, p. 277201.

Petti, D., Tacchi, S. and Albisetti, E. (2022). Review on magnonics with engineered spin textures, *J. Phys. D: Appl. Phys.* **55**, 29, p. 293003.

Romero Vivas, J., Mamica, S., Krawczyk, M. and Kruglyak, V. V. (2012). Investigation of spin wave damping in three-dimensional magnonic crystals using the plane wave method, *Phys. Rev. B* **86**, p. 144417.

Rosi, G., Auffray, N. and Combescure, C. (2020). On the failure of classic elasticity in predicting elastic wave propagation in gyroid lattices for very long wavelengths, *Symmetry* **12**, 8.

Ross, C. A., Berggren, K. K., Cheng, J. Y., Jung, Y. S. and Chang, J.-B. (2014). Three-dimensional nanofabrication by block copolymer self-assembly, *Adv. Mater.* **26**, 25, pp. 4386–4396.

Rychły, J., Gruszecki, P., Mruczkiewicz, M., Kłos, J. W., Mamica, S. and Krawczyk, M. (2015). Magnonic crystals – prospective structures for shaping spin waves in nanoscale, *Low Temp. Phys.* **41**, 10, pp. 745–759.

Rychły, J., Tkachenko, V. S., Kłos, J. W., Kuchko, A. and Krawczyk, M. (2018). Spin wave modes in a cylindrical nanowire in crossover dipolar-exchange regime, *J. Phys. D: Appl. Phys.* **52**, 7, p. 075003.

Sadovnikov, A., Talmelli, G., Gubbiotti, G., Beginin, E., Sheshukova, S., Nikitov, S., Adelmann, C. and Ciubotaru, F. (2022). Reconfigurable 3D magnonic crystal: Tunable and localized spin-wave excitations in CoFeB meander-shaped film, *J. Magn. Magn. Mater.* **544**, p. 168670.

Sahoo, S., Mondal, S., Williams, G., May, A., Ladak, S. and Barman, A. (2018). Ultrafast magnetization dynamics in a nanoscale three-dimensional cobalt tetrapod structure, *Nanoscale* **10**, pp. 9981–9986.

Scherer, M. R. J., Cunha, P. M. S. and Steiner, U. (2014). Labyrinth-induced faceted electrochemical growth, *Adv. Mater.* **26**, 15, pp. 2403–2407.

Schindler, F., Cook, A. M., Vergniory, M. G., Wang, Z., Parkin, S. S. P., Bernevig, B. A. and Neupert, T. (2018). Higher-order topological insulators, *Sci. Adv.* **4**, 6, p. 0346.

Schneider, T., Serga, A. A., Leven, B., Hillebrands, B., Stamps, R. L. and Kostylev, M. P. (2008). Realization of spin-wave logic gates, *Appl. Phys. Lett.* **92**, 2, p. 022505.

Schoen, A. H. (1970). Infinite periodic minimal surfaces without self-intersections, Tech. rep., NASA Electronics Research Center, Cambridge.

Schoen, A. H. (2012). Reflections concerning triply-periodic minimal surfaces, *Interface Focus* **2**, 5, pp. 658–668.

Segalman, R. A. (2005). Patterning with block copolymer thin films, *Mater. Sci. Eng. R Rep.* **48**, 6, pp. 191–226.

Serga, A. A., Chumak, A. V. and Hillebrands, B. (2010). YIG magnonics, *J. Phys. D: Appl. Phys.* **43**, 26, p. 264002.

Serha, R. O., Bozhko, D. A., Agrawal, M., Verba, R. V., Kostylev, M., Vasyuchka, V. I., Hillebrands, B. and Serga, A. A. (2022). Low-damping spin-wave transmission in YIG/Pt-interfaced structures, *Adv. Mater. Interfaces* **9**, 36, p. 2201323.

She, M.-S., Lo, T.-Y. and Ho, R.-M. (2013). Controlled ordering of block copolymer gyroid thin films by solvent annealing, *Macromolecules* **47**, p. 175–182.

Sheka, D. D. (2021). A perspective on curvilinear magnetism, *Appl. Phys. Lett.* **118**, 23, p. 230502.

Sheka, D. D., Pylypovskyi, O. V., Volkov, O. M., Yershov, K. V., Kravchuk, V. P. and Makarov, D. (2022). Fundamentals of curvilinear ferromagnetism: Statics and dynamics of geometrically curved wires and narrow ribbons, *Small* **18**, 12, p. 2105219.

Shindou, R., Matsumoto, R., Murakami, S. and Ohe, J. I. (2013a). Topological chiral magnonic edge mode in a magnonic crystal, *Phys. Rev. B* **87**, p. 174427.

Shindou, R., Ohe, J. I., Matsumoto, R., Murakami, S. and Saitoh, E. (2013b). Chiral spin-wave edge modes in dipolar magnetic thin films, *Phys. Rev. B* **87**, p. 174402.

Simon, P. F. W., Ulrich, R., Spiess, H. W. and Wiesner, U. (2001). Block copolymer-ceramic hybrid materials from organically modified ceramic precursors, *Science* **261**, 5126, pp. 1299–1303.

Singh, M., Callaway, J. and Wang, C. S. (1976). Calculation of g and g' for iron and nickel, *Phys. Rev. B* **14**, 3, pp. 1214–1220.

Skjærvø, S. H., Marrows, C. H., Stamps, R. L. and Heyderman, L. J. (2020). Advances in artificial spin ice, *Nat. Rev. Phys.* **2**, pp. 2522–5820.

Skoric, L., Sanz-Hernández, D., Meng, F., Donnelly, C., Merino-Aceituno, S. and Fernández-Pacheco, A. (2020). Layer-by-layer growth of complex-shaped three-dimensional nanostructures with focused electron beams, *Nano Lett.* **20**, 1, pp. 184–191.

Stenning, K. D., Xiao, X., Holder, H. H., Gartside, J. C., Vanstone, A., Kennedy, O. W., Oulton, R. F. and Branford, W. R. (2023). Low-power continuous-wave all-optical magnetic switching in ferromagnetic nanoarrays, *Cell Rep. Phys. Sci.* **4**, 3, p. 101291.

Sunada, T. (2008). Crystals that nature might miss creating, *Not. Am. Math. Soc.* **55**, 2, pp. 208–215.

Szulc, K., Tacchi, S., Hierro-Rodriguez, A., Diaz, J., Gruszecki, P., Graczyk, P., Quiros, C., Marko, D., Martin, J. I., Vélez, M., Schmool, D. S., Carlotti, G., Krawczyk, M. and Álvarez-Prado, L. M. (2022). Reconfigurable magnonic crystals based on imprinted magnetization textures in hard and soft dipolar-coupled bilayers, *ACS Nano* **16**, 9, pp. 14168–14177.

Szulc, K., Zelent, M. and Krawczyk, M. (2024). Reconfigurable spin-wave platform based on interplay between nanodots and waveguide in hybrid magnonic crystal,.

Tacchi, S., Flores-Farias, J., Petti, D., Brevis, F., Cattoni, A., Scaramuzzi, G., Girardi, D., Cortes-Ortuno, D., Gallardo, R. A., Albisetti, E., Carlotti, G. and Landeros, P. (2023). Experimental observation of flat bands in one-dimensional chiral magnonic crystals, *Nano Lett.* **23**, 14, pp. 6776–6783.

Tacchi, S., Gruszecki, P., Madami, M., Carlotti, G., Kłos, J. W., Krawczyk, M., Adeyeye, A. and Gubbiotti, G. (2015). Universal dependence of the spin wave band structure on the geometrical characteristics of two-dimensional magnonic crystals, *Sci. Rep.* **5**, 1, p. 10367.

Tai, J.-S. B. and Smalyukh, I. I. (2018). Static hopf solitons and knotted emergent fields in solid-state noncentrosymmetric magnetic nanostructures, *Phys. Rev. Lett.* **121**, p. 187201.

Tanaka, G., Yamane, T., Héroux, J. B., Nakane, R., Kanazawa, N., Takeda, S., Numata, H., Nakano, D. and Hirose, A. (2019). Recent advances in physical reservoir computing: A review, *Neural Netw.* **115**, pp. 100–123.

Tang, N., Liyanage, W. L. N. C., Montoya, S. A., Patel, S., Quigley, L. J., Grutter, A. J., Fitzsimmons, M. R., Sinha, S., Borchers, J. A., Fullerton, E. E., DeBeer-Schmitt, L. and Gilbert, D. A. (2023). Skyrmion-excited spin-wave fractal networks, *Adv. Mater.* **35**, 33, p. 2300416.

Tselikas, Y., Hadjichristidis, N., Lescanec, R. L., Honeker, C. C., Wohlgemuth, M. and Thomas, E. L. (1996). Architecturally-induced tricontinuous cubic morphology in compositionally symmetric miktoarm starblock copolymers, *Macromolecules* **29**, 10, pp. 3390–3396.

Turner, M., Saba, M., Zhang, Q., Cumming, B., Schröder-Turk, G. and Gu, M. (2013). Miniature chiral beamsplitter based on gyroid photonic crystals, *Nat. Photonics* **7**, pp. 801–805.

van den Berg, A., Caruel, M., Hunt, M. and Ladak, S. (2023). Combining two-photon lithography with laser ablation of sacrificial layers: A route to isolated 3D magnetic nanostructures, *Nano Res.* **16**, 1, pp. 1441–1447.

Vasileiadis, T., Varghese, J., Babacic, V., Gomis-Bresco, J., Navarro Urrios, D. and Graczykowski, B. (2021). Progress and perspectives on phononic crystals, *J. Appl. Phys.* **129**, 16, p. 160901.

Vignolini, S., Yufa, N. A., Cunha, P. S., Guldin, S., Rushkin, I., Stefik, M., Hur, K., Wiesner, U., Baumberg, J. J. and Steiner, U. (2012). A 3D optical metamaterial made by self-assembly, *Adv. Mater.* **24**, 10, pp. OP23–OP27.

Wagner, K., Kákay, A., Schultheiss, K., Henschke, A., Sebastian, T. and Schultheiss, H. (2016). Magnetic domain walls as reconfigurable spin-wave nanochannels, *Nat. Nanotechnol.* **11**, 5, pp. 432–436.

Wang, X.-G., Nie, Y.-Z., Xia, Q.-L. and Guo, G.-H. (2020). Dynamically reconfigurable magnonic crystal composed of artificial magnetic skyrmion lattice, *J. Appl. Phys.* **128**, 6, p. 063901.

Wang, X. S., Zhang, H. W. and Wang, X. R. (2018). Topological magnonics: A paradigm for spin-wave manipulation and device design, *Phys. Rev. Appl.* **9**, p. 024029.

Wannier, G. H. (1950). Antiferromagnetism. the triangular Ising net, *Phys. Rev.* **79**, pp. 357–364.

Wohlgemuth, M., Yufa, N., Hoffman, J. and Thomas, E. L. (2001). Triply periodic bicontinuous cubic microdomain morphologies by symmetries, *Macromolecules* **34**, 17, pp. 6083–6089.

Wu, J.-S. and Smalyukh, I. I. (2022). Hopfions, heliknotons, skyrmions, torons and both abelian and nonabelian vortices in chiral liquid crystals, *Liquid Crystals Reviews* **10**, 1-2, pp. 34–68.

Yamamoto, K., Thiang, G. C., Pirro, P., Kim, K.-W., Everschor-Sitte, K. and Saitoh, E. (2019). Topological characterization of classical waves: The topological origin of magnetostatic surface spin waves, *Phys. Rev. Lett.* **122**, p. 217201.

Yan, C., Hao, L., Hussein, A. and Raymont, D. (2012). Evaluations of cellular lattice structures manufactured using selective laser melting, *Int. J. Mach. Tools Manuf.* **62**, pp. 32–38.

Yan, P. and Bauer, G. E. (2013). Magnon mediated domain wall heat conductance in ferromagnetic wires, *IEEE Trans. Magn.* **49**, 7, pp. 3109–3112.

Yánez, A., Herrera, A., Martel, O., Monopoli, D. and Afonso, H. (2016). Compressive behaviour of gyroid lattice structures for human cancellous bone implant applications, *Mater. Sci. Eng. C* **68**, pp. 445–448.

Yu, H., Xiao, J. and Schultheiss, H. (2021). Magnetic texture based magnonics, *Phys. Rep.* **905**, pp. 1–59.

Zhuo, F., Kang, J., Manchon, A. and Cheng, Z. (2023). Topological phases in magnonics, *Adv. Phys. Res.*, p. 2300054.

Chapter 7

Theoretical Advances in Understanding the Physics of Core–Shell Spin Transition Nanoparticles

Nicolas di Scala, Nour El Islam Belmouri, Yogendra Singh, and Kamel Boukheddaden
Université Paris-Saclay, UVSQ, GEMAC, UMR CNRS 8635, 45 Avenue des États Unis, 78035 Versailles, France
kamel.boukheddaden@uvsq.fr

7.1 Introduction

Spin-crossover (SCO) solids have been studied for their fascinating properties, exhibiting first-order phase transitions and macroscopic bistabilities, accompanied by significant magnetic, structural, and optical changes. These exceptional properties make these materials promising for applications as high-density information storage and optical switches. Recently, the critical progress made in chemistry allowed the design of spin-crossover nanocomposites, combining the properties of two types of spin-crossover solids having different properties, like different lattice parameters, bulk moduli, transition

2D and 3D Nanostructures: Structure, Properties, and Applications
Edited by Jean-Claude Serge Levy

ISBN 978-981-5129-97-7 (Hardcover), 978-1-003-74871-7 (eBook)
www.jennystanford.com

DOI: 10.1201/97810037487-7

temperatures, ligand fields, etc. Despite these advances, several fundamental questions related to the mechanism of switching in SCO nanostructures remain open. For example, the role of the size of the nanoparticles, their shape as well as the nature of the coupling at the interface are still largely misunderstood. In addition, little theoretical work has been done in this field to understand how elastic stresses control magnetic properties, and even less has been devoted to the search for thermal behavior scaling laws for these nanoswitchable materials.

In the present theoretical contribution, we examine the equilibrium and non-equilibrium properties of spin-crossover (SCO) nanostructures, focusing on two-dimensional core–shell nanoparticle systems composed of an active core SCO material surrounded by an inactive shell which plays the role of the matrix. Together, the two components form a nanocomposite, whose thermodynamic properties crucially depend on the difference of elastic properties of the two moieties. Thus, when the lattice parameters of the core and the shell are different, which happens when the core undergoes a transition between the low-spin (LS) and the high-spin (HS) states, an intrinsic elastic frustration does take place at the core–shell interface, leading to complex distribution of stress inside the core. The latter, strongly disturbs the nucleation of the spin states along the thermal spin transition by causing a spatial distribution of the ligand-field energy, which is highly sensitive to local pressure effects. This leads to the emergence of new types of nucleation and growth mechanisms of the spin states in 3D systems, as well as unexpected multistep transitions along which self-organized HS–LS structures are revealed in the case of both SCO active core and shell moieties. These results highlight the unpredictable nature of the magnetoelastic properties in switchable SCO nanoparticles whether for bare particles or core–shells.

In this theoretical work, we use a 2D version of an electroelastic model to describe both bare and core–shell 2D SCO nanoparticles in which spin states couple through springs allowing the lattice volume change along the phase transition. The simulations employ a Metropolis Monte Carlo algorithm to allow spin change within the nanoparticle core, while lattice position dynamics is achieved through a molecular dynamics algorithm at 0 K to enable mechanical relaxation of site positions in the whole nanoparticle, including core and shell moieties. Through the results of these simulations, we investigate how mechanical stresses and elastic interactions influence the spin transitions in these nanoparticles. One of the objectives is to gain insights into the nucleation and growth mechanisms of spin phases within the core, as well as the impact of the surrounding shell, which induces elastic frustration effects. We also analyze the thermal and structural properties of these systems as a function of nanoparticle size and shell thickness around the core, detailing the hysteresis regimes and critical behaviors observed in each case.

7.1.1 Spin-Crossover Systems and Nanoparticles

Spin-crossover (SCO) molecular systems are fascinating objects with rich physical properties ranging from optics, magnetism, elasticity, statistical mechanics to spintronics [1–9]. They exhibit various aspects of condensed matter, including first-order phase transitions in the solid state with molecular-scale bistability, mechanical properties of materials due to changes in the crystal lattice volume accompanying the transformations, thermochromism, and many more. These properties open up a range of potential applications in the field of switchable molecular solids, such as high-density reversible information storage [6, 10–12], actuators [13–15], pressure sensors [16–18] or display screens [19, 20] to name but a few. For several years, scientists have been particularly interested in nanomaterials or nanocomposites with spin-crossover, which possess physical and chemical properties at the nanoscale that depend on their size [21–25]. The potential applications of these nanomaterials are intriguing in many areas, including high-density memories that would become accessible at the scale of nanoparticles or small microcrystals. Unlike "powder" systems, it should be easier to switch the entire nanoparticle using a pulsed laser. Other potential applications include the use of these nanoparticles in drug delivery [26], pigments [27], cosmetics and so on. The synthesis of SCO materials in the form of nanoparticles allows them to be processed in various forms, ranging from colloidal suspensions [28] and thin films [29] to other nanometric assemblies. There is a significant interest in the literature regarding the effect of nanoparticle size on the properties of SCO—in particular, the hysteresis property, which is expected to be affected by the decrease in the number of molecules constituting the nanoparticle. Indeed, as the number of metal centers that can undergo the transition decreases, it is expected that this decrease in cooperativity in the system could lead to a decrease in bistability. Even though other parameters can also come into play to explain these changes in the properties of SCO nanomaterials [30]. In particular, researchers have designed core–shell SCO nanoparticles consisting of two different materials: one for the core and one for the shell of the nanoparticle, combined in a single system. Some studies have focused on the presence of an active SCO core and an active SCO shell [25, 31–33], with different properties (such as different ligand fields, transition temperatures, or hysteresis widths between the core and the shell), while others have preferred to approach the nanocomposite as consisting of an active SCO core and an inert shell [21, 24]. The introduction of SCO compounds into an active core within a polymer matrix has proven to be promising for technological applications, whether in organic polymers, biopolymers, or mesoporous silica matrices [34–36]. Indeed, the presence of the shell offers the possibility of designing and controlling the spin transition in a more subtle way than what occurs naturally in a free spin transition nanoparticle. The field of SCO nanoparticles has seen many advances in recent years, whether through the synthesis of new materials at the nanoscale, the use of specific experimental techniques to measure and characterize the materials (electron and atomic force microscopy [37], X-ray crystallography [38], etc.), or the theoretical approach to describe and understand the inherent properties of these materials. Several models have been developed or used to describe and understand the physical properties of core–shell nanoparticles. In addition to Ising-type models solved numerically by Monte Carlo simulations [39, 40], models that take into account the coupling between the change in the spin state of the molecule and the change in the local molecular volume during the spin transition have been used to try to understand the link between the first-order phase transitions and the electroelastic behavior of core–shell nanoparticles [41–44]. For example, some studies have addressed the case of nanoparticles consisting of an active SCO core and an inert shell [45], with an emphasis on the properties of the SCO core as a function of the elastic properties of the shell (rigidity, lattice parameter mismatch, and system size).

7.1.2 Spin-Crossover in Nanoparticles, Thin Films and Nanocomposites

The appealing properties of thin films [46] or nanoparticles [21, 40, 47–53] confer a promising market for industry while simultaneously bridging the gap between the understanding of single molecules and bulk macroscopic materials behaviors. When individual materials and technologies achieve their limits in terms of application needs, new technological strategies are introduced to substitute or turn their value in a different manner. One of the ways to overcome the drawbacks of a given material is to make it work in combination with other materials in composites and nanocomposites. This obviously requires a more complex, interdisciplinary approach. Basically, a composite material is defined as a material made from two or more constituents with different properties which, when combined, produce a material with different characteristics from the individual components. Regarding this aim, the manufacture of thin films, nanoparticles and other nano-objects together with a nanoscale understanding of the material properties is of central interest [39, 52, 54–62]. The evolution in the understanding of spin transition at the nanoscale level has been stimulated through instrumental improvements in nanoscale manipulation [63],measurement methods [64]. Simultaneously, a continuous boost in computing power has permitted the modeling of the materials and their properties at the nanoscale level. New attractive progress has been made in the area of SCO complexes with infinite one-, two- or three-dimensional (1D, 2D, 3D) networks, the so-called coordination polymers [65]. The major goals of that approach are to enhance and fine-tune cooperative properties by the strong covalent links between the metallic cores in the polymers. In recent literature, several highly cooperative polymer systems have been reported displaying hysteretic behavior (thermal and piezo) even, in some cases, at room temperature. It is generally accepted that the SCO phenomenon depends on the molecular environment (anion, solvent, packing) [66]. Among the "exotic" SCO nanoparticles system, one can cite the nanoball [67], a $Fe(II)(NCS)_2Cu(I)(Tp^{4-py})(CH_3CN)$ association that exhibit thermally induced, light-induced, and guest-induced (by complexation with an external, "guest" molecule) spin transition. 1D nanowires or nanotubes of SCO systems have not been successfully produced [68] before 2012 (although enhanced shape anisotropy by confined growth and patterning of nanocrystals [69, 70], into 1D-like systems were known), when the first example of the template assembly of a mononuclear SCO molecule into nanowires was reported [71]. Some surprising developments have come from an unexpected exchange with biology [72], where a biomembrane has been used to grow $[Fe(ptz)_6](BF_4)_2$ nanocrystals, using it as a natural stencil to transfer the patterns on Si wafers by conformational contact. The practical applications range from moisture sensors to digital displays [73].

7.1.3 Spin-Crossover as MEMS Actuators

Microelectromechanical systems (MEMS) are microscale devices able to transform a mechanical signal into an electrical one and vice versa. The long-range order of switching molecules results in a collective or cooperative volume shift at the macroscopic level, which can be easily exploited for actuation purposes [74–76]. The combination of SCO materials and MEMS opens up new possibilities for MEMS actuation as well as the study of SCO mechanical characteristics. Although SCO compounds that act at room temperature are now widely available [77], their deposition on delicate mechanical parts and reliable performance remain a significant issue. The potential of SCO materials as actuators requires a thorough understanding of their mechanical properties, as well as their effective integration into practical devices and appropriate transduction mechanisms. Recently, scientists have been able to achieve MEMS devices in which the SCO molecules are not only used to tune the device mechanical properties but also perform controlled and reversible macroscopic work under ambient conditions [13, 78–83]. In the future, nanosystems will provide new opportunities for the selective and sensitive recognition of a wide range of measuring capabilities. Furthermore, the development of new stable and processable compounds as well as a new dedicated procedure has allowed for significant progress, opening up new technological perspectives for SCO compounds and reviving interest in the field. For instance, miniaturized sensors for detecting diverse physical contact such as tiny weights, pressure, radiation (electromagnetic or simple light), and chemical and biological substances may be realized with nanosystems or nano-elements. There is no doubt that the trend is to create nanosystems, and in this context, systems that are superior in energetic efficiency, easily integrable, environmentally friendly, and cost-efficient. However, these real-world applications have just not yet reached industry, primarily because there are still scientific and technological barriers [9]. The parameter space where bistability can be reached is still too small compared to industrial needs, and triggered effects such as LIESST (Light-induced Excited Spin State Trapping) happen at too low temperatures. It could be a possibility that a pure SCO system might not be able to satisfy all the requirements [84]. However, more exotic systems can be devised to make use of the high sensitivity to the environment of SCO centers, for example, in molecular sensing by guest-dependent SCO in an SCO framework [85]. Research applications, notably in the field of non-linear optics [84], are also considered.

7.1.4 Spin-Crossover: A Molecular Complex

An SCO molecule is a coordination complex where a central metal ion is bound to six donor atoms (N, O, S, etc.) [3, 66, 86] in an octahedral symmetry. These inorganic molecular complexes, widely studied in the literature for their molecular-scale bistability, have an electronic configuration between $3d^4$ and $3d^7$ and mainly concern the first-row transition metal complexes: chromium (II), manganese (III), iron (II, III), and cobalt (II). The case of Fe(II)-based SCO materials, with $3d^6$ configuration, is the most common in the literature. The central transition metal ion undergoes a lifting of the degeneracy of its five *d* orbitals due to the presence of a ligand field from the surrounding nitrogen atoms, with a strength $\Delta = E(t_{2g}) - E(e_g)$. The lifting of the degeneracy splits the orbitals into three weakly bonding t_{2g} orbitals and two antibonding e_g orbitals. Depending on the value of Δ relative to the electronic repulsion energy Π between neighboring molecules, a competition between two spin states emerges. If $\Delta \gg \Pi$, the electrons will tend to occupy the low-energy t_{2g} orbital and the metal complex will adopt a low-spin diamagnetic state ($S = 0$, denoted LS), while if $\Delta \ll \Pi$, the filling of the d orbitals will follow Hund's rule by filling all the t_{2g} and e_g levels, thus maximizing the total spin and the metal complex will adopt a high-spin paramagnetic state ($S = 2$, denoted HS). The transition between these two LS and HS states becomes thermally accessible when $\Delta \sim k_B T$, where T is the temperature and k_B is the Boltzmann constant. This spin transition is accompanied locally by an expansion of the volume of the molecular coordination sphere (by about 30%) [87] which can be interpreted by the redistribution of electrons between the t_{2g} and e_g orbitals during the spin transition of the metal ion. This leads to an increase in the Fe-N bond lengths from ~2.0 Å in the LS state to ~2.2 Å in the HS state [66, 88]. These local volume expansions occur at different locations in the material's lattice, but the overall expansion of the entire lattice is relatively small, around 3% to 5% [1], compared to the volume expansion of the molecular coordination sphere: a large part of the volume expansion is absorbed by the molecular structure of the lattice via ligand rotations or changes in molecular packing. The spin transition, achievable when Δ and Π have equivalent energy values, can be controlled by various external parameters, which makes it very interesting from an application point of view, as mentioned previously. Indeed, we can cite as control parameters: external pressure [89–91], magnetic fields [92–94], temperature variation [95–97] or even light radiation [25, 75, 98] (via the LIESST effect, meaning Light-Induced Excited Spin State Trapping, which consists of trapping a metastable HS state at low temperature by light). Once the transition between the LS and HS states is achieved, it leads to many changes in the material's properties, including magnetic, structural, optical and vibrational changes that are measurable during the transition. It is common in the SCO literature to characterize the state of the system, and in particular to characterize the cooperativity of the SCO system, by the high-spin fraction denoted n_{HS}, which is the fraction of molecules in the HS state for a given set of parameters (usually temperature or pressure). Depending on the intensity of the elastic interactions between the molecules, the literature reports a wide variety of spin transitions. In particular, in the case of thermally induced spin transitions, a continuous progressive transition is observed in weakly cooperative SCO materials [66, 99], corresponding to a Boltzmann population of two degenerate states, while a sharp transition or a first-order transition with hysteresis can be observed in strongly cooperative systems [88, 100]. The origin of the hysteresis phenomenon from a physical point of view, and the desire to understand its link with the degree of cooperativity of the system, has led to the development of various theoretical models. From mean-field approaches using phenomenological interaction parameters acting similarly for all molecules [66, 101], or Ising-like approaches in which interactions are written under a form of an exchange term [102, 103], to elastic models where interactions between spin state of molecules and the lattice are introduced, taking into account the lattice volume change at the transition (mechanoelastic model [104, 105], anharmonic model [106, 107] or electroelastic model [108, 109]).

7.1.5 A Brief Survey on Core–Shell Spin-Crossover Nanoparticles: Experiment and Theory

During the last decade, compounds with multistep spin conversion remained quite rare and attracted increasing attention in theoretical and experimental studies [33, 110–113] due to their possibility of building up three-byte electronics. Multistep SCO behavior results from structural ordering or/and the existence of multistability in the molecule itself, like in binuclear SCO systems [114, 115]. In contrast, some multistep SCO systems consist of an asymmetric unit containing two or more non-equivalent sites [116, 117], having different local environments. At the macroscopic scale, the competition or the interplay between the two types of SCO sites manifests through the existence of an intermediate phase associated with partial conversion from HS to LS states. It is worth noticing that the processing of SCO materials with two non-equivalent sites, although possible, is hardly controllable, and most of the time, the behavior of the high-spin fraction is hardly predictable. That is why

(either photoinduced or thermoinduced) and ferromagnetic shells were conducted, which discussed about the role strain played in shifting the magnetization of core entities [140] (Co^{II} Fe^{III} pairs), towards higher temperatures which were closer to that of shell (CsNiCr) entities. This synergy between core and shell was explained through magnetic dipolar interactions between the core and the shell entities (CsNiCr@$Co^{III}Fe^{II}$@CsNiCr particles, 10 nm photoactive shell sandwiched between a 12 nm ferromagnetic core and 5 nm external ferromagnetic shell). However, almost no photo-transformation was observed [141] for larger particles ($Rb_{0.5}Co^{III}Fe^{II}_{0.8}$@$Rb_{0.2}NiCr_{0.7}$, 45 nm core and 26 nm shell), for a similar misfit range, which was again explained due to strong coupling between the core and shell and the ratio between the core and shell thicknesses (shell layers affecting the expansion of core). Although, studies conducted on larger heterostructures (greater than 200 nm, of $Rb_{0.2}Co^{II}Fe^{III}$@RbMCr, where M = Ni, Co, and Cr), with low misfits ($\sim$0.9%) [138, 141–143], shows a 90 K decrease in magnetization, due to CTIST induced switching in core from HS to LS state during the cooling regime. A strained layer with a thickness of 24 nm for KCoCr and 43 nm for KNiCr was proposed [32] see Fig. 7.2, above which a bulk behavior could be considered. This highlights the importance of the volume ratios between the core and the shell in optimizing these mechanical effects, for both SCO core's impact on the shell and vice versa. This is a simple demonstration of mechanical coupling caused by the CTIST's volume transition. The most significant argument is that, even though CsNiCr shell exerts strain over the core, SCO transition is sustained at a slightly lower transition temperature (though without hysteresis) while the core remains contracted. These examples show how synergies can be obtained by elastically binding the center to a magnetic shell.

Aside from PBA-based core–shells, an interesting illustration was given on iron triazole-based core–shells [33], which combine a $[Fe(NH_2trz)_3](BF_4)_2$ core and a $[Fe(Htrz)_2\ (trz)](BF_4)$ shell with the aim of coupling different SCO behaviors. The rods (crystallites that make up the shell) on the upper/lower sides of the core are shorter than the rods rising on the core sides (see Fig. 7.3).

The authors propose a mixed-composition shell with $[Fe(NH_2\ trz)_{3z}(Htrz)_{3-3z}](BF_4)_2$ instead of pure $[Fe(Htrz)_2(trz)]\ (BF_4)$. The total SCO behavior indicates the existence of three SCO transitions

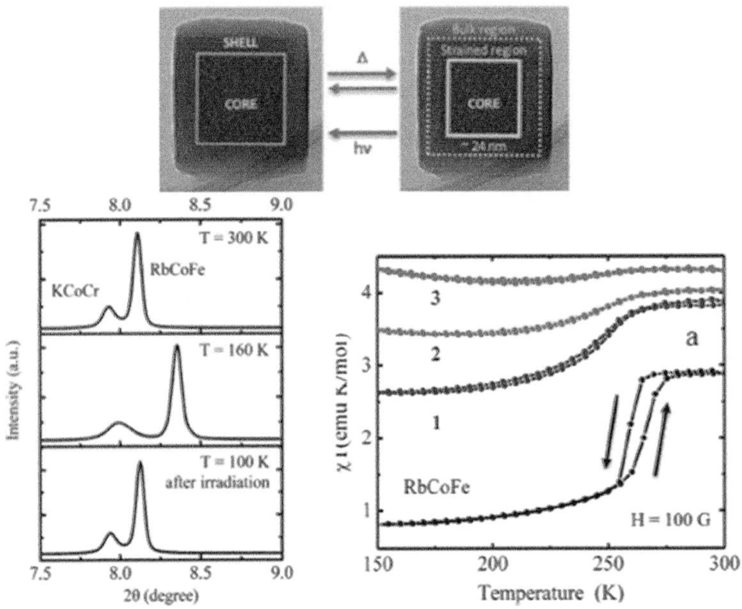

Figure 7.2 RbCoFe@KCoCr heterostructures and scheme showing the presence of a strained region released by light. (Bottom left) XRD diagrams showing the contraction of the core and strained shell at 160 K and the released strain under light. (Bottom right) $\chi T = f(T)$ plots for heterostructures comprising a 137 $\pm$ 12 nm core and thicknesses of 11 nm (1), 23 nm (2), and 37 nm (3) showing SCO of the core at the same temperature. Adapted from Ref. [32].

(core, shell, and interface layer) on the ascending branch, while the $[Fe(NH_2trz)_{3z}(Htrz)_{3-3z}](BF_4)_2$ shell has intermediate temperatures, resulting in a very gradual conversion around 285 K.

7.1.7 Focus on Some Theoretical Descriptions

In a pioneering work [144], it was the first time it was point out the value of surface relaxations, demonstrating that the hysteresis width decreases with size reduction in a different manner depending on the nanoparticle form. At the same time, the equilibrium temperature remains unchanged (Fig. 7.4). However, experimental observations of phase stability in some coordination nanoparticles have revealed a clear size dependence, resulting in a downshift

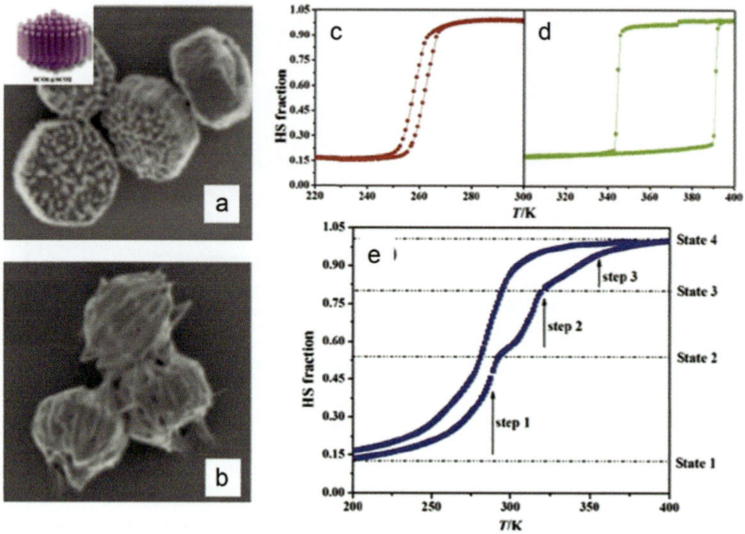

Figure 7.3 SEM images of the SCO@SCO particles at an early stage (a) and final stage (b). $n_{HS} - f(T)$ plots for the core alone (c), the shell alone (d), and the resulting core–shells (e). Adapted from Ref.[33].

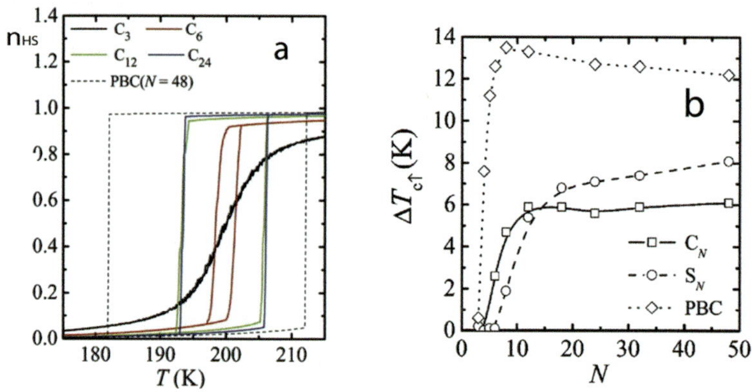

Figure 7.4 (a) The fraction of the unit in the HS state in an equilibrium state, n_{HS}, for various sizes of cubic particles. For comparison, the result at $N = 48$ with the PBC is also shown. (b) Size dependence of the width of the thermal hysteresis DTc for cubic particles (CN), spherical ones (SN) and the PBC case.

of the equilibrium temperature and the existence of incomplete transitions [53, 131]. Surface chemical reactions, structural disorder, and/or ligand modification by the external environment may all be blamed for various physical and chemical properties at the surface. Such observations have been taken into account by adding specific boundary conditions (surface molecules fixed in the HS state [39] or a weakening of the ligand field at the surface [145, 146]). An investigation of local surface effects using the Ising-like model has shown that the HS surface sheet acts as a "negative pressure" [39], but pressure effects can be grasped through an effective modification of the model.

Since the thermodynamic method can allow direct comparisons with experimental observations [147], the existence of surface physical properties different from those of the nanoparticle core has been integrated in terms of surface energies in a simple nano-thermodynamic model [62]. The so-called core–shell approach is used to create spherical SCO nanoparticles with a bulk (core-like) and a surface (shell-like). As the nanoparticle radius is reduced, the size reduction causes either a downshift or an upshift of the transition temperature (Fig. 7.6) and follows the same decay as presented in [39] (Fig. 7.5). Incomplete transitions and the inactivity of molecules at the surface can also be predicted using thermodynamic models, so the fixed boundary conditions suggested in [146] are retrieved; however, the possibility of fixing the surface in the LS state is not ruled out. Main conclusions derived thus far from these models are that the surface energy plays a significant role as the driving force of the spin transition at the nanoscale, and downshifting of the transition temperature with the occurrence of a residual HS fraction at low temperature appears to be much more prevalent than upshifting of the transition temperature with the occurrence of a residual LS fraction at low temperature [147].

Continuation is essential to keep enriching the future prospects of any field. Similarly, here spin-phonon models [148–151] and the extension of these models accounting for the volume change during spin transition designed by several groups [104, 106, 152] leads to the electroelastic model [41], which takes into account the coupling between the spin state change of the molecule (from LS to HS) and the change of the local molecular volume in the course of

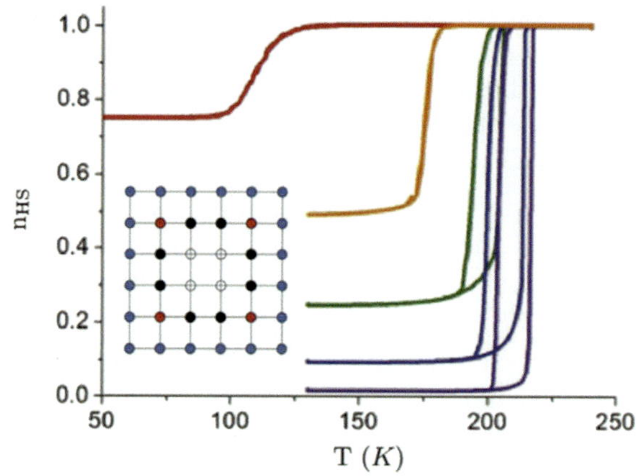

Figure 7.5 Calculated temperature dependence of the HS fraction for square-shaped nanoparticles of different sizes in the framework of the Ising-like model with fixed boundary conditions (HS state fixed at the surface). From left to right, L = 4, 7, 10, 40, 200 (quasi-infinity). Insert: Lattice configuration in the case L = 6. Blue filled circles are HS-fixed (edge) atoms. All inside atoms are active: filled red (black) circles stand for atoms having two (one) inactive HS as the nearest neighbors, and open circles stand for atoms surrounded by only active sites.

the spin transition and other models including the role of surface relaxations [153, 154] and the particle environment [45, 155, 156]. Scientists conducted Monte Carlo simulations using the electro-vibrational model [41], with a nanoparticle embedded in an external environment whose elastic properties and structure were chosen to be close to the HS phase ("soft" matrix). The strain and local pressure mappings show that raising the thickness of the surrounding medium causes the elastic strain at the nanoparticle/matrix interface to be released and the transition temperature to be lowered (Fig. 7.7). This work results in predicting the thermal hysteresis, which moves according to the atomic surface/volume ratio and is in excellent agreement with the available experimental data [53]. Since this pioneering work, the electroelastic model has been extended to incorporate the case of core–shell SCO nanoparticles, made of an active SCO core and an inert shell, where SCO core is investigated in detail as a function of the elastic properties (rigidity, lattice parameter misfit, and size) of the shell [56, 133, 157–159].

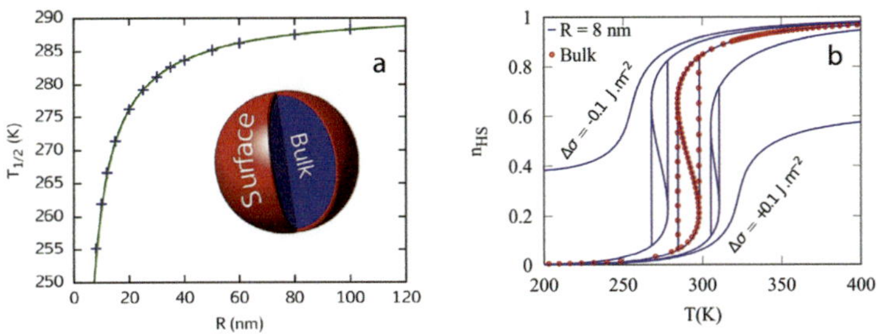

Figure 7.6 (a) Evolution of the equilibrium temperature with the size reduction. Insert: schematic representations of a spherical core–shell nanoparticle; (b) simulation of spin transition curves with the nano-thermodynamic model. The symbols display the spin transition for the bulk material. The lines show the transition for an 8 nm nanoparticle with different surface energy density differences between the HS and LS phases $\Delta\sigma = \sigma_{Hs} - \sigma_{Ls}$.

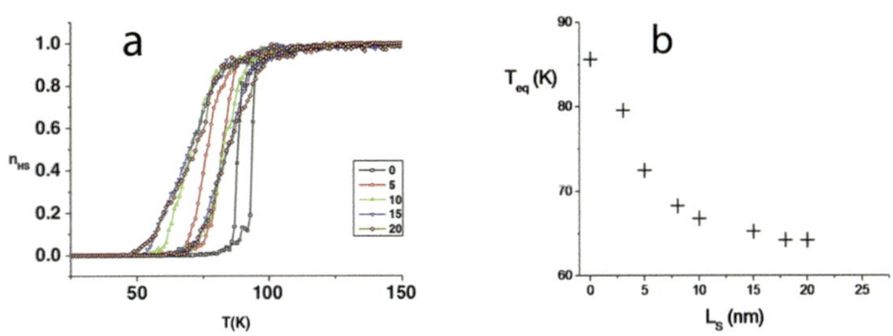

Figure 7.7 (a) Thermal dependence of the HS fraction for square-shaped nanoparticles at constant core size surrounded by a various number of HS shell layers (0-20). The hysteresis shifts downwards as a result of the negative elastic pressure induced by the shell. (b) Evolution of the equilibrium temperature for different shell thickness (LS).

All of these experimental and theoretical findings of the physical properties of hybrid nano-objects and hollow nanoparticles pique enough curiosity and interest to warrant further research for both functional and fundamental reasons, as these structures depict the fundamental role of the external environment and interfacial interactions on phase stability at the nanoscale.

Scientists used a similar method to show that if the matrix/nanoparticle couplings are strong enough, the external matrix

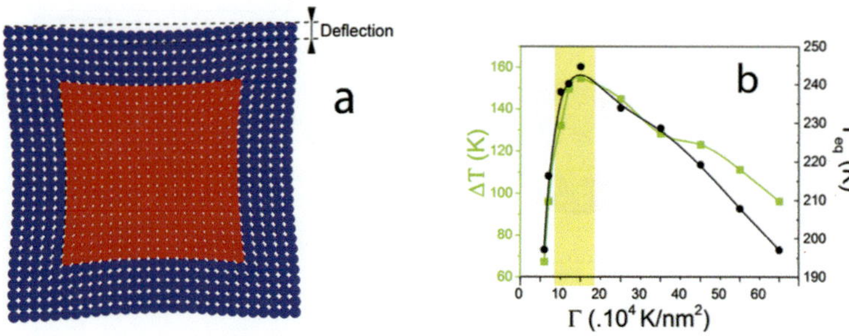

Figure 7.8 (a) Example of a simulated squared core–shell nanoparticle using the electroelastic model. The elastic misfit between the core and the shell induces a strain in the heterostructure, which the deflection calculation can quantify. (b) Effect of the shell stiffness on the thermoinduced spin transition of the core–shell nanoparticle: hysteresis width (green line) and equilibrium temperature (black line).

will retain the bistability property in the SCO nanoparticle [158]. To simulate SCO heterostructures, a further extension of the electroelastic model was used [45]. Monte Carlo simulation was used to examine the thickness and rigidity effect of the surrounding shell on the spin transition properties of the core. The SCO core's thermal hysteresis is found to be heavily influenced by the shell rigidity. Furthermore, when the shell rigidity is similar to that of the core, resonance activity is observed (Fig. 7.8).

Some studies [146] used the spin-phonon model and Monte Carlo simulations to simulate core–shell nanoparticles. The interfacial energy is zero when the core and shell have identical elastic properties, and no change in phase stability shift is observed when compared to the corresponding nanoparticle or hollow nanoparticle. The elastic misfit, on the other hand, causes a compressive (respectively tensile) stress on the SCO core, resulting in an upshift (respectively downshift) in the transition temperature. Despite the fact that elastic models provide an origin for the core–shell coupling process, establishing a relation between Hamiltonian parameters and experimentally accessible quantities remains difficult. As a result, a continuum medium mechanism-based analytical method for modeling a core–shell nanoparticle with a coherent interface has been developed [160]. To optimize the elastic coupling between the core and the shell, a quantitative analysis of the effect of the interfacial elastic energy on phase stability can be carried out, and thus by injecting the interfacial energy density into the model, the spin transformation of the core–shell nanoparticle can be simulated. Since the elastic interaction between molecules is not explicitly taken into account in the model [161], and the elastic interfacial stress cannot be coupled with the distortions occurring during the spin-state transition, the effect of the elastic misfit on collective behavior and bistability cannot be grasped with this model. This is the key disadvantage of this method. Elastic models are a step forward in understanding the behavior of cooperative SCO materials and providing a full explanation of the SCO material's interaction with its environment. They are more accurate and functional than mean-field or Ising-like models for simulating phenomenological behaviors of bulk SCO in all of their various yet convergent ways, vastly improving previous models. One of the most challenging tasks is to thoroughly monitor and evaluate the design of the interface in ultrasmall nanosystems (defects, types of bonds, etc.). Computational methods will be crucial in evaluating certain surface/interface energies and guiding chemists in nanostructure optimization. Experimental investigations in this field have been performed by several groups [33, 34, 162–164] among whom [33] (see Fig. 7.3), in particular, have succeeded to grow high-quality spin-crossover nanocomposites, which have different SCO properties since the former has a transition temperature $T_1 = 260$ K and a hysteresis width of 4 K, while the latter shows a transition temperature $T_2 = 368$ K and an associated thermal hysteresis width of 46 K. Thus, these two materials have different ligand fields and most likely different strengths of cooperativity between the molecules. In the experiment, the core is constituted by the material which has the smaller ligand field, which was then covered by the second material (stronger ligand field) in rather heteroepitaxial conditions. The magnetic studies of the formed nanocomposite, reported by the authors [33, 163], showed the occurrence of clearly two transition temperatures on heating, located at 288 K and 314 K. Thus, the transition temperature of the first constituent (the core) increased by 28 K while that of the second material (the shell) decreased by 54 K, due to the elastic interplay between the core

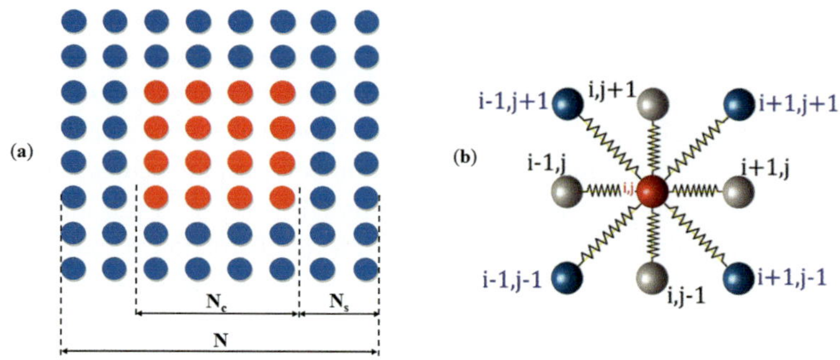

Figure 7.9 (a) Schematic structure of the SCO nanocomposite. Blue and red dots are two spin-crossover sites belonging to different materials. (b) The configuration of the elastic interaction in the two-dimensional square model was considered in this previous study.

and the shell nanostructures. Furthermore, a very gradual and third transition involving a small fraction of molecules occurs at 347 K. It is also interesting to notice that the 2D version of the electroelastic model was used to describe core–shell SCO nanoparticles with an active SCO core and an inert shell in [25, 41] as well as in the case of ideal situations of active core–shell SCO nanoparticle, made of two SCO constituents able to convert between LS and HS states [43, 44]. These situations are sketched in Fig. 7.9, which presents the schematic view of the elastic lattice.

Without going into the details of these previous investigations, we briefly summarize some of the important results.

In the case of an inert soft shell, having the same lattice parameter as that of the HS phase, it is found that the spin transition of the core shifts downwards with an increase in the thermal hysteresis width, attributed to the existence of a negative local pressure on the core part. On the other hand, in the case of active shell (able to convert between HS and LS states) an interplay between the transitions of the two constituents was demonstrated, leading to multistep transitions stabilized by elastic frustration effects.

This chapter is organized as follows: in Section 7.2, we present the 2D electroelastic model used and the simulation method; Section 7.3, present the results obtained from these simulations for the core–shell nanostructure, and in Section 7.4, we conclude and outline some possible developments of this work.

7.2 Electroelastic Model and Numerical Implementation

7.2.1 2D Electroelastic Model

Although SCO nanoparticles are usually 3D materials, whose structures are hardly measurable by conventional X-ray diffraction techniques, it has been well demonstrated using TEM (transmission electron microscopy) that their shape are quite regular, even if controlling their size is often challenging. In some cases, as for Fe(pyrazine)Pt$(CN)_4$ [51], nice platelets SCO nanoparticles are obtained, with a strong 2D character. In the same way, new low-energy pulsed laser deposition techniques are now used to grow molecular spin transition nanoparticles [165–167], replacing Langmuir–Blodgett techniques which, although they allow layer-by-layer growth, lead to very non-cooperative materials. Thus, we chose here to study the size and elastic constants effects in SCO nanoparticles using a two-dimensional electroelastic model, which is a microscopic model coupling the electronic and elastic properties of SCO lattices, already successfully employed in previous publications to obtain results close to experiments [108, 109]. The case of 3D nanoparticles is certainly very interesting, but it requires the use of a large number of physical parameters for the core and the shell, which makes the studies very time consuming. It will be dealt with in a separate work.

Thus, here we will focus on the case of 2D nanoparticles defined by an active core of SCO spins, and an inert shell that surrounds the latter. The spins located in the core of the nanoparticle can thermally switch between LS and HS states: we allow a volume expansion of the nanoparticle during this spin transition. We are interested here in 2D square-shaped lattices of different sizes with free boundary conditions in order to allow the lattice to expand or contract in all directions of space. These macroscopic deformations of the lattice result from the reversals of the spins of the nanoparticle

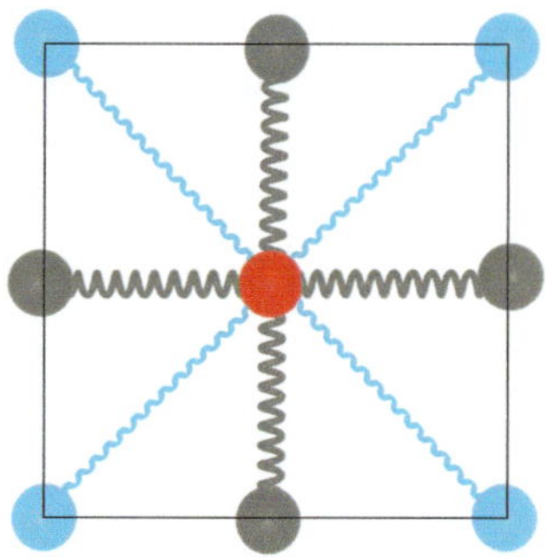

Figure 7.10 Schematic view of the 2D elastic lattice within an elementary simulation cell for a test site represented in red, connected to its neighbors by springs represented in grey for the first neighbors (1n), in blue for the second neighbors (2n) The 1n neighbors are located along the edges of the cell, the 2n neighbors are located along the diagonals of the cell.

core which will locally lead to volume (here surface) changes during their transition between the LS $\leftrightarrow$ HS state. The fact of having free boundary conditions will also allow us to reveal possible surface effects expected in the case of low-dimensional systems as it is the case here.

In all of these simulations, each site (or node) of the lattice represents an SCO molecule described by a fictitious spin S having two possible electronic states, $S = +1$ (respectively -1) associated with the HS (respectively LS) state of the molecule. The sites of the lattice are all connected to each other by springs to represent the elastic interactions between the molecules, the equilibrium distances of these springs depending on the state of the spins of the neighboring connected sites. As illustrated in Fig. 7.10, the first neighbors (1n) are located along the edges of the simulation unit cell, and the second neighbors (2n) are found along the diagonals of the cell. The 2nd neighbors are considered to avoid a shear of the system and thus ensure mechanical stability to the lattice. The equilibrium distance between two neighboring HS sites is naturally greater than that between two LS sites because of the volume change accompanying the spin transition from LS to HS. The deformation of the elementary unit cell can be isotropic or anisotropic, with in the latter case expansion/contraction rates depending on the direction

of space during the transition from the LS to HS state [109, 168, 169]. However, in the present investigations, we will restrict ourselves to the case where the deformation of the simulation unit cell remains isotropic.

The total effective Hamiltonian of the system is given by:

$$H = H_{elec} + H_{elast} \tag{7.1}$$

The first temperature-dependent term in Eq. 7.1 corresponds to the effective electronic contribution of the system's "ligand field":

$$H_{elec} = \sum_{i} \left(\Delta - k_B T \ln g\right) S_i \tag{7.2}$$

With Δ representing the energy gap of the ligand field arising from the nitrogen atoms surrounding the SCO metal center, T being the temperature, and k_B the Boltzmann constant. The entropy term $(-k_B T \ln g)$ is an entropic contribution arising from the difference in vibrational properties between the HS and LS states (spin, orbit, phonons, and intramolecular), with g being the degeneracy ratio between the HS and LS states.

The second term of Eq. 7.1 corresponds to the elastic contribution of the system defined by the elastic energy of the lattice via the potential energies of all neighboring molecules:

$$\begin{aligned} H_{elast} &= V_{1n}\left(|\vec{r}|\right) + V_{2n}\left(|\vec{r}|\right) \\ &= \underbrace{A_{1n} \sum_{(i,j)} \left(r_{ij} - R_{1n}(S_i, S_j)\right)^2}_{1n \text{ neighbors}} \\ &\quad + \underbrace{B_{2n} \sum_{(i,k)} \left(r_{ik} - R_{2n}(S_i, S_k)\right)^2}_{2n \text{ neighbors}} \end{aligned} \tag{7.3}$$

With elastic constants connecting a site to its neighbors which are defined by A_{1n} for first neighbors (1n), B_{2n} for second neighbors (2n). These elastic coefficients are chosen to be isotropic and do not depend on a specific direction in space or the spin state of neighboring sites, although it has been observed experimentally that the LS state is generally more rigid than the HS state [170] even if in rare cases the opposite has been observed [171]. The Euclidean distances are given by r_{ij} (resp. r_{ik}) between the (1n) (resp. (2n)) sites i and j (resp. i and k), write $r_{ij} = \sqrt{(x_j - x_i)^2 + (y_j - y_i)^2}$, and similarly for r_{ik}, by exchanging the index j by k in the previous expression.

The equilibrium distance between two 1n sites is given by $R_{1n}(S_i, S_j)$, and because of the volume change accompanying the spin flip, the equilibrium distance is considered to depend on the spins S_i and S_j of the connected sites. Its expression is then written as

$$R_{1n}(S_i, S_j) = R_0 + R_1(S_i + S_j), \tag{7.4}$$

Where $R_0 = \frac{(R_{HH} + R_{LL})}{2}$ is the average lattice bond length and $R_1 = \frac{\delta_R}{4}$ with $\delta_R = (R_{HH} - R_{LL})$ the lattice misfit between HS and LS. In the formula (7.4), we considered that the distance between a HS and a LS neighboring sites is exactly located at the center of mass of those of HS–HS and LS–LS distances. By noting R_{LL}, R_{HH}, R_{HL} the equilibrium distances between the (1n) sites associated with the respective electroelastic configurations LS–LS, HS–HS and HS–LS, this gives the following relations for the equilibrium distances as a function of the spin state of the neighboring connected sites:

$$\begin{aligned} R_{LL} &= R_{1n}(-1, -1) \\ R_{HH} &= R_{1n}(+1, +1) \\ R_{HL} = R_{1n}(+1, -1) &= R_{1n}(-1, +1) \end{aligned} \tag{7.5}$$

For simplicity, we assume in this study that $R_{HL} = \frac{(R_{HH} + R_{LL})}{2}$ and we choose for the next-nearest-neighbor elastic constants a value of $B_{2n} = 0.3A_{1n}$ in order to keep consistency with our previous studies, and to compare the studies carried out with the same model [108, 109]. The equilibrium distance between two 2n sites denoted by $R_{2n}(S_i, S_k)$ simply corresponds to the Euclidean distance along the diagonals of the unit cell. Its expression is then: $R_{2n}(S_i, S_k) = \sqrt{2}R_{1n}$. We summarize in Table 7.1 the numerical values of the equilibrium distances and elastic constants used in the simulations, for better readability.

The other values of the thermodynamic parameter used in all simulations carried out in this study are: $\Delta = 450$ K for the ligand-field energy and $g = 150$ for the HS/LS degeneracy ratio, assumed

Table 7.1 Values of the nearest-neighbor equilibrium lattice parameters and elastic constants used in the simulations.

	Distances (nm)			Elastic constants (K.nm^{-2})
$1n$ distances	$R_{HH} = 1.20$	$R_{HL} = 1.10$	$R_{LL} = 1.00$	$A_{1n} = 1.5 \cdot 10^4$
$2n$ distances	$\sqrt{2}R_{HH} \approx 1.70$	$\sqrt{2}R_{HL} \approx 1.56$	$\sqrt{2}R_{LL} \approx 1.41$	$B_{2n} = 0.3A_{1n}$

here as temperature independent. This leads to an equilibrium transition temperature of the Ising-like model $T_{eq} = \dfrac{\Delta}{k_B \ln g} \approx 90$ K and an entropy variation of $\Delta S = R \ln g \approx 42$ J.K^{-1}.mol^{-1}, where R is the perfect gas constant. The value of the elastic constant is taken here to be equal to $A_{1n} = 15{,}000$ K.nm^{-2}, allowing us, as it will be seen later, to obtain hysteretic phase transition with thermal hysteresis width of about 30–40 K depending on the nanoparticle size. Previous studies, which we have been carried out on bare particles, have shown that such values lead to many interesting results on the dynamical spatiotemporal aspects [43–45] of the spin transition phenomenon, the main results of which will be summarized below. Within the fixed values of the elastic constant and the lattice spacing, the isostatic elasticity modulus (alias bulk modulus), is roughly estimated to be $E \approx \dfrac{A_{1n}}{R_{1n}} \approx 8$ GPa, which is rather in good agreement with those reported in the literature [172, 173].

7.2.2 Numerical Ingredients

To numerically simulate the behavior of an SCO nanoparticle, we use a computing code developed in CUDA (Compute Unified Device Architecture) and running on an Nvidia graphics card with an RTX A5000 GPU based on an Ampere architecture with 8192 computing cores, 24 GB GDDR6 memory and a single-precision performance of up to 27.8 TFLOPS. In order to solve the Hamiltonian Eq.7.1 describing our physical system, we implement a two-step approach on the GPU in a parallelized way depending on whether we are dealing with the electronic degrees of freedom (spins) or the structural degrees of freedom (lattice positions). The evolution of spins will be performed using a Monte Carlo Metropolis simulation, while the evolution of lattice positions will be done through a molecular dynamics simulation allowing for mechanical relaxation of positions.

The Monte Carlo Metropolis stochastic algorithm is well-known for updating spin values following this procedure: on the two-dimensional lattice of size $(N_x \times N_y)$, a site i is randomly selected from all the sites in the lattice (with variable $S_i = \pm 1$ and given position r_i), a new value $S_i^{'} = -S_i$ is proposed for this site where we flip its spin value at unchanged position r_i. This spin change is accepted or rejected by the usual Metropolis criterion. Regardless of the result (acceptance or rejection of the proposed new spin value), the entire lattice is then mechanically relaxed by the molecular dynamics algorithm (where the lattice sites are displaced 300 times), with a fixed spin configuration (there are therefore no real magnetic interactions in the model where the spins would interact through the elastic lattice). A new site is then randomly selected, and so on: lattice relaxation is repeated for each spin flip. When all sites in the lattice have been visited for the spin flip proposal, which will be the case after $(N_x \times N_y)^2$ times, we define this as the time unit of our simulations, and call it the Monte Carlo time step (MCS): each lattice site is therefore updated $300 \times (N_x \times N_y)^2$ times. We use an amount of 100 MCS as a waiting time to reach the stationary states for each temperature during the thermal transition' study. Then, over the next 50 MCS, statistics are made to determine average physical quantities, such as the HS fraction n_{HS}, the average distance between sites, etc. Each averaged measurement is carried out after performing, at first, three independent runs (each run corresponding to $300 \times (N_x \times N_y)^2$ times sites updates). This is realized to ensure that the different configurations used for the determination of the averages are statistically independent, so as to obtain uncorrelated equilibrium states. Thereby, the physical quantities of interest (site positions, spin configurations, etc.) are recorded at each temperature step along the thermal cycling after a waiting time corresponding to 150 MCS.

Molecular dynamics algorithm for mechanical relaxation of the lattice to reach a stable mechanical state is performed using classical mechanics by solving Newton's equations at 0 K while imposing a high viscosity to avoid site oscillations. We do not directly introduce a numerical parameter in our calculation code for the damping factor, but instead we choose to impose a velocity $v_i = 0$ for all sites of the lattice after each time step. Indeed, to do this, the potential energy gradient of each site is calculated to evaluate the force vector $F_i = -\nabla H_{elast}$ acting on each site. The forces obtained are then normalized: by noting these normalized forces f_i, we simply divide the forces F_i acting on each site by the largest force value obtained (denoted F_{max}) and the normalization relationship is given by $f_i = \frac{F_i}{F_{max}}$. These normalized forces are then multiplied by the chosen simulation time step, dt, to calculate the velocities by the simple relation $v_i = f_i \times dt$, and which are then used to determine the new positions of all particles via the relation $r_i = (v_i \times dt) + r_0 = (f_i \times dt^2) + r_0$, where $dt^2 = 0.001$, r_0 corresponds to the position of the sites at the previous time step, and the discretization step value for the finite differences during the calculations is set to 10^{-5}. This procedure corresponds to the well-known principle of the steepest descent algorithm, which is useful for finding the local minimum of an elastic energy landscape. This whole procedure is then repeated until mechanical equilibrium is reached as explained above. This can be controlled by visualizing the time dependence of the total elastic energy of the system which reaches a minimum average value when mechanical equilibrium is reached. In our simulations, we repeat this procedure 300 times to ensure sufficient mechanical relaxation after each spin flip, and so that the relaxation of the lattice positions does not interfere with that of the spin lattice. Molecular and Monte Carlo dynamics are therefore well separated, and the lattice dynamics is assumed to be faster than the relaxation of the spin states: experimentally, it is observed that the relaxation of the lattice is on the picosecond timescale, while the relaxation of the HS and LS states of the molecule is in the micro-nanosecond range as observed in some Mössbauer experiments of non-cooperative SCO materials [174–176]. Of course, it may be useful to recall that the use of 0 K molecular dynamics simulation can lead the lattice to fall into a metastable state, however as we allow spin-state fluctuations during the Monte Carlo process which affect the lattice energy due to the spin-distortion interaction, it is always possible to leave these metastable states in a reasonable finite time, except when the simulations are done at very low temperature, which obviously increases the lifetime of metastable states exponentially.

From a numerical implementation standpoint on the GPU, when the code is executed, CUDA generates a grid of threads that are organized in a three-dimensional hierarchy. Each grid is organized into blocks of threads, where each block can contain up to 1,024 threads. While the size of each thread block is ideally a multiple of 32 due to the warp structure of the GPU architecture, the total number of thread blocks does not need to be a multiple of 32, as the multiprocessors can manage and schedule threads in groups of 32 parallel threads, called warps. For more details on the algorithm and the implementation of the calculation code, we refer the reader to the original article describing this in detail for the three-dimensional SCO simulations [108]. In our simulations, we then choose "reasonable" square-shaped sizes for our 2D lattices, of spins size ($N \times N$) in the two directions of space, in order to keep an acceptable computation time. Indeed, we have taken different sizes grouped as follows $N = 4, 6, 7, 8, 10, 12, 14, 16, 24, 32, 48, 64, 80, 96, 112, 128$ in our simulations, thus providing small systems of 16 spins to larger ones of 16,384 spins.

7.3 Results

7.3.1 Core–Shell Nanostructure Model

To simulate 2D SCO nanoparticles, we consider a square-shaped spin lattice consisting of an active spin-crossover core of spins that can thermally switch between LS and HS states, surrounded by an inert shell (without spin) with a fixed lattice parameter as described in Fig. 7.11.

We perform a thermal cycle on the nanoparticle thus described, starting from zero temperature $T = 0$ K up to a high temperature of $T = 200$ K (well above the transition temperature of the core), and then cool it back down to the initial temperature of $T = 0$ K. During this cycle, we analyze the thermal behavior of the HS fraction, n_{HS},

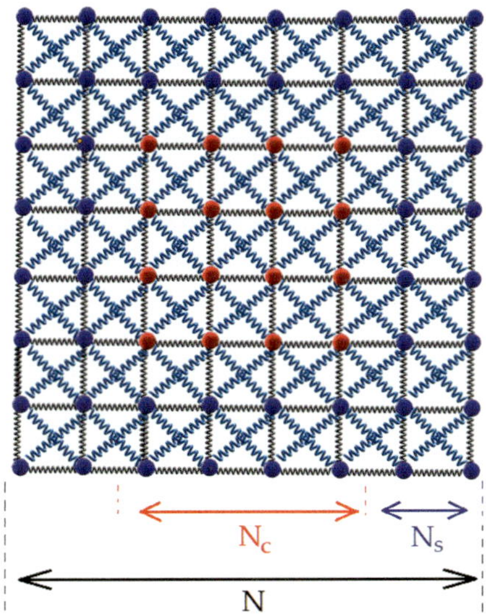

Figure 7.11 Schematic representation of the structure of the 2D core–shell SCO nanoparticle. Two types of sites are represented: core sites (in red) and shell sites (in blue). The springs represent elastic interactions as described previously in the text: grey ones for the first neighbors (1n) and blue ones for the second neighbors (2n) to be consistent with Fig. 7.10.

and several other physical quantities of interest: site positions, spin configurations, average nearest-neighbor distance, spatiotemporal distribution of the local pressure field, etc.

To facilitate the construction of the network and the implementation of this problem, we have considered that the matrix (the shell) has the lattice parameter of the core in the HS state. In this way, it becomes possible to consider the matrix as an extension of the core in the HS state, thus imposing an (1n) equilibrium distance equal to R_{HH} but with a fixed spin-state configuration $S_i = +1$ ($\forall i \in$ shell), even though the elastic network of the shell can adapt to core deformations.

There are two ways of describing the hysteresis cycle. Either we start from low temperatures, with the LS core surrounded by the shell, and thus increase the temperature, or we start from high temperature with a HS core and decrease the temperature. Naively, starting from the high-temperature phase with a HS core and a shell having the same lattice parameter ensures to be at mechanical equilibrium. However, on decreasing the temperature, the core will convert to LS which tend to decrease all its (1n) distances, causing lattice frustration in both moieties and notably at their interface. This frustration reaches its peak when the core switches totally in the LS state. This raises the legitimate question of whether the initial state influences the shape and transition temperatures of the obtained thermal cycles.

If we start the simulations from the low-temperature phase, with core sites with the spin configuration $S = -1$, it imposes (1n) core equilibrium distances equal to R_{LL}, while those of the shell are equal to R_{HH}. Thus, initially one can prepare the whole nanoparticle with (1n) distances equal to R_{LL} or to R_{HH}. In the former, the elastic frustration is put in the shell while in the latter, it is injected in the core.

It should be noted that whatever the choice is made for the initial low-temperature lattice distances, leading to a frustrated shell or core, we have verified for different lattice sizes that this initially "frustrated" configuration converges to the same stationary lattice state after mechanical relaxation and provides the same results, in terms of the thermal dependence of the n_{HS} fraction and average bond length. Moreover, starting the simulations from the high-temperature phase (HS core) where shell and core have the same equilibrium distances, also leads to the same results. These results ensure that, from a structural point of view, we are following stationary states of the elastic network during the thermal cycle.

In a manner similar to what we have done in 2D studies of nanoparticles [43–45], we choose a square-shaped core size (red dots in Fig. 7.11) containing ($N_c \times N_c$) sites, and a shell (blue dots in Fig. 7.11) surrounding the core in epitaxial form, with a width of N_s sites.

The total number of sites in the whole square-shaped nanoparticle is given by $N^2 = (N_c + 2N_s)^2$, the number of sites solely in the core is straightforward and equals N_c^2, and finally, the number of sites in the shell is given by $N_{shell} = N^2 - N_c^2 = 4N_s(N_c + N_s)$. These relations, particularly the last one, are useful for analytically calculating the number of sites present solely in the shell or the core. This allows us to determine and subsequently verify numerically the expected residual fraction n_{HS} at low temperature in the core of the nanostructure as a function of the shell size N_s, for example.

7.3.2 Thermal Properties of the Uncoated Core

To begin with, we first examine the core lattice alone, without any surrounding shell decoration (called later as "uncoated core"), for different square-shaped system sizes during a thermal cycle. Our objective here is to investigate whether there are scaling laws related to the cooperative nature of these core particles by studying the behavior of thermal hysteresis, in terms of position, width, shape, etc. as a function of size. Furthermore, we will attempt, as much as possible, to separate size effects from surface effects, which clearly interfere in systems subjected to long-range interactions. The thermal properties of the nanoparticle are measured by monitoring as function of temperature, the following macroscopic parameters: (i) the HS fraction defined by the relation $n_{HS} = \frac{(1 + \langle S_i \rangle)}{2}$, where $\langle \ldots \rangle$ represents thermal averages (S_i are the core spins) and the average nearest-neighbor distance, denoted $\langle d_{1n} \rangle$.

7.3.2.1 Thermal spin transition: Hysteresis and transition temperature

This section is devoted to the analysis of the thermal evolution of the HS fraction an uncoated 2D nanoparticle as a function of its size, N. Figure 7.12 illustrates the thermal evolution of this fraction n_{HS} for different square-shaped system sizes N = 16, 24, 32, 48, 64, 80, 96, 112, 128 during a thermal cycle. According to the values of the elastic constants chosen for the core, given in Table 7.1, hysteretic thermally induced first-order transitions are obtained regardless of the simulated nanoparticle size. The shape and width of the hysteresis depend on the simulated size, with the width becoming more pronounced as the size increases. It is also noted that at smaller sizes, the abrupt nature of the first-order transition is more pronounced during cooling around the lowest transition temperature than during heating around the highest transition temperature. Conversely, for larger sizes, the abruptness of the transition softens at low temperatures, and the

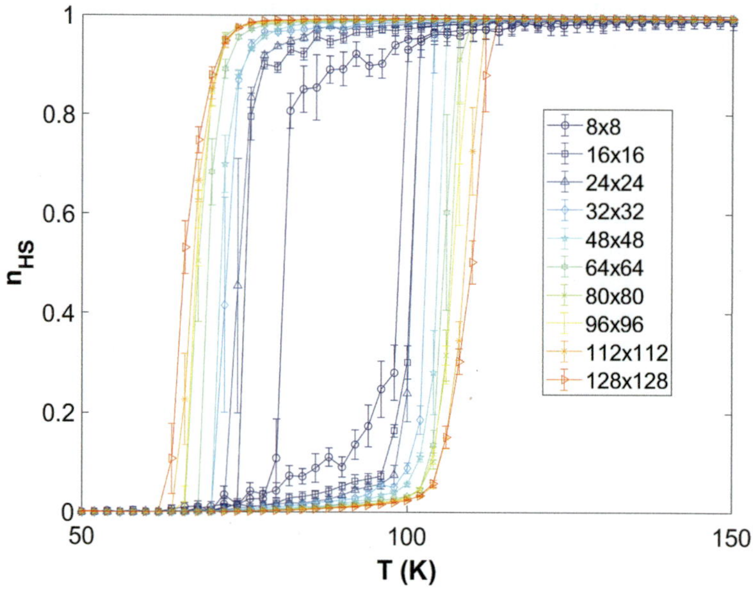

Figure 7.12 Thermal dependence of the HS fraction n_{HS} represented for different square-shaped sizes, $N = 8, 16, 24, 32, 48, 64, 80, 96, 112, 128$, of the uncoated core. Error bars represent the standard deviation on the average n_{HS} for each temperature. Remark: each presented point has an error bar; however, the latter are visible only near the HS–LS transition, while above and below this region, the sizes of the error bars are smaller than that of the points.

hysteresis cycle becomes more squared at high temperatures. The error bars displayed on the n_{HS} versus temperature curve represent the standard deviation calculated from the average of n_{HS} for each temperature. Each presented point has an error bar, but they are only visible near the HS–LS transition; above and below, the size of the error bars is smaller than that of the points. This confirms the choice of parameters defining our MCS steps to accurately calculate the statistical averages of our physical quantities.

The width of the hysteresis, represented by ΔT, is defined as the temperature interval between the transition temperature (T_{up}) on heating and that on cooling (T_{down}), at the value $n_{HS} = \frac{1}{2}$. We can express this relationship as $\Delta T = T_{up} - T_{down}$, whose evolution

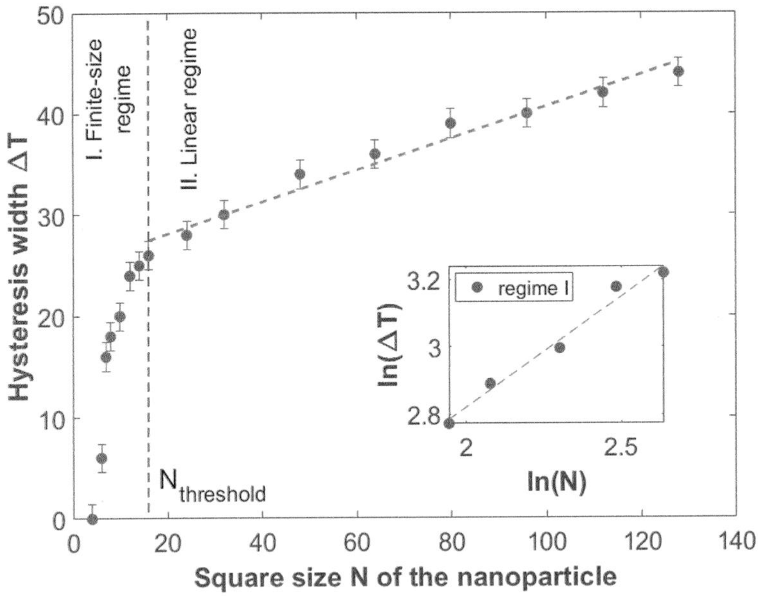

Figure 7.13 Nanoparticle's size dependence of the thermal hysteresis' width, ΔT, of the HS fraction, derived from data of Fig. 7.12, showing two distinct regimes represented for different square-shaped sizes $N =$ 4, 6, 7, 8, 10, 12, 14, 16, 24, 32, 48, 64, 80, 96, 112, 128 of the uncoated core. A first regime (I) for $N < N_{threshold} = 16$ in which ΔT drops abruptly to zero and a second regime (II) for $N \geq N_{threshold}$ where ΔT follows an almost linear behavior with N. Error bars represent the absolute uncertainty associated with ΔT. The red dashed curve in regime II represents a linear regression with a slope given by (0.16 ± 0.01). See text for more explanations. (Insert) In log-log scale, ΔT is plotted as a function of N for the points of regime I, from the main figure, having a value close to $N_{threshold}$ (i.e., the points $N = 7, 8, 10, 12, 14$). ΔT fits with the linear regression (red dashed line) here with a slope $\alpha = 0.66 \pm 0.06$.

as a function of size is presented in Fig. 7.13. It is observed that the dependence of the hysteresis width on size exhibits two distinct regimes: a first non-linear regime for small sizes (denoted as regime I), for $6 \leq N \leq 16$ where the value of ΔT decreases rapidly to zero as the size decreases, and a quasi-linear growth beyond a threshold nanoparticle size value of $N_{threshold} = 16$ (denoted as regime II). Below the size value $N = 6$, only gradual transitions are obtained.

For the simulated sizes, when no shell is present and only the core exists, the hysteresis width thus increases with the square-shaped size of the nanoparticle; this behavior has already been observed in studies of 2D core–shell nanoparticles [43]. It is noteworthy that, as shown in both 2D and 3D simulations of SCO materials [105, 108, 133, 177], the stiffness of the system's elastic constants governs the spin transition regime. Indeed, for low values of the elastic constant A_{1n}, below a critical threshold, a homogeneous and gradual spin transition takes place without the presence of macroscopic LS/HS domains, consisting in a homogenous multidroplet nucleation of the sin states everywhere in the lattice forming a ramified structure, which is consistent with the weakly cooperative nature of elastic interactions compared to the ligand-field energy. Conversely, for high values of the elastic constant A_{1n} above a critical value, the spin switching phenomenon at a particular site has a low probability of occurring due to the presence of a large elastic energy barrier that must be overcome in the system for a spin to flip. The existence of this macroscopic elastic barrier generally leads to a cluster formation process that primarily occurs at the surface and preferentially at the corners and edges of the lattice, where the nucleation energy is lower due to the lack of neighboring sites, compared to those located in the bulk.

Let us begin by explaining the behavior of regime II, where the hysteresis width follows a quasi-linear law, as indicated by the linear regression represented in Fig. 7.13 by a red dashed curve, which fits the data with a slope of (0.16 ± 0.01). This linear character can be explained by the fact that the hysteresis arises from the existence of a macroscopic elastic energy barrier, mainly stored in the HS/LS interface region. This energy excess is due to intermolecular cooperative interactions, resulting from the interference between local volume changes accompanying each spin flip. When the transition proceeds in one or two single domains nucleation, the length of the interface at which the elastic energy barrier reaches its maximum, and from which the lattice instability occurs (the collective switching between the two states), varies linearly with crystal size. In other words, the elastic energy is inherently an extensive quantity since it is proportional to the size of the system, which explains why the hysteresis width also increases quasi-linearly as the size of the nanoparticle increases [178]. Paradoxically, for larger sizes, it's surface effects that drive physical behavior in these cooperative bare nanoparticles.

Note that in the previous discussion, the effect of entropy was not mentioned, because in single-domain propagation, the configurational entropy remains very low except at the interface, which represents a very small fraction of the spin lattice.

Furthermore, it is easily understood that hysteresis disappears below a certain simulated system size in regime I, based on the physical arguments stated above. Indeed, the width of the thermal hysteresis of the HS fraction is closely related to the elastic energy barrier present in the system. Since this barrier depends on the size of the simulated system, smaller nanoparticle sizes correspond to smaller barriers. Thus, in small nanoparticles below the threshold size $N_{threshold}$, thermal fluctuations can easily overcome the barrier, thus favoring the disappearance of hysteresis because the elastic energy barriers are no longer sufficient to maintain the metastable states. As a result of this low barrier energy, several metastable domains can grow up, leading to interface proliferation. In this case, interface length is no longer proportional to lattice size and depends little on lattice shape. Instead, it depends on "volume," leading to a deviation from linearity as a function of size, as observed in Fig. 7.13.

In the inset of Fig. 7.13, we have plotted ΔT as a function of the size N on a log-log scale for the points in regime I, which have values close to $N_{threshold}$ (i.e., the points $N = 7, 8, 10, 12, 14$). Here, ΔT fits a linear regression (red dashed line) with a slope of $\alpha = 0.66 \pm 0.06$, suggesting the power law $\Delta T \sim N^{\alpha}$ in the first regime I. This exponent is close to that of the mean-field universality class, as found for the elastic model of spin-crossover materials in 2D systems [179], with $\nu^* = 2/3$.

As a final remark, the value of the slope of ΔT as a function of N in regime II likely depends on the value of the elastic constant A_{1n} (and thus on B_{2n}) and the lattice mismatch $\delta_R = R_{HH} - R_{LL}$. It would be interesting to investigate this in order to examine the universality or non-universality of the quasi-proportionality between the hysteresis width and the nanoparticle size. We leave this work for future study and will not focus on this issue here.

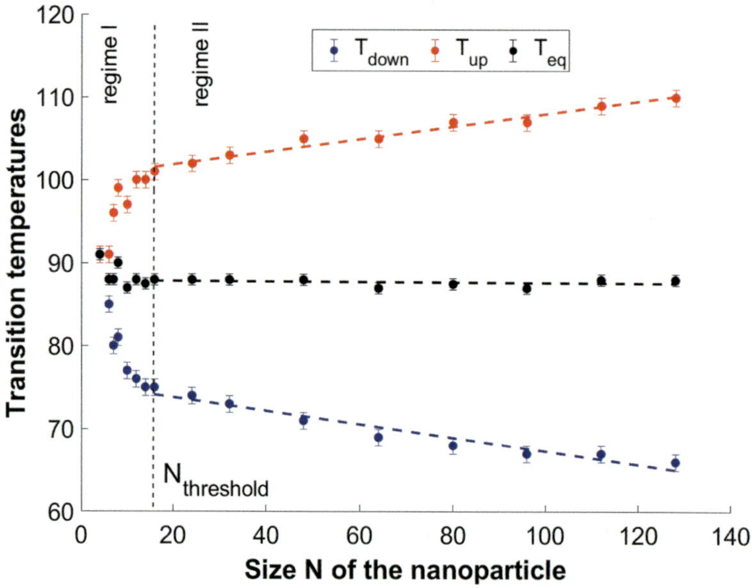

Figure 7.14 Nanoparticle's size N dependence of the transition temperature of the heating, T_{up} (red dots), and cooling, T_{down} (blue dots), branches as well as that of the equilibrium temperature T_{eq} (black dots). Error bars represent the absolute uncertainty associated with each temperature. The blue (respectively red) dashed line in regime II represents a linear regression to help the reader visualize the nearly linear behavior of T_{down} with a slope of (-0.08 ± 0.02) (respectively T_{up} with a slope of (0.08 ± 0.01)) with N. Note the non-dependence of the equilibrium temperature, T_{eq}, on N.

The numerically calculated equilibrium temperature, denoted as T_{eq}, is given by the relation $T_{eq} = \frac{T_{down} + T_{up}}{2}$. Figure 7.14 shows the evolution of the heating T_{up} and cooling T_{down} transition temperatures, as well as the deduced equilibrium temperature T_{eq}, as a function of the size, N, of the nanoparticle. We consistently find again the two previous regimes described in Fig. 7.13, with the convergence of the two temperatures T_{down} and T_{up} into a single value T_{eq} as the size decreases in regime I, and a quasi-linear behavior of T_{down} and T_{up} as the nanoparticle size increases in regime II. It is important to mention that below regime I, i.e., for $N <$ 6, the HS fraction of the nanoparticle leads to a gradual transition ($\Delta T = 0$) for which the equilibrium transition temperature is simply given by that of the Ising-like model, $T_{eq} = \dfrac{\Delta}{k_B \ln g} \approx 90$ K, calculated in Section 7.2.1. Thus, a horizontal straight line exists in the phase diagram of Fig. 7.14, corresponding to $\Delta T = 0$, with the presence of a bifurcation point at $N \simeq 6$ above which the two more or less symmetrical branches of the thermal hysteresis emerge.

The appearance of regime I is very abrupt, and the switching temperatures T_{up} and T_{down} deviate sharply from T_{eq} following a non-linear behavior with system's size and lastly stabilize in regime II above $N_{threshold} \simeq 16$, above which T_{up} and T_{down} behave linearly with N. Moreover, it is interesting to note that in regime II, the equilibrium temperature T_{eq} remains almost constant regardless of the nanoparticle size, with a value of approximately 89 K. Therefore, this means that despite the increase in the hysteresis width as the nanoparticle size increases, the equilibrium temperature remains unchanged. Furthermore, the slopes of the linear regressions characterizing the N-dependence of T_{up} (with a value of 0.08 ± 0.01) and T_{down} (with a value of -0.08 ± 0.02) are identical in absolute value, implying that the thermal bistability window expands almost symmetrically around T_{eq} without shifting as a function of N.

7.3.2.2 Structural properties of the spin transition

In this section, we explore the structural aspects of the lattice to better understand how it expands or contracts during the thermal spin transition. For this purpose, Fig. 7.15(a) collects the thermal dependence of the average nearest-neighbor distance (1n), noted $\langle d_{1n} \rangle$, for various sizes, $N = 16, 24, 32, 48, 64, 80, 96, 112, 128$, of square-shaped nanoparticles. Several observations can be made from the evolution of these curves.

First of all, and consistently with the corresponding behavior of the HS fraction, thermal hysteresis is also obtained for this structural parameter, although some differences in the error bar intensities indicate the existence of additional information compared to the electronic response. Thus, at low (respectively high) temperatures, the average distance $\langle d_{1n} \rangle$ approaches 1.00 nm (1.20 nm), which is consistent with the equilibrium distance for the LS (HS) state R_{LL} (R_{HH}). The error bars in Fig. 7.15(a) represent the standard

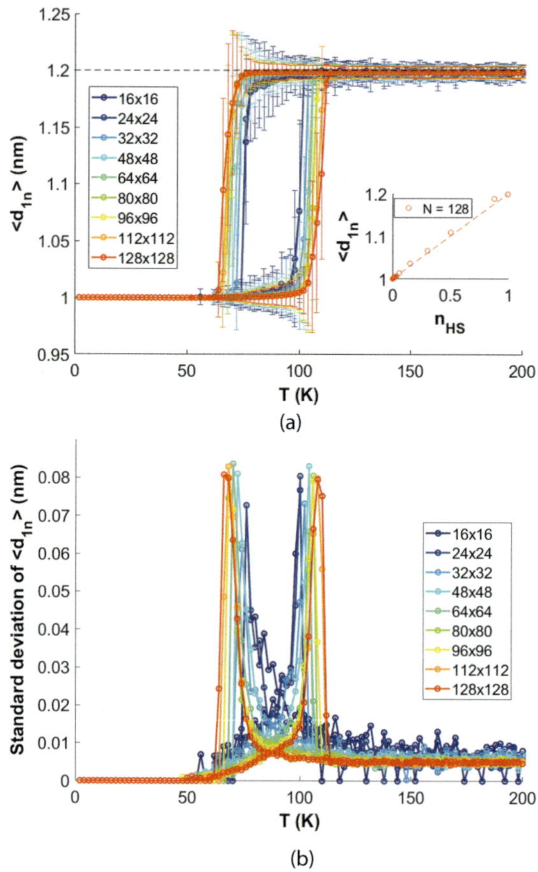

Figure 7.15 (a) Thermal dependence of the average (1n) distance, $\langle d_{1n} \rangle$, represented for different sizes of the uncoated square-shaped SCO nanoparticle with: $N = 16, 24, 32, 48, 64, 80, 96, 112, 128$. Error bars represent the standard deviation on the average $\langle d_{1n} \rangle$ for each temperature. The dashed horizontal line shows the average equilibrium (1n) lattice spacing in the HS state, $R_{HH} = 1.20$ nm. (Insert) Average (1n) distance $\langle d_{1n} \rangle$ as a function of the HS fraction, n_{HS}, along the heating branch of the hysteresis loop for the uncoated nanoparticle of size $N = 128$ (identical results are found for other sizes and branches). The colored dashed line represents a linear regression between the two quantities, helping to better illustrate the relatively good proportionality between them. (b) Thermal evolution of the amplitude of error bar values of Fig. 7.15(a), showing that they peak at the transition temperatures T_{down} and T_{up}, respectively, during cooling and heating.

deviation of the average $\langle d_{1n}\rangle$ at each temperature. These error bars indicate the presence of residual stresses in the system, reflecting local positional disorder within the lattice due to the persistence of unreleased mechanical stresses generated as the transformation front advances. This behavior can also be observed in Fig. 7.15(b), which shows the evolution of the error bar values from Fig. 7.15(a) as a function of temperature. From this curve, we can observe that the intensity of the error bars peaks twice, well correlated with the transition temperatures T_{down} (respectively T_{up}) during the cooling (heating) phases. Moreover, it is noted that the half-width of these two peaks increases with the nanoparticle size N, which is linked to the widening of the thermal hysteresis for larger values N of the nanoparticle sizes.

Furthermore, the transition temperatures from the structural LS state to the HS state (and vice versa), which can be identified numerically at $\langle d_{1n}\rangle = 1.10$ nm in Fig. 7.15(a), coincide with those measured in the previous Section 7.3.2.1 in Fig. 7.14. This correspondence between the structural and electronic aspects of the lattice is illustrated by the inset of Fig. 7.15(a), which shows the variation of $\langle d_{1n}\rangle$ as a function of n_{HS} during the heating phase for a given nanoparticle size ($N = 128$). This inset reveals a fairly good linear relationship between the structural (average distances (1n)) and the electronic (average HS fraction) degrees of freedom: the same behavior was observed for other nanoparticle sizes and for other heating or cooling phases.

All these results indicate the existence of a strong correlation between the electronic and structural properties of the uncoated core of the nanoparticle during the thermal spin transition. Thus, when all core spins switch from $S = -1$ to $S = +1$ (or vice versa), the average lattice parameters align with the expected equilibrium distance values for the HS (LS) state, resulting in complete mechanical relaxation of the lattice.

Moreover, in complete agreement with our previous observations, and as mentioned above, the hysteresis width in Fig. 7.15(a) for the average (1n) distance exhibits a behavior similar to that shown in Fig. 7.13 for the HS fraction. To illustrate this point quantitatively, Fig. 7.16 summarizes the evolution of the hysteresis width ΔT associated with $\langle d_{1n}\rangle$ as a function of the square-shaped

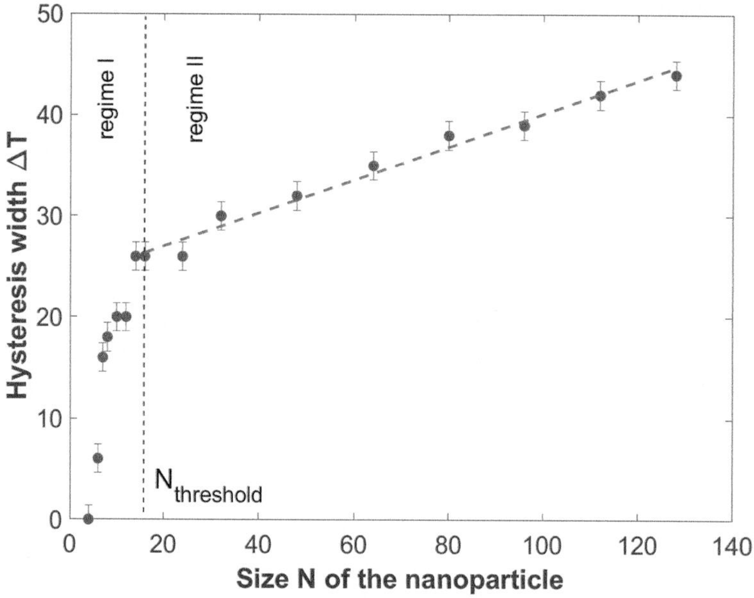

Figure 7.16 Nanoparticle size, N, dependence of the hysteresis width, ΔT of $\langle d_{1n} \rangle$, corresponding to Fig. 7.15(a), showing the two same regimes as in Fig. 7.13: for $N < 16$, a finite-size regime (regime I) where ΔT suddenly collapses, followed by a linear behavior (regime II) beyond. Error bars represent the absolute uncertainty associated with ΔT. The red dashed curve in regime II represents a linear regression with a slope given by (0.16 ± 0.01), the same value as in Fig. 7.13.

nanoparticle size, N. The two regimes previously identified in Fig. 7.13 are clearly present and separated by the same threshold size value, $N_{threshold} = 16$. Additionally, ΔT collapses to zero for $N < N_{threshold}$, while for $N \geq N_{threshold}$, an almost linear behavior is also observed, with the same slope of (0.16 ± 0.01) as for the electronic properties n_{HS}.

On the other hand, in Fig. 7.17, we have plotted the thermal evolution in a two-scale graph for both $\langle d_{1n} \rangle$ and $\langle E_{elast} \rangle$, which corresponds to the total elastic energy of the lattice, for the thermal cycle of a nanoparticle of size $N = 64$. The elastic energy is zero in the HS and LS phases up to fluctuations. As the transition approaches, the growing of a new phase increases the elastic energy due to the formation of H–S–LS interfaces and $\langle E_{elast} \rangle$ peaks

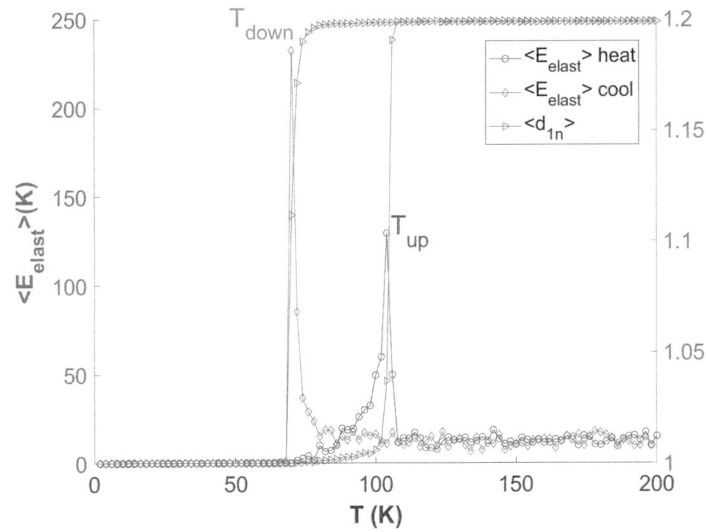

Figure 7.17 Two-scale plot showing the thermal dependence of the average (1n) distance, $\langle d_{1n}\rangle$ (in red), as well as the thermal dependence of the total elastic energy of the lattice $\langle E_{elast}\rangle$ for the heating and cooling branches during the corresponding hysteretic cycle for the nanoparticle size $N = 64$ (in bluish). The transition temperatures T_{down} and T_{up} are indicated to allow the reader to correctly identify the good thermal correlation between $\langle d_{1n}\rangle$ and $\langle E_{elast}\rangle$. We obtained identical results with the other nanoparticle sizes considered previously in the chapter.

exactly around the transition temperature T_{up} (respectively T_{down}) on heating (cooling). Identical results have also been measured for all other sizes studied. This confirms that when the lattice switches collectively from LS (respectively HS) to HS (LS), the total elastic energy of the system gradually increases until it reaches a maximum of approximately 130 K (230 K) at the transition temperatures T_{up} (T_{down}) and then decreases sharply once this temperature is exceeded. These results are consistent with those obtained in 3D SCO systems [108, 109], where the measured elastic energy of the system is primarily localized around the transformation front during its propagation throughout the thermal spin transition. The width of the $\langle E_{elast}\rangle$ peaks is slightly more significant in the heating branch compared to that in the cooling branch: this can be correlated with the behavior of the hysteresis cycle, which is somewhat wider and less abrupt during heating compared to cooling. Conversely, the value of $\langle E_{elast} \rangle$ is nearly zero at low temperatures, while it asymptotically approaches zero at high temperatures. This indicates that the lattice still experiences stresses when at high temperatures, and that with the defined simulation parameters, it would be necessary to wait longer in terms of MCS steps or to increase further the temperature to achieve a fully relaxed situation. This phenomenon also explains the asymptotic thermal growth of $\langle d_{1n} \rangle$ towards R_{HH} at elevated temperatures in Fig. 7.15(a).

7.3.2.3 Spatiotemporal aspects of the spin transition

In this part, we aim to study the dynamics of the spin transition, by illustrating, in particular, the spatiotemporal properties of the first-order phase transition involving an intimate coupling between the electronic and the elastic degrees of freedom. For that, various snapshots of the spin and lattice configurations of an uncoated core of size of $N = 64$, are presented in Fig. 7.18. For convenience reasons, we choose to display the spin-lattice configurations obtained near the region of the transition temperature, T_{down}, during the cooling process, from the HS state to the LS state, in the six subfigures, Fig. 7.18a–f, where yellow (blue) spheres represent HS (LS) spin sites. The configurations obtained during the heating process from the LS state to the HS state are very similar, up to thermal fluctuations since $T_{up} > T_{down}$, and likewise, analogous results are observed regardless of the studied size of the system.

For each snapshot, we have indicated the corresponding (n_{HS}, T) values to allow the reader to locate them on the corresponding thermal hysteresis in Fig. 7.12. In agreement with our previous discussions, we observe a growth of the transformation front characterized by a clustering process of the LS state within a lattice that is initially in the HS state at high temperatures (Figs. 7.18a)-b)). As explained above, these clusters appear independently at the equivalent four corners of the lattice (Fig. 7.18.c) due to the previously discussed energetic reasons (i.e., the nucleation energy is lower in these regions because the number of neighboring sites is lower compared to those located in the bulk and at the edges). Initially, these clusters exhibit a shape close to a quarter circle in 2D,

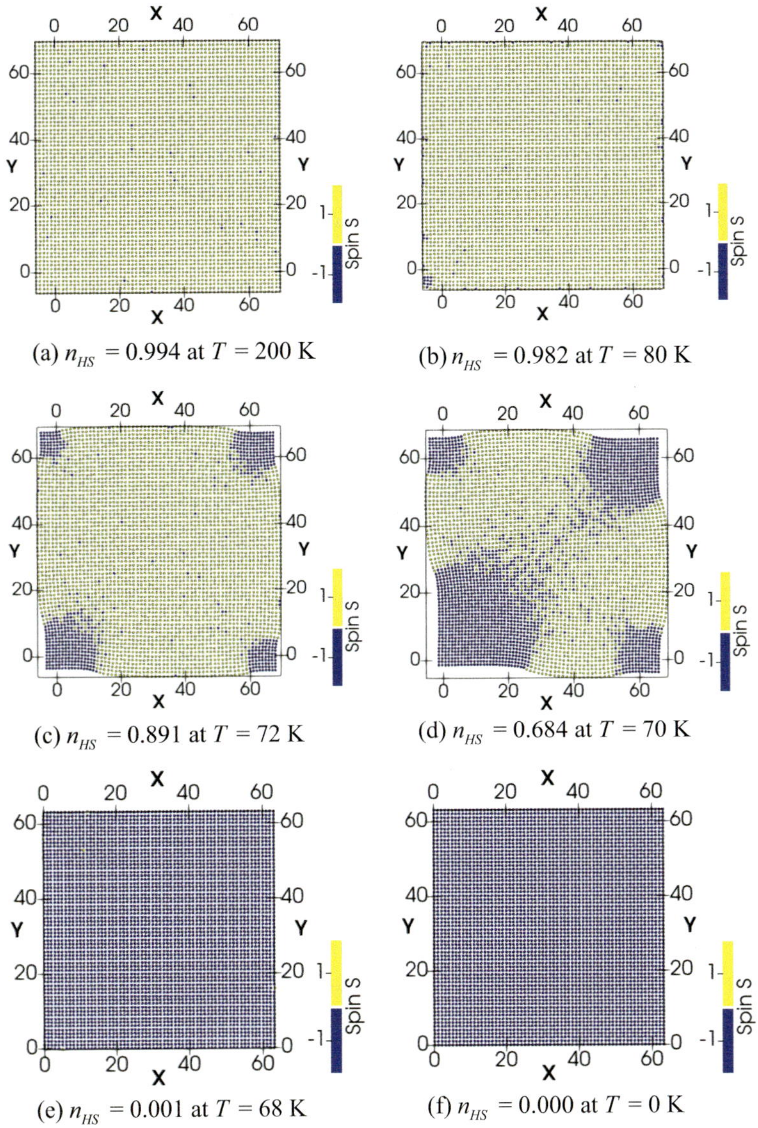

(a) $n_{HS} = 0.994$ at $T = 200$ K

(b) $n_{HS} = 0.982$ at $T = 80$ K

(c) $n_{HS} = 0.891$ at $T = 72$ K

(d) $n_{HS} = 0.684$ at $T = 70$ K

(e) $n_{HS} = 0.001$ at $T = 68$ K

(f) $n_{HS} = 0.000$ at $T = 0$ K

Figure 7.18 Selected snapshots showing the spatiotemporal configurations of the lattice of the uncoated core for $N = 64$ during the cooling branch at various n_{HS} and T values during the thermal transition from HS to LS. Yellow (blue) spheres represent HS (LS) sites. Remark the macroscopic domain character of the HS → LS transformation.

since the circularly-shaped interface is the one that minimizes the interfacial elastic energy. This first process is stochastic and emerges between 80 and 72 K (Fig. 7.18c), where the LS domains become apparent. It is interesting to note the overall deformation of the network that accompanies the appearance of the LS domains. This deformation also influences the mode of propagation and growth of these domains through a subtle feedback effect: the spin flip contracts the bonds, deforming the surface and influencing the volume distances due to elastic interactions. Interestingly, although the four corners of the lattice are equivalent, as preferential nucleation points, as indicated above, the random character of the MC procedure breaks the symmetry between them. Indeed, as can easily be seen in Fig. 7.18.c, the LS domains in the top right and bottom left corners are larger than the other two, whose progression will be smaller during the transition. As the temperature decreases, their size increase during their propagation (Fig. 7.18.d) until they coalesce at the center of the lattice (Fig. 7.18e), resulting in a nearly complete transition of the lattice to the LS state, accompanied with the lattice contraction and a regular surface from which the previous deformations vanished. Further cooling down to very low temperatures (Fig. 7.18.f) only allows the system to complete the conversion of the remaining HS sites, ultimately leading to a single-phase LS lattice configuration.

7.3.3 Core–Shell Spin-Crossover Nanoparticles

From this point onward, we will focus on examining the lattice of the whole nanoparticle, with its core–shell structure. This implies that from now on, the core will be surrounded by a matrix (referred to as the shell in the following text), as explained in Section 7.3.1. To recall, this shell has no spin and an equilibrium nearest-neighbor distance equal to R_{HH}. However, the elastic lattice of the shell can adapt to the deformations induced by the core during its LS to HS conversion and vice versa. We have chosen that, for a given nanostructure (including core and shell) of fixed size N, when the shell thickness, N_s increases by adding more sites surrounding the core, the core size N_c decreases accordingly, following the relations defined: $N = N_c + 2N_s$, in such a way that the total size N of the whole nanoparticle remains constant. This means that within this approach, the total size N of the nanoparticle remains constant while N_s and N_c vary relative to each other. This approach slightly differs from what is typically found in other 2D studies [43, 45], where the core size remains constant, and thus the total size N of the nanoparticle increases as the shell thickness increases.

For the present investigations, we consider a nanoparticle of size, $N = 64$, in which we inject from the surface a shell whose thickness, N_s, will vary from 0 to 30. Similarly to what was done for the bare core, the thermal properties of the core–shell nanoparticle will be measured by monitoring, as a function of temperature, the following macroscopic parameters: (i) the HS fraction of the core defined by the relation $n_{HS}^{core} = \frac{(1 + \langle S_i \rangle)}{2}$, ($S_i$ are the core spins), (ii) the average nearest-neighbor distances of the core and shell, denoted $\langle d_{1n} \rangle$, and (iii) the mechanical properties of the lattice. Here, too, the core's spatiotemporal behavior will be analyzed, the elastic interplay between the core and the shell will be elucidated, and scaling laws for the behavior of the whole nanoparticle will be deduced.

7.3.3.1 Thermoinduced spin transition

We will now analyze the thermal evolution of the HS fraction of the core part of the whole square-shaped core–shell nanoparticle, during a thermal cycle. We focus exclusively on the core sites, as only the core sites of the nanoparticle have the ability to switch between the LS and HS states. This evolution will be examined for different shell thicknesses, N_s, ranging from 0 to 30 in increments of 1. For clarity, we have chosen to illustrate this behavior in Fig. 7.19(a) for a limited set of N_s values, where we plot the thermal dependence of the HS fraction, n_{HS}^{core}, for shell's thickness values, $N_s = 1, 5, 10, 15, 20, 25, 27, 30$. The displayed error bars represent the standard deviation calculated from the average of n_{HS}^{core} for each temperature.

Thermally induced first-order transitions with hysteresis are obtained regardless of the simulated N_s thickness. As observed previously for the uncoated core, the shape and width of the hysteresis now depend on the shell's thickness. Specifically, the hysteresis

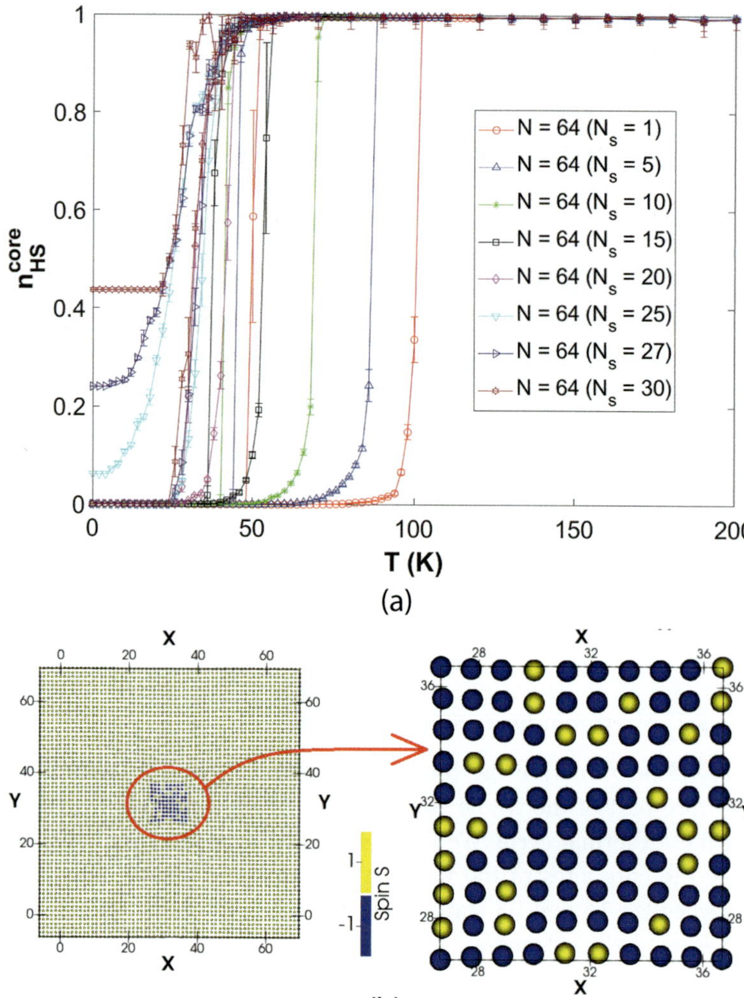

Figure 7.19 (a) Temperature dependence of the core HS fraction n_{HS}^{core} plotted for square-shaped size of the spin lattice $N = 64$ and with different shell thicknesses N_s: $N_s = 1, 5, 10, 15, 20, 25, 27, 30$. The shell has the same equilibrium lattice parameter as that of the HS core. (b) Left side: snapshot of the spatiotemporal configurations of the lattice of the core–shell nanoparticle for $N = 64$ with $N_s = 27$, and $N_c = 10$, during the cooling branch at $T = 10$ K (corresponding to $n_{HS}^{core} = 0.25$). Right side: only the core of the nanoparticle is displayed for better visibility. Yellow (blue) spheres represent HS (LS) sites.

width decreases as the shell thickness increases. Regarding the hysteresis shape, it remains similar to that of the uncoated core, with a roughly rectangular cycle from the smallest thickness $N_s = 1$ up to relatively large values (e.g., $N_s = 20$). In this range of N_s values, we observe sharp transitions at the transition temperatures T_{up} during heating and T_{down} during cooling.

However, for thicker shell thickness (e.g., for $N_s = 25, 27, 30$) to which correspond smaller core sizes (e.g., $N_c = 14, 10, 4$), a change in behavior occurs in the evolution of the HS fraction, n_{HS}^{core}, of the core, particularly during the low-temperature transition from the HS to LS state. While the LS to HS transition in the core remains more or less abrupt for the heating branch at T_{up}; along the cooling branch, we observe at very low temperatures the spontaneous emergence of a plateau with a non-zero residual value in the n_{HS}^{core} curve. This means that a fraction of the core sites remain trapped in the HS state, even at very low temperatures, where the LS state is expected to be the fundamental state. In addition, as the shell thickness increases, the magnitude of the residual HS fraction also increase to reach more than 40% for $N_s = 27$. Add to this the fact that the growth of this residual fraction is also accompanied by the widening of the plateau of the intermediate state appearing at low temperature, thus reinforcing its thermal stability range. This raises the question of the origin of this low-temperature plateau and the organization of HS sites in the core in this particular state.

As shown in Fig. 7.19(b), where we present the spatial distribution of the spin states inside the core at $T = 10$ K for the particular case where $N_s = 27$ and $N_c = 10$, it appears clearly that this phenomenon results from the emergence of a spontaneous self-organization of HS and LS spin states, taking place around 10 K. This configuration remains stable when the temperature is further decreased down to $T = 0$ K. Thus, beyond a critical shell thickness N_s, the core of the nanoparticle is no longer able to fully switch to the LS state at low temperature. Additionally, as shown in Fig. 7.19(a), the transition temperatures T_{up} and T_{down} shift towards lower temperatures as the shell thickness increases. In the simple Ising-like view, presented above, this is equivalent to lowering the effective ligand-field energy. Indeed, these observations suggest that the core part experiences a negative pressure effect.

These phenomena can be explained by the fact that, beyond a certain shell thickness N_s, the core sites which are initially at high temperature in the HS state, separated by distances equal to R_{HH}, are no longer able during cooling to sufficiently "pull" on the thick shell, which behave as a wall, to achieve an average nearest-neighbor distance close to the equilibrium distance R_{LL}. Indeed, since the LS state is thermodynamically stable at low temperatures, due to entropic effects, the core naturally tends to convert from the HS to the LS state, which requires a reduction in the average nearest-neighbor distances, shifting from R_{HH} to R_{LL}. Indeed, the tendency of the core to contract results in tensile stress inside the shell exerted by the core sites all around the core–shell interface. Thus, to convert the whole core from HS to LS, its surface must decrease by the amount $(N_c - 1)^2 \left(R_{HH}^2 - R_{LL}^2\right)$ and the shell must also contract by the same amount. However, beyond a critical shell thickness, these forces become insufficient to overcome the internal forces within the shell structure, resulting in a structurally frustrated core state whose total surface remains higher than $(N_c - 1)^2 R_{LL}^2$. In this state, the core attempts to adopt the average distance R_{LL} but cannot fully achieve it, leading to self-organization to best optimize the spin configuration according the system's constraints. This is confirmed by the curve of the HS fraction at low temperatures for large shell thickness values (e.g., $N_s = 30$), where the residual non-zero value of n_{HS}^{core} tends to stabilize around the value 0.5.

Similar to what we presented for the uncoated core case, we now illustrate in Fig. 7.20 the size evolution of the respective heating and cooling transition temperatures T_{up} and T_{down} measured at $n_{HS}^{core} = 0.5$, as well as the equilibrium temperature, $T_{eq} = \frac{T_{up} + T_{down}}{2}$, using the same average relation as before. However, this time, the x-axis corresponds to the ratio of the shell thickness to the core size, i.e., N_s/N_c. A general decrease in all transition temperatures is observed as the N_s/N_c ratio increases. In particular, for low ratio values, typically $N_s/N_c \ll 1$, the decrease is more pronounced and sharper than for higher ratios. Moreover, the temperature T_{up} drops much faster than T_{down}. The understanding of these behaviors requires the analysis of the structural part, particularly the evolution of the average (1n) distances $\langle d_{1n} \rangle$ of core and shell

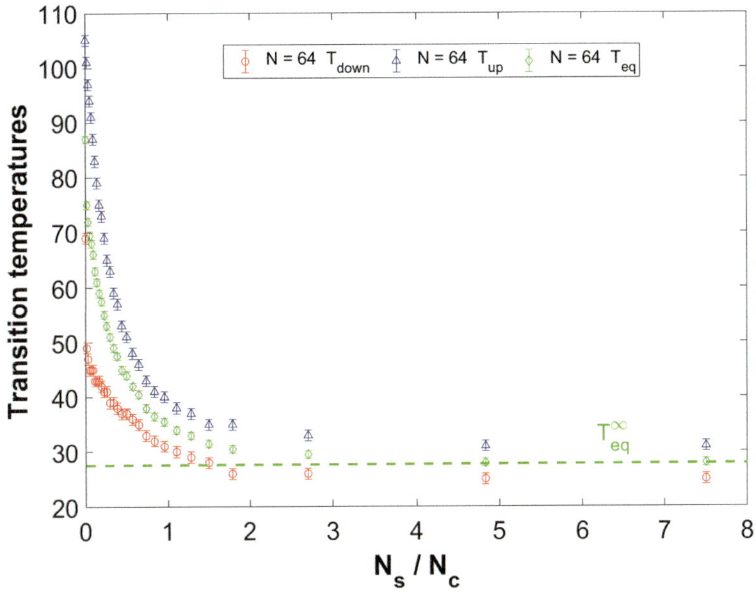

Figure 7.20 Evolution of transition temperatures as a function of the ratio N_s/N_c for the core–shell spin-crossover (SCO) nanoparticle, with a constant total size, $N = 64$. The shell thickness (resp. core), N_s (resp. N_c) values are ranging from $N_s = 0$ to $N_s = 30$ (resp. $N_c = N - 2N_s$). The error bars represent the absolute uncertainty associated with each temperature. The horizontal dashed green line illustrates the temperature $T_{eq}^{\infty} \approx 27.5$ K obtained for large values of the N_s/N_c ratio.

for the different sizes. Finally, it is important to point out that, when the ratio increases beyond $N_s/N_c \simeq 1$, the decrease in transition temperatures seems to asymptotically reach a constant value close to $T_{eq}^{\infty} \approx 27.5$ K for $N_s/N_c \gg 1$. However, in reality, this value strongly depends on the total size of the nanoparticle, and it is therefore expected to reach 0 for a core surrounded by a shell having an infinite thickness.

A meticulous analysis of the data from Fig. 7.20 shows that the representative curve of T_i (with $i = \{\text{down, up}\}$) as a function of the N_s/N_c ratio can be refined using a stretched exponential function of the form: $T_i = T_i^{\infty} + a\ e^{-(x/x_0)^c}$, where a has the dimension of a temperature, $x = N_s/N_c$, x_0 corresponds to a characteristic value of N_s/N_c, and c is an exponent ($\neq 1$) reflecting the non-linear nature of the curve. To validate the relevance of this evolution law, we plotted, in the insert of Fig. 7.21(a), the quantity $\ln\left[-\ln\frac{(T-T^{\infty})}{a}\right]$ as a function of $\ln\left(\frac{N_s}{N_c}\right)$ for a core–shell nanoparticle with shell thickness varying from $N_s = 1$ to $N_s = 30$. We plotted in red (blue) the evolution of this quantity for T_{down} (T_{up}), with $T^{\infty}_{down} \approx 24.5$ K and $T^{\infty}_{up} \approx 30.5$ K.

A straight line is expected, with a slope equal to c and an intercept at the origin equal to $-c \ln x_0$. A rather good linear regression is obtained for each case, represented by a corresponding continuous colored line. From these regressions, we extract the coefficients of the stretched exponential regression: $a_{down} = 23.82 \pm 1.10$, $c_{down} = 0.83 \pm 0.05$, $x_{0\,down} = 0.78 \pm 0.08$ for the temperature transition T_{down}, and $a_{up} = 78.94 \pm 2.91$, $c_{up} = 0.69 \pm 0.04$, $x_{0\,up} = 0.38 \pm 0.04$ for the temperature transition T_{up}. These values of x_0 are relatively small and represent the characteristic scale that defines the N_s/N_c ratio beyond which the transition temperature changes significantly. Additionally, $c_{up} < c_{down}$, indicating a slower decrease in T_{up} compared to T_{down} as the N_s/N_c ratio increases.

Since both T_{up} and T_{down} as a function of N_s/N_c follow a stretched exponential behavior, the behaviors of both T_{eq} and ΔT should be linear combinations of these functions. Therefore, we plotted in the main panels of Figs. 7.21(a) and 7.21(b) the respective evolution of $T_{eq} - T^{\infty}_{eq}$ and $\Delta T - \Delta T^{\infty}$ ($\Delta T^{\infty} \approx 5$ K) as a function of the N_s/N_c ratio. The green (resp. red) continuous line represents the regression obtained by performing a linear combination of the two stretched exponentials, following the relation $T_{eq} = \frac{T_{up} + T_{down}}{2}$ (resp. $\Delta T = T_{up} - T_{down}$), and using the previous coefficients derived from the fitting. A very good agreement is obtained between the data extracted from the simulation and the refinement performed by the stretched exponential laws.

All these elements seem to confirm the stretched exponential relationship characterizing the behavior of the upper and lower transition temperatures, T_{up} and T_{down} of the thermal hysteresis as function of the N_s/N_c ratio, recalling the well-known Avrami laws [180–182] which describe the behavior of heterogeneous media.

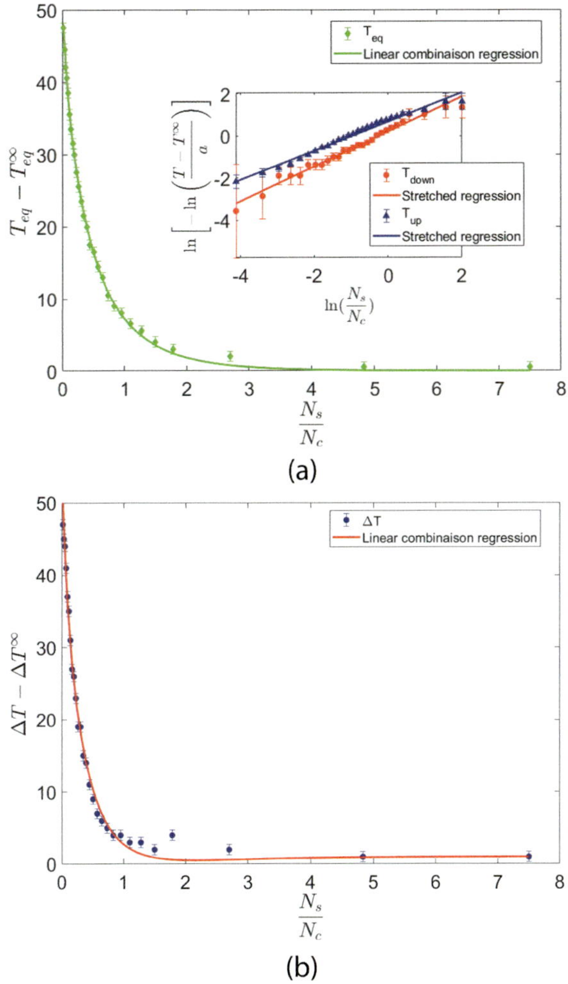

Figure 7.21 Evolution of (a) $T_{eq} - T_{eq}^{\infty}$ and (b) $\Delta T - \Delta T^{\infty}$ ($\Delta T^{\infty} \approx 5$ K), as a function of the N_s/N_c ratio. The solid green (resp. red) curve in Fig. 7.21(a) (resp. 7.21(b)) represents the regression obtained by performing a linear combination of the two stretched exponentials $T_i = T_i^{\infty} + a\ e^{-(x/x_0)^c}$ for T_{down} and T_{up}, following the relation $T_{eq} = (T_{up} + T_{down})/2$ (resp. $\Delta T = T_{up} - T_{down}$), and using the coefficients derived from the fitting of the insert of Fig. 7.21(a). See the text for further explanations. Insert of panel (a): evolution of the quantity $\ln\left[-\ln\frac{(T_{\pm} - T^{\infty})}{a}\right]$ vs $\ln\left(\frac{N_s}{N_c}\right)$ for a core–shell nanoparticle with shell thickness, $1 \leq N_s \leq 30$, where $T_- = T_{down}$ (red curve) and $T_+ = T_{up}$ (blue curve). A linear regression is shown for each temperature, represented by a solid line of the corresponding color.

The two exponents $0.69 \leq c \leq 0.83$ found for the regressions of T_{eq} and ΔT are both less than 1 and their values are quite close, up to standard deviations, indicating a slower decrease of these parameters as a function of the ratio N_s/N_c compared to a standard exponential decay. Furthermore, this value likely suggests variability in the two parameters from one region to another within the 2D spin lattice. Indeed, corner, edges and bulk regions of the core lattice respond differently to the elastic strain imposed by the shell, resulting in an inhomogeneous response of the lattice to external perturbations. This spatial aspect of stress distribution will be discussed later in the section on pressure field distribution in the lattice network during transition.

As was the case for the uncoated core, it would be interesting to investigate in more detail the universality (or lack thereof) of this non-linear relationship when varying the value of the elastic constant A_{1n} as well as the lattice mismatch $\delta_R = R_{HH} - R_{LL}$. This analysis is left for future work.

7.3.3.2 Structural properties of the spin transition in core–shell nanoparticles

We now take a look at the structural aspects of the lattice of the whole square-shaped nanoparticle during a thermal cycle to gain a better understanding of the interplay between the electronic and the elastic degrees of freedom during the thermal spin transition. To this end, Fig. 7.22 presents the thermal dependence of the average nearest-neighbor distance (1n), $\langle d_{1n} \rangle$, for a nanoparticle of size, $N = 64$. Specifically, we have plotted $\langle d_{1n} \rangle$ as a function of temperature T for (i), the whole nanoparticle in panel (a); (ii) exclusively the core part, in panel (b) and (iii) for only the shell in panel (c). For each of these three figures, the evolution of $\langle d_{1n} \rangle$ vs T for core–shell nanoparticles is calculated for various thicknesses N_s. For clarity, we have restricted ourselves to presenting only the following cases where, the thickness N_s equals 0, 5, 10, 15, 20, 25. The $N_s = 0$ case is shown as a reference for comparison with the case of bare nanoparticles, presented in Fig. 7.15(a).

Consistent with the thermal behavior of the HS fraction of the core, n_{HS}^{core}, thermal hysteresis is also observed for the structural

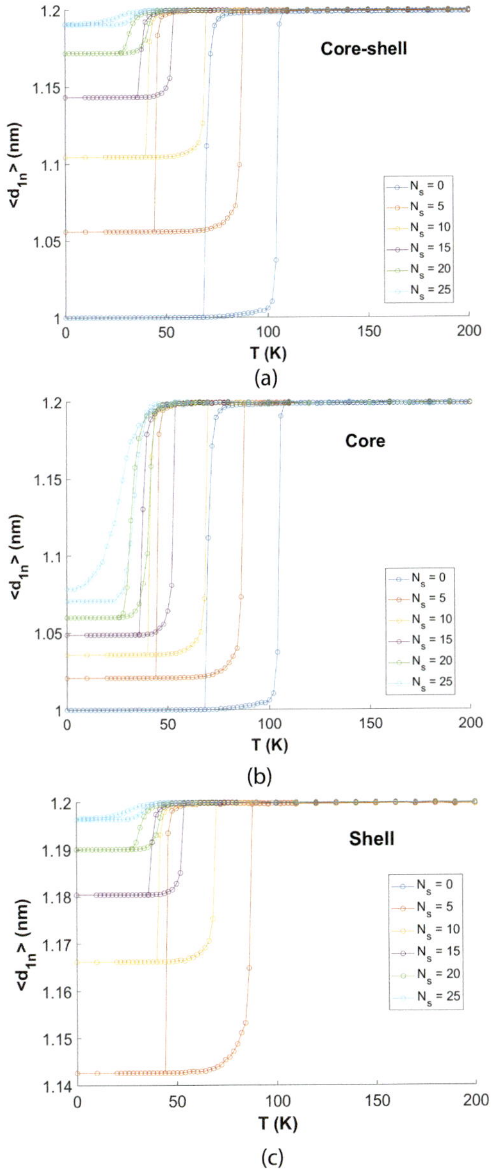

Figure 7.22 Thermal dependence of the average first neighbors distance, $\langle d_{1n} \rangle$, of (a) the whole core–shell system, (b) only the core and (c) only the shell, plotted for the nanoparticle size $N = 64$ for various shell thickness, N_s, values, indicated in the figures.

parameter $\langle d_{1n} \rangle$. For the whole nanoparticle, the core only, and the shell only, it can be seen that at high temperatures, the average distance $\langle d_{1n} \rangle$ consistently approaches 1.20 nm, regardless of the shell thickness $N_s > 0$. This result is in line with the choice made from the outset of a lattice parameter of the shell equal to that of the core, R_{HH}, in the HS state. Thus, in the HS state, the nanoparticle has only bond lengths equal to R_{HH}, and is in a state of mechanical equilibrium. In contrast, in the low-temperature phase, regardless of the shell thickness N_s studied, except for uncoated core, the average value of the core lattice parameters, $\langle d_{1n} \rangle$, exceed that of the LS state, $R_{LL} = 1.00$ nm and increase with N_s. For the whole nanoparticle (Fig. 7.22(a)), this residual value approaches R_{HH} for very thick shell ($N_s \gg N_c$), leading to a reduction in width and amplitude of the corresponding thermal hysteresis of $\langle d_{1n} \rangle$.

If we look at the structural properties of the core (Fig. 7.22(b)), we notice that the average bond length value at low temperatures tends towards $R_{HL} = 1.10$ nm as N_s increases sufficiently. Furthermore, at very low temperatures (e.g., $T = 0$ K), for large N_s values, the thermal cycle of the structural parameter of the core opens during the cooling branch, reaching an average nearest-neighbors distance, $\langle d_{1n} \rangle_{LT}$, being slightly higher than that of the true LS state on heating. The obtained structural phase is then under tensile stress (negative internal pressure), which favors the emergence of HS species inside the lattice. This observation is consistent with the findings in Fig. 7.19(b), where a fraction of the core sites remained trapped in the HS state when the nanoparticle's shell thickness N_s was substantial, resulting in a plateau with a non-zero residual value in n_{HS}^{core}. It's worth mentioning that continuing the simulations from this low-temperature state by increasing the temperature will lead to a relaxation of the structural stress stored in the network to reach the first heating branch. Finally, the analysis of the structural changes in the shell moiety (Fig. 7.22(c)), the average (1n) bond length at low-temperature, $\langle d_{1n} \rangle$, starts slightly above R_{HL} for thin shell thickness N_s and tends towards R_{HH} for thicker shells. The case of very thin shell at low temperature is quite straightforward, since at the interface core–shell sites have from the one side R_{LL} equilibrium distance (core) and R_{HH} from the other side. The situation becomes similar to that of LS core surrounded by a thin HS layer. As a result, the outer layer will have a lattice parameter equal to R_{HL}. The behavior of thick shells at low temperatures is also well understood in terms of elastic energy. Indeed, the contraction of the core from the HS state to LS state tends to contract the shell. However, the increasing thickness of the shell prevents its deformation, thus acting as an increasingly strong barrier for the core sites, preventing their own contraction, which is necessary to switch into the LS state. Consequently, these core sites are unable to exert sufficient influence on the thick shell to have a significant impact on the average nearest-neighbor distance (1n). Thus, as the shell thickness increases, the average distance within it inevitably stays around R_{HH}.

Now, we aim to conduct a quantitative study on the evolution of the low-temperature lattice parameter $\langle d_{1n}\rangle_{LT}$, for both core and shell parts, as a function of the N_s/N_c ratio, using data obtained from Fig. 7.22 at $T = 0$ K. This is an important parameter which provides information on the degree of elastic stress remaining in the lattice at low temperature, due to the misfit of lattice parameters between core and shell. It is observed that the data can once again be refined using a stretched exponential function of the form:

$$\langle d_{1n}\rangle_{LT} = \langle d_{1n}\rangle_{LT}^{\infty} + a\, e^{-(x/x_0)^c}, \tag{7.6}$$

where a now has the dimension of a length. Here, $x = N_s/N_c$, x_0 corresponds to a characteristic value of the ratio N_s/N_c, and c is an exponent reflecting the non-linearity of the curve. The value of $\langle d_{1n}\rangle_{LT}^{\infty} \approx R_{HH}$ is obtained for large values of N_s/N_c in Fig. 7.22, as previously explained.

Employing a similar approach as before, we plot in Fig. 7.23 the quantity $\ln\left[-\ln\frac{(\langle d_{1n}\rangle_{LT} - \langle d1n\rangle_{LT}^{\infty})}{a}\right]$ as a function of $\ln\left(\frac{Ns}{N_c}\right)$ for the core–shell nanoparticle, with various shell thicknesses. This quantity has been plotted for the entire core–shell nanoparticle (in blue), the core (in red) and shell (in green). In these representations, the shell thickness is varied from $N_s = 1$ to $N_s = 30$ for the whole core–shell and the shell part plots, whereas for the core, the data are restricted for N_s going from 1 to 21. This choice results from the fact that, beyond $N_s = 21$, the lattice exhibits trapped HS sites at $T = 0$ K, and the average distance between (1n) sites saturates, as previously observed.

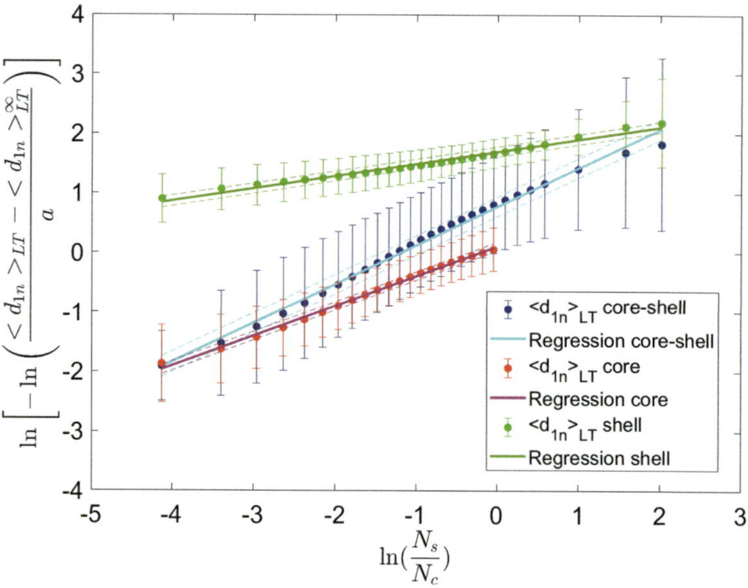

Figure 7.23 Evolution of the quantity $\ln\left[-\ln\frac{(\langle d_{1n}\rangle_{LT}-\langle d_{1n}\rangle_{LT}^{\infty})}{a}\right]$ as a function of $\ln\left(\frac{N_s}{N_c}\right)$ for the core–shell nanoparticle, for various shell thicknesses. The low-temperature lattice parameter value $\langle d_{1n}\rangle_{LT}$ is derived from Fig. 7.22 at $T = 0$ K, where $\langle d_{1n}\rangle_{LT}^{\infty} \approx R_{HH}$ for large value of ratio N_s/N_c. This quantity is plotted here for the whole core–shell nanoparticle (in blue), for the core (in red) and for the shell (in green). A linear regression is shown for each ones, represented by a solid line of the corresponding color. The quasi-linear character of the data confirms the stretched exponential behavior of low-temperature (1n) bond length vs $\frac{N_s}{N_c}$. The colored dotted lines indicate the 95% confidence prediction bounds for each regression.

A linear regression is shown for each of these three average lattice parameters, represented by a solid line of the corresponding color. The colored dotted lines indicate the 95% confidence prediction bounds for each refinement. A satisfactory linear regression is obtained in each case, from which we can extract the following key parameters: $c_{core--shell} = 0.65 \pm 0.02$, $x_{0,core--shell} = 0.30 \pm 0.02$, $c_{core} = 0.50 \pm 0.01$, $x_{0,core} = 0.80 \pm 0.04$, $c_{shell} = 0.21 \pm 0.01$, and $x_{0,shell} = 0.03 \pm 0.01$.

Based on the previously obtained results concerning the average nearest-neighbor distance (1n) at $T = 0$ K, we plot in Fig. 7.24 the evolution of the residual HS fraction $n_{HS\,LT}^{core}$ as a function of $\langle d_{1n}\rangle_{LT}^{core}$ in the core at $T = 0$ K. For this, we considered shell thicknesses ranging from $N_s = 1$ to 30. The curve exhibits two distinct regimes. In the first regime, corresponding to N_s values below 21, the low-temperature HS fraction, $n_{HS\,LT}^{core}$, is nearly zero as long as the average core bond length is between $\langle d_{1n}\rangle_{LT}^{core} = 1.0$ nm and slightly over 1.05 nm. This implies that for approximately the first 25% of the possible misfit increase between the LS and HS states of the lattice, $n_{HS\,LT}^{core} \approx 0$, and thus the core remains in the LS state. Then, in the second regime, which appears above $N_s \simeq 21$, as starting spontaneously at around $\langle d_{1n}\rangle_{LT}^{core} \approx 1.05$ nm and continuing up to nearly $\langle d_{1n}\rangle_{LT}^{core} \approx 1.15$ nm, the value of $n_{HS\,LT}^{core}$ increases almost linearly (as indicated by the red dashed line in Fig. 7.24), until it reaches the previously explained saturation value $n_{HS\,LT}^{core} \approx 0.5$. It is interesting to understand the physical origin of this linear relationship between $n_{HS\,LT}^{core}$ and $\langle d_{1n}\rangle_{LT}^{core}$, as the ratio $\frac{N_s}{N_c}$ increases. Since we deal with average values, the average distance, $\langle d_{ij}^{core}\rangle$ between two nearest-neighboring core sites, i and j, relates to the average connected fictitious magnetizations, $\langle s_i\rangle$ and $\langle s_j\rangle$ through $\langle d_{ij}^{core}\rangle = R_{HL} + \frac{\delta R}{4}\left(\langle s_i\rangle + \langle s_j\rangle\right)$. In the hypothesis of a homogeneous medium, one has $\langle s_i\rangle + \langle s_j\rangle = 2\langle s\rangle = 4n_{HS} - 2$. Thus, one arrives to the simple linear relation, $n_{HS}^{core} = \frac{1}{2} + \frac{1}{\delta R}\left(\langle d_{ij}^{core}\rangle - R_{HL}\right)$, which leads to a slope $\frac{1}{\delta R} = 5$. Despite the limitations surrounding this equation, which does not take into account matrix frustration effects and leads to linear behavior for the full range of $\langle d^{core}\rangle$ values, its asymptotic behavior seems very reasonable. Indeed, a linear regression in the second regime of Fig. 7.24 yields an estimated slope of (6.06 ± 1.58), which is not far from the above prediction, although based on a rough approximation. It is interesting to notice that this slope value of $n_{HS\,LT}^{core}$ as a function of $\langle d_{1n}\rangle_{LT}$ in the core at $T = 0$ K in the second regime is consistent with that found in Fig. 7.13 and Fig. 7.16, which respectively represent the behavior of the thermal hysteresis width of the HS fraction and of the average distance

$\langle d_{1n} \rangle$ for the uncoated core as a function of nanoparticle size, which ties in with the analytical prediction made earlier, assuming a homogeneous system. Indeed, by swapping the two axes in Fig. 7.24, and plotting $\langle d_{1n} \rangle_{LT}$ as a function of $n_{HS\ LT}^{core}$, a slope of (0.16 ± 0.03) is measured, in fair agreement with the value $1/5$ found in the above simple analytical calculations. This suggests that a universal behavior seems to characterize the thermal hysteresis of both the HS fraction and the average nearest-neighbor distance in the core of the nanoparticle, whether coated or not. In particular, we can attempt to establish a connection with classical 2D percolation theory [183–185]. The formation of infinite clusters in the 2D percolation model has analogies with Ising-like model: the emergence of ferromagnetic clusters at the critical threshold mirrors the collective behaviors observed during percolation, highlighting the universal properties shared by these two systems during phase transitions. The exponent β, which characterizes the growth of infinite cluster density from the percolation threshold, is $5/36 \approx 0.1389$ [185]. This value is universal for 2D percolation and is quite close (within uncertainties) to what we have found for the thermal hysteresis behavior of the HS fraction and the average nearest-neighbor distance (1n) in the core of our 2D core–shell nanoparticles (coated or uncoated), except that the interactions in Ising's model are short-range, whereas in this case, the elastic interactions generate correlations that are often directional at both long and short ranges, mixing ferroelastic interactions in one direction with antiferro interactions in other directions, making the establishment of universality laws very difficult. In the end, the 0.160 slope value deduced from the simulations is well ranged between the value ~0.139 of the Ising model and the value, 0.2, derived from the mean-field view of long-range elastic interactions. However, studies of the electroelastic model with different misfit parameters are probably still needed before any definitive conclusions can be drawn on this point.

7.3.3.3 Spatial organization of the spin states

This section is devoted to the investigations of the spatiotemporal properties of the core–shell nanostructure along the first-order phase transition. It is restricted to examining the behavior of the

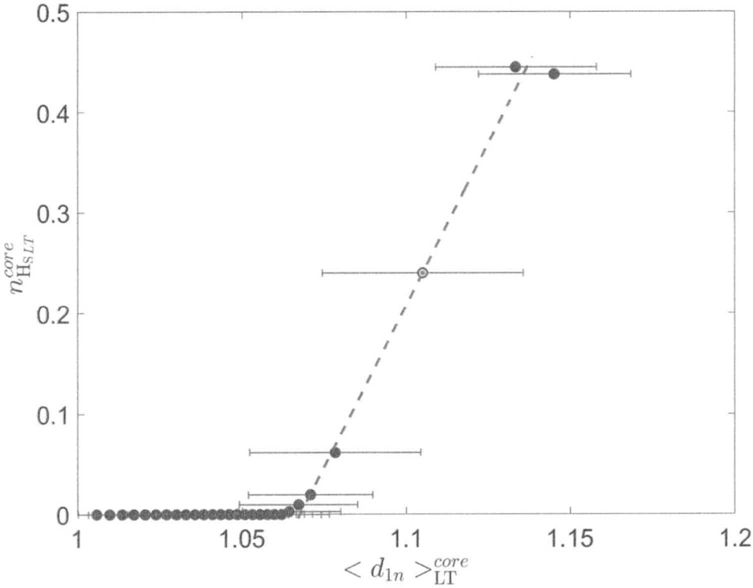

Figure 7.24 Evolution of the residual HS fraction $n^{core}_{HS\ LT}$ of the core at $T = 0$ K as a function of the corresponding distance, $\langle d_{1n} \rangle^{core}_{LT}$, for various shell thicknesses, N_s, ranging from 1 to 30, indicating the existence of a threshold $\langle d_{1n} \rangle^{core}_{LT}$ value above which a residual HS fraction subsists at low-temperature inside the core. The error bars represent the previously calculated uncertainty on the values of $\langle d_{1n} \rangle^{core}_{LT}$, while those of $n^{core}_{HS\ LT}$ are negligible. The red dashed curve represents a linear regression. See text for more explanations.

spin lattice part. A separate section on the mechanical properties will follow.

To illustrate these behaviors, we have displayed in Fig. 7.25 several snapshots of the spin and lattice configurations of a nanoparticle of size $N = 64$ on heating for different shell thicknesses N_S. To avoid overloading the figure, we have selected the following cases: $N_S = 0, 1, 2, 4, 8, 12, 16$, and 24, respectively, shown in Figs. 7.25(a), 7.25(b), 7.25(c), 7.25(d), 7.25(e), 7.25(f), 7.25(g), and 7.25(h). We have chosen to display configurations obtained just below LS to HS transition temperature, T_{up}, where the HS fraction of the core has a value around 0.15. In these figures, yellow (blue) spheres still represent HS (LS) spin sites. In good agreement with

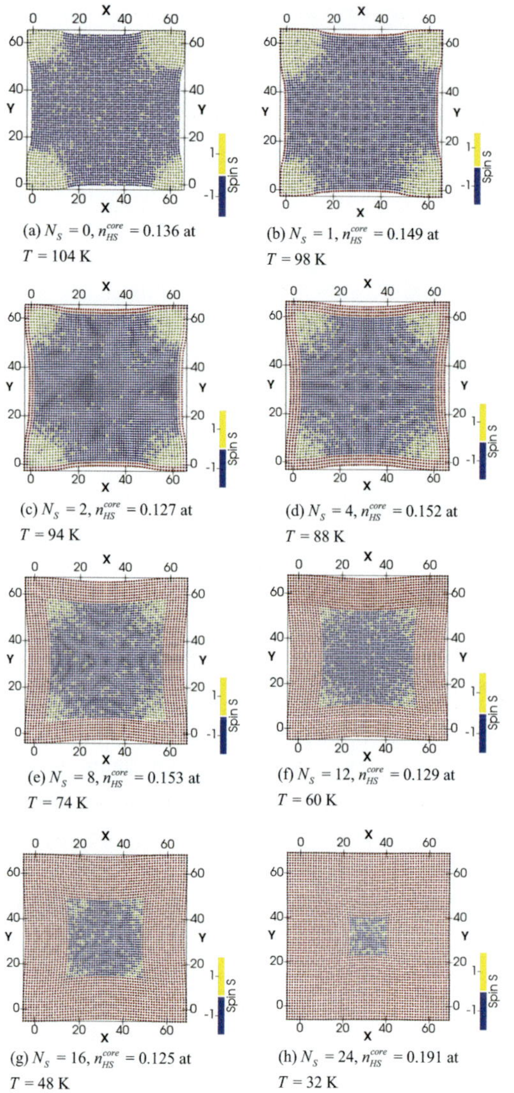

(a) $N_S = 0$, $n_{HS}^{core} = 0.136$ at $T = 104$ K

(b) $N_S = 1$, $n_{HS}^{core} = 0.149$ at $T = 98$ K

(c) $N_S = 2$, $n_{HS}^{core} = 0.127$ at $T = 94$ K

(d) $N_S = 4$, $n_{HS}^{core} = 0.152$ at $T = 88$ K

(e) $N_S = 8$, $n_{HS}^{core} = 0.153$ at $T = 74$ K

(f) $N_S = 12$, $n_{HS}^{core} = 0.129$ at $T = 60$ K

(g) $N_S = 16$, $n_{HS}^{core} = 0.125$ at $T = 48$ K

(h) $N_S = 24$, $n_{HS}^{core} = 0.191$ at $T = 32$ K

Figure 7.25 Selected snapshots showing the spatial distribution of the spin states inside the whole core–shell nanoparticle for $N = 64$ during the heating branch at various N_s thicknesses, while T values approach the thermal transition from LS to HS. Yellow (blue) spheres represent HS (LS) sites, while the shell sites are represented by the red spheres. Note the progressive disappearance of macroscopic nucleation domains in the four corners of the square as the thickness increases.

our earlier findings, for the reference case where $N_S = 0$, the system forms HS clusters emerging from the four equivalent corners of the LS core lattice, as the spin transition approaches. It has not been shown here for readability reasons, but naturally, as the temperature increases, the size of the clusters grows until they merge at the center of the lattice, leading to the complete transformation of the LS core to the HS state.

As the shell thickness increases and takes on low values (for example, $N_S = 1, 2, 4$), the lattice experiences a stressed structure due to the restoring forces of the shell, which prevent the expansion of the core. This strain reaches its maximum at the interface due to the lattice mismatch between the LS core (with R_{LL} lattice spacing) and the shell (with R_{HH} lattice spacing). This incompatibility of lattice bond lengths results in the contraction of the part of the shell all around the core and the slight expansion of the core part near the shell. For this situation of thin shells ($N_S \leq 4$), the propagation of macroscopic clusters occurs through big single-domain nucleation, since their size increases until they merge at the center of the lattice. However, we start to notice, as Figs. 7.25(b), 7.25(c), and 7.25(d) demonstrate, a certain diffusivity, of the HS states from the four clusters, across the lattice towards its center. As the shell thickness, N_S, increases, this phenomenon becomes more pronounced.

In particular, at higher $N_S = 8, 12, 16, 24$, the system forms fewer clusters at the four corners of the lattice as the spin transition progresses from the LS to the HS state. Instead, as the thickness N_S increases, it shows, as seen in Figs. 7.25(f)–7.25(h) a more random flipping of the core's spins located near-shell. In addition, as N_S increases, the lattice shows less overall structural stress and regains its square shape. This phenomenon can be explained by the previously mentioned fact that when the shell, which has the lattice parameter of the HS state, becomes very thick, its deformation becomes very hard, which forces the core to switch in the LS state from the electronic point of view with bond lengths close to those of the HS state. As a result, a negative pressure does take place inside the core, causing the decrease of lattice contribution to the effective ligand-field acting on each spin, although the latter is not homogeneously distributed inside the core due to shape effects. From a simple Ising-like perspective, this is equivalent to lowering the effective ligand-field energy, Δ, for the core's spins. From a physical standpoint, the energy of the ligand field must be moderated to avoid creating significant energy differences between the HS and LS spin states. Thus, the more the ligand-field energy is reduced, the smaller the energy gap becomes between the HS and LS spin states. Given the constant cooperative nature of the elastic interactions, the system should then undergo a gradual transition rather than a sharp first-order transition. This is precisely what we observe for large shell thickness values N_S, as indicated above. Stated differently, in terms of barrier energy and lifetime of metastable states, a very thick layer favors small changes in the core's lattice parameters at the HS–LS transition, where elastic interaction energy is proportional to the quantity $(\langle r\rangle_{HT} - \langle r\rangle_{LT})^2$, where HT (resp. LT) denote high- (resp. low-) temperature limit, indicating that this quantity collapses faster than the simple lattice mismatch $(\langle r\rangle_{HT} - \langle r\rangle_{LT})$ between high- and low-temperature phases. In this particular case, the elastic energy barrier facing each spin flip decreases as N_S increases, which lowers the interaction energy and favors the emergence of gradual transitions. This is in agreement with the observed significant change in the growth mode of the HS fraction of the core. Indeed, the transition of the core does not lead to the usual nucleation of a single domain, even though this transition is of first order. Thus, the nucleation sites within the core are made easier to form by the shell, which permits nucleation to begin anywhere in the core and forms a ramified HS structure. As such, the shell prevents the growth of macroscopic HS domains, changing the growth mechanism to that of homogeneous ramified structures.

7.3.3.4 Mechanical properties of the lattice

As mentioned earlier, it is important to analyze the spatial distribution of local pressure within the core and shell lattices to investigate how internal stresses self-organize. Depending on the elastic stress applied to a site in the lattice, the local pressure field can be either positive or negative. Thus, compressive deformation causes positive pressure at the site, whereas tensile deformation results in negative pressure. This makes the local pressure field an interesting quantity, which provides information on the areas of the crystal where elastic energy excess is stored.

The local pressure, P_i at site i, is calculated using the following expression:

$$P_i = -A_{1n}\sum_j \left(r_{ij} - d_{1n}(S_i, S_j)\right) - B_{2n}\sum_k \left(r_{ik} - d_{2n}(S_i, S_k)\right) \quad (7.7)$$

In Fig. 7.26, we have plotted the distribution of the local pressure field across the lattice of the whole core–shell nanoparticle for a size $N = 64$ during the heating branch, corresponding to the spatiotemporal configurations of the lattice represented in Fig. 7.25. A complex distribution of pressure is observed within the lattice, depending on the shell thickness, N_S.

Indeed, for the uncoated core ($N_s = 0$), shown in Fig. 7.26(a), a relatively uniform and very weak distribution of internal pressure, close to zero, is observed within the lattice, except in the four clusters where compressive deformation (positive pressure) appears near the HS–LS interface, storing a sizable amount of elastic energy as a result of the volume misfit.

For moderate values of the shell thickness (e.g., $N_S = 1, 2, 4$), as seen in Figs. 7.26(b)–7.26(d), a positive pressure region emerges in the shell, resulting from the core's expansion, while a negative pressure region appears in the core. As the shell thickness N_S increases, so does the intensity of the negative pressure inside the core. Near the clusters, it appears clearly that in comparison to sites nearer the core center, the sites at the HS–LS interface on the core side undergo much greater tensile strain (negative pressure). On the other hand, although less intensely than the sites in the shell, the sites inside the HS clusters undergo compression (positive pressure). In the end, the lattice exhibits an inhomogeneous distribution of pressure: at the shell level, the contraction effect leads to positive pressure regions (compressive stresses) at the edges of the lattice, while the bending effect maximizes negative pressure regions (tensile stresses) in the core near the corners.

When the shell thickness increases further and becomes substantial (e.g., $N_S = 8, 12, 16, 24$), as seen in Figs. 7.26(e)–7.26(h), its effect tensile effect becomes dominant: the negative pressure distribution within the core tends to homogenize as N_S increases, despite the presence of a few residual HS clusters (compression

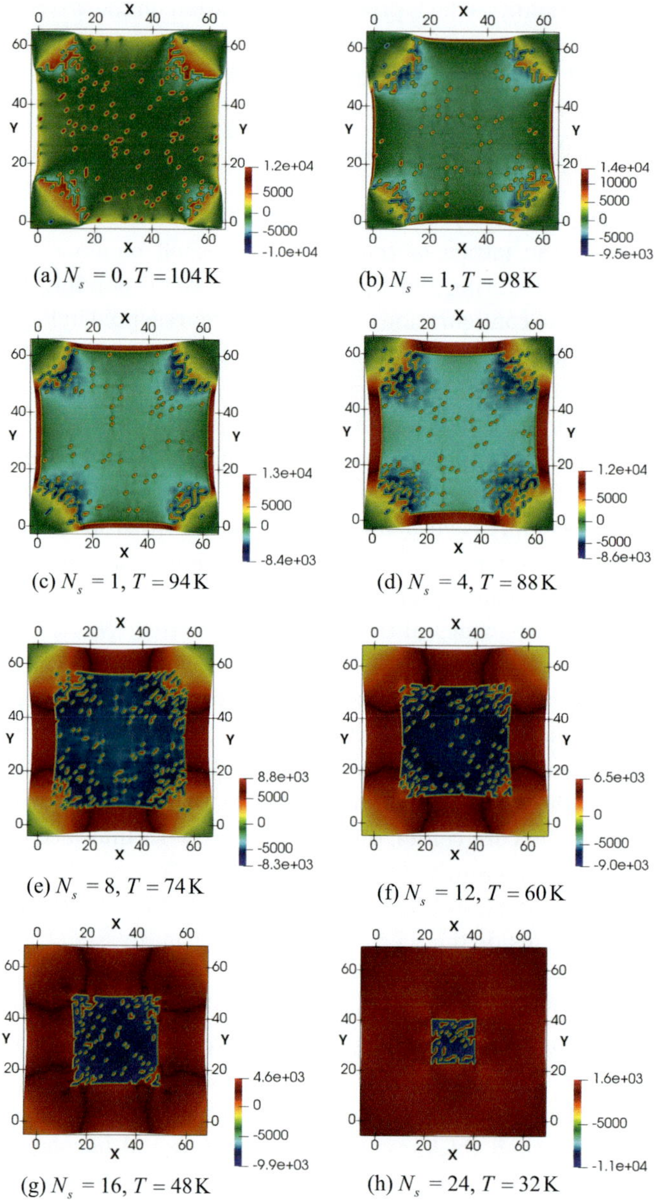

Figure 7.26 Selected snapshots showing the spatial distribution of the local pressure field through the lattice of the whole core–shell nanoparticle for $N = 64$ during the heating branch corresponding to Fig. 7.25.

islands) within the LS core (in tension). Similar to this, as the shell thickness N_S increases, the positive pressure distribution likewise, becomes more homogeneous throughout the shell, producing a quasi-uniform pressure field for thick shells.

This inhomogeneous distribution of pressure in the whole core–shell nanoparticle, with compressive stresses in the shell and tensile stresses in the core, results from the retroaction of the shell to the spin conversion of the core. In other words, it is the result of the subtle elastic interplay between the two components of this switchable nanostructure.

7.4 Conclusions

In this contribution, we examined the equilibrium and non-equilibrium properties of square-shaped core–shell nanostructures made of an active SCO core material surrounded by an inactive shell that serves as the matrix. The latter has a lattice spacing equal to that of the core in the HS phase. When the lattice parameters of the core and the shell differ, such as during the spin transition of the core, on cooling from the HS to LS, an intrinsic elastic frustration takes place at the core–shell interface. This leads to a complex distribution of stress within the core and the shell, which significantly disrupts the nucleation of the core's spin states along the thermal spin transition. These features make the magnetoelastic properties of switchable SCO nanoparticles unpredictable, for both core and shell structures.

Initially, we investigated the case of uncoated nanoparticles, with the aim of understanding the nucleation mechanisms of the HS and BS phases in a system where the interactions between spins are of elastic nature and transitions are accompanied by significant volume changes in the lattice.

For sufficiently high elastic constants, the thermal properties of the uncoated core revealed a thermally induced first-order phase transition, accompanied by a thermal hysteresis, whose size and width depend on the simulated nanoparticle's size, N. We found that the width of the hysteresis, ΔT, increases with the size of the nanoparticle, whose evolution exhibits two distinct regimes.

Finite-size effects are responsible for the non-linear growth of ΔT, following a power of the form $\Delta T \sim N^{\alpha}$ with $\alpha = 0.66 \pm 0.06$.

Beyond a threshold size, $N_{threshold} = 16$, a nearly linear growth of ΔT is obtained, with a slope of 0.16 ± 0.01, regardless of the lattice size. This behavior is linked to the vanishing of the elastic surface energy compared to that of the bulk, as well as to the extensive nature of the macroscopic elastic energy barrier, present in the system. The study of the respective switching temperatures, T_{down} and T_{up}, obtained during cooling and heating, reveals that the thermal bistability window is almost symmetrically centered around the equilibrium value $T_{eq} = (T_{down} + T_{up})/2$. This equilibrium temperature remains nearly constant regardless of the simulated lattice size, with a value of $T_{eq} \approx \dfrac{\Delta}{k_B \ln g} = 90$ K, corresponding to the equilibrium transition temperature, predicted by the Ising-like model.

We have looked into the structural changes in the lattice, along the thermal spin transition of square-shaped nanoparticles of different sizes. It is found that the average nearest-neighbor distance, $\langle d_{1n} \rangle$, exhibits thermal hysteresis, which well correlates with the electronic HS fraction response of the spin states. The statistical analysis of the thermal dependence of the error bars showed that they peaked around the transition temperatures, indicating the presence of residual mechanical stresses during the transition, which increase with nanoparticle size. A strong correlation between the structural (lattice expansion/contraction) and electronic (average HS fraction) properties is observed, showing that the elastic network is slaved to the behavior of the HS fraction at each temperature, so that the values of the lattice parameters relax to their expected equilibrium values in the HS and LS phases.

We confirmed that the hysteresis width of the average nearest-neighbor distance $\langle d_{1n} \rangle$ behaves similarly to that of the HS fraction, showing two distinct regimes based on nanoparticle size N, with a threshold at $N_{threshold}$ and an almost linear behavior with the same slope of 0.16 ± 0.01 as for electronic properties. The total elastic energy of the lattice $\langle E_{elast} \rangle$ peaks around the transition temperatures T_{up} and T_{down}, reflecting the formation of HS–LS interfaces. The energy rises during the transition and decreases sharply afterward, consistent with 3D spin-crossover systems. Stresses persist at high temperatures, explaining the asymptotic growth of $\langle d_{1n} \rangle$.

We have also examined the dynamics of the spin transition, focusing on the spatiotemporal aspects of the first-order phase transition. Snapshots of a nanoparticle's uncoated core during the cooling process, from the HS to LS state, show nucleation starting at the lattice's corners, where elastic energy is minimized. LS clusters grow and coalesce as temperature decreases, leading to a nearly complete transformation of the lattice into the LS state, accompanied by lattice contraction and surface smoothing as the transition progresses.

Next, we turned out our attention to the whole nanoparticle, including the core and shell moieties: the shell, which plays the role of a matrix (without spin) surrounding the core, adapts elastically to the core's deformations during its spin transition. As the shell thickness N_s increases, the core size N_c decreases, maintaining a constant total nanoparticle size N.

The thermal evolution of the HS fraction in the core part was analyzed, focusing on varying shell thicknesses. A hysteretic thermally induced first-order transition is found once more. The width of which decreases as the shell thickens. Furthermore, at larger shell sizes ($N_s \geq 25$), a low-temperature intermediate plateau appears where a significant residual HS fraction of core sites remain trapped in the HS state. This phenomenon is attributed to the tensile mechanical stresses induced by the lattice parameter misfit between the large shell and the core, causing its elastic frustration, and preventing the switching of the HS species to the LS state.

The transition temperatures T_{up} and T_{down} of core–shell nanostructure decrease as the ratio of shell thickness to core size (N_s/N_c) increases, with T_{up} decreasing faster than T_{down}. A meticulous analysis of the size dependence of these switching temperatures revealed that they first follow a stretched exponential law ($\sim e^{-\left(\frac{x}{x_c}\right)^c}$) before saturating at constant values for large shell thicknesses. Regression analysis of the equilibrium temperature, $T_{eq} = \dfrac{T_{up} + T_{down}}{2}$ and the thermal hysteresis ΔT led to exponents $0.69 \leq c \leq 0.83$, close to each other, and both less than 1, indicating a slower decrease of these parameters as a function of the ratio N_s/N_c compared to a standard exponential decay. These results confirm that the transition temperatures and hysteresis width follow stretched exponential behaviors, reminiscent of Avrami laws for heterogeneous systems.

In order to disentangle the interplay between the electronic and the structural elastic degrees of freedom during the thermal spin transition, we next examined the structural features of the nanoparticle. For all studied sizes of core (N_c) and shell (N_s), we found that the average nearest-neighbor distance, $\langle d_{1n} \rangle$, at high temperatures, approaches 1.20 nm, thus aligning with the lattice parameter of the core in the HS state. On the other hand, $\langle d_{1n} \rangle$, increases with N_s at low temperatures and exceeds the equilibrium bond length value, $R_{HL} = 1.0$ nm, of the LS state. For the core, as N_s increases, the average bond length approaches $R_{HL} = 1.10$ nm, indicating tensile stress that promotes the emergence of metastable HS species.

The analysis of the structural behavior of the shell, at low-temperature, reveals that the average bond length varies from R_{HL} for thin shells to R_{HH} for thick shells, highlighting the influence of the core contraction on the shell, as well as that of the shell thickness on the structural properties of the whole nanoparticle. We conducted a quantitative analysis of the low-temperature lattice parameter $\langle d_{1n} \rangle_{LT}$, as a function of the shell thickness at $T = 0$ K. Key parameters were extracted from the relationship, which we described using a stretched exponential function. This showed a satisfactory regression for the entire nanoparticle, the core and the shell.

We then investigated the spatiotemporal properties of the core–shell nanostructure during the first-order phase transition, focusing on the spin-lattice subsystem. Various snapshots of the spin and lattice configurations of a nanoparticle ($N = 64$) were presented for different shell thicknesses (N_s) near the LS to HS transition temperature. The results proved that for thin shells, macroscopic HS clusters nucleate from the four corners of the core and propagate towards the center, causing a sharp first-order transition due to long-range cooperative interactions, driven by the significant volume change accompanying the spin transition of the core. In contrast, for thicker shells, fewer clusters form, and spins near the shell flip randomly. Increased shell thickness prevents the volume change of the core, resulting in the reduction of overall structural stress, thus restoring the lattice's square shape and promoting a gradual transition rather than a sharp one. This gradual transition is attributed to reduced energy differences between HS and LS states, facilitating nucleation throughout the core and leading to the formation of ramified HS structures, with the shell inhibiting the growth of large HS domains.

To understand the organization of internal stresses, we lastly looked—along the heating regime—at the spatial distribution of local pressure within the core and shell lattices of the core–shell nanoparticle. Based on elastic deformations, local pressure can be either positive (compressive) or negative (tensile). For an uncoated core, the internal pressure is relatively weak and uniformly distributed over the lattice, with positive pressure near HS–LS interfaces due to compressive deformation. As shell thickness increases, positive pressure regions develop in the shell due to core expansion, while negative pressure intensifies in the core, particularly in the vicinity of the HS–LS interface. Finally, we have shown that thick shells intensify homogenization of the pressure field in both the core and the shell, which respectively experience negative and positive pressures, finally leading to a largely inhomogeneous distribution of the pressure field on the nanoparticle scale.

Among the possible developments of the present work, one can identify the following extensions. The first one consists in investigating the impact of the shell lattice parameter on the core properties. Thus, instead of using R_{HH} for the shell, use R_{LL}, for example, which means that the core–shell nanoparticle is grown at a low temperature with a core in LS state. Incidentally, the study can also be carried out for a shell with lattice spacing equal to R_{HL}, but in this case, the nanoparticle will be frustrated all the time. Second point: extending the current study to include shells that are themselves switchable, with a ligand field distinct from that of the core, This will result in a double effect of competitive interactions, bearing in mind that the core is completely surrounded by the shell, while the shell has a surface area that allows it to relax its stresses during transition-induced volume changes. Third point: modifying the nanostructure configuration to a sushi-like arrangement, alternating between SCO sites and the shell (e.g., SCO/shell/SCO/shell) to explore the possibility of achieving multistep transitions and possible spatial self-organization of the spin stresses and spin states. Another interesting extension relates to examining the effects of shape on the nanostructures by utilizing circular cores and shells, which would eliminate the shape effects observed at the corners of the square lattice. Finally, extending the present work to three dimensions to better align with experimental realities and analyze whether the behaviors and finite-size laws obtained at 2D remain valid. This last point, transitioning from 2D to 3D nanostructures, is currently underway and will be the subject of a forthcoming publication.

Another possible development, which would require a more substantial modification of the current electroelastic model, involves including exchange interactions between SCO species. Due to the volume change accompanying the spin transition, the exchange interaction should depend on the distance between connected spins. This additional interaction term would enable an investigation of: (i) the effect of elastic properties on the magnetic order, on the one hand, and that of (ii) the magnetic ordering on the emergence of the HS fraction, on the other. The interplay between these two phenomena opens the way to a better understanding of local spin-distortion coupling, and to the study of magnetostrictive systems, which deform when their magnetization is reoriented, and vice versa.

Acknowledgments

We gratefully acknowledge Universities of Versailles and Paris-Saclay, ANR (Agence Nationale de la Recherche Scientifique) Grant No. Mol-CoSM ANR-20-CE07-0028-02. This project has received financial support from the CNRS ("Centre National de la Recherche Scientifique") through the MITI interdisciplinary and exploratory research program. This work was also supported by ANR PIA funding: ANR-20-IDEES-0002. The authors declare neither conflict of interest nor competing financial interest.

References

1. Philipp Gütlich. Spin crossover in iron(II)-complexes. In *Metal Complexes*, pp. 83–195, Berlin, Heidelberg, 1981. Springer Berlin Heidelberg.
2. S. Decurtins, P. Gütlich, C.P. Köhler, H. Spiering, and A. Hauser. Light-induced excited spin state trapping in a transition-metal complex: The hexa-1-propyltetrazole-iron(II) tetrafluoroborate spin-crossover system. *Chem. Phys. Lett.*, 105(1):1–4, 1984.
3. C.P. Köhler, R. Jakobi, E. Meissner, L. Wiehl, H. Spiering, and P. Gütlich. Nature of the phase transition in spin crossover compounds. *J. Phys. Chem. Solids*, 51(3):239–247, 1990.
4. W. Haase. Oliver Kahn: Molecular Magnetism. VCH-Verlag, Weinheim, New York 1993. ISBN 3-527-89566-3. *Berichte der Bunsengesellschaft für physikalische Chemie*, 98(9):1208–1208, 1994.
5. F Palacio. Magnetic phenomena in molecular materials. In *Molecular Magnetism: From Molecular Assemblies to the Devices*, pp. 5–63. Springer, 1996.
6. O. Kahn and C. Jay Martinez. Spin-transition polymers: From molecular materials toward memory devices. *Science*, 279(5347):44–48, 1998.
7. K Boukheddaden, J Linares, H Spiering, and F Varret. One-dimensional Ising-like systems: An analytical investigation of the static and dynamic properties, applied to spin-crossover relaxation. *Eur. Phys. J. B*, 15:317–326, 2000.
8. P Gütlich, HA Goodwin, and H Spiering. Elastic interaction in spin-crossover compounds. *Spin Crossover in Transition Metal Compounds III*, pp. 171–195, 2004.
9. Keith S. Murray. *The Development of Spin-Crossover Research*, chapter 1, pp. 1–54. John Wiley and Sons, Ltd, 2013.
10. Olivier Fouché, Jérôme Degert, Gediminas Jonusauskas, Nathalie Daro, Jean-François Létard, and Eric Freysz. Mechanism for optical switching of the spin crossover $[Fe(NH_2\text{-trz})_3](Br)_2 3H_2O$ compound at room temperature. *Phys. Chem. Chem. Phys.*, 12:3044–3052, 2010.
11. Torben Jasper-Toennies, Manuel Gruber, Sujoy Karan, Hanne Jacob, Felix Tuczek, and Richard Berndt. Robust and selective switching of an Fe^{III} spin-crossover compound on $Cu_2N/Cu(100)$ with memristance behavior. *Nano Lett.*, 17(11):6613–6619, 2017.
12. Francisco Javier Valverde-Muñoz, Maksym Seredyuk, Manuel Meneses-Sánchez, M. Carmen Muñoz, Carlos Bartual-Murgui, and José A. Real.

Discrimination between two memory channels by molecular alloying in a doubly bistable spin crossover material. *Chem. Sci.*, 10:3807–3816, 2019.

13. Maria D. Manrique-Juarez, Sylvain Rat, Fabrice Mathieu, Daisuke Saya, Isabelle Séguy, Thierry Leïchlé, Liviu Nicu, Lionel Salmon, Gábor Molnár, and Azzedine Bousseksou. Microelectromechanical systems integrating molecular spin crossover actuators. *Appl. Phys. Lett.*, 109(6):061903, 08 2016.

14. Mario Piedrahita-Bello, José Elias Angulo-Cervera, Alejandro Enriquez-Cabrera, Gábor Molnár, Bertrand Tondu, Lionel Salmon, and Azzedine Bousseksou. Colossal expansion and fast motion in spin-crossover@polymer actuators. *Mater. Horiz.*, 8:3055–3062, 2021.

15. Y. Singh, K. Affes, N.-I. Belmouri, and K. Boukheddaden. Clamping of spin-crossover solid inducing crystal bending and spatial spin organization. *Mater. Today Phys.*, 27:100842, 2022.

16. Jorge Linares, Epiphane Codjovi, and Yann Garcia. Pressure and temperature spin crossover sensors with optical detection. *Sensors*, 12(4):4479–4492, 2012.

17. Catalin Jureschi, Jorge Linares, Aurelian Rotaru, Marie Hélène Ritti, Michel Parlier, Marinela M. Dîrtu, Mariusz Wolff, and Yann Garcia. Pressure sensor via optical detection based on a 1D spin transition coordination polymer. *Sensors*, 15(2):2388–2398, 2015.

18. Kamel Boukheddaden, Marie Hélène Ritti, Guillaume Bouchez, Mouhamadou Sy, Marinela M. Dîrtu, Michel Parlier, Jorge Linares, and Yann Garcia. Quantitative contact pressure sensor based on spin crossover mechanism for civil security applications. *J. Phys. Chem. C*, 122(14):7597–7604, 2018.

19. Yann Garcia, Petra J. van Koningsbruggen, Epiphane Codjovi, René Lapouyade, Olivier Kahn, and Louis Rabardel. Non-classical Fe^{II} spin-crossover behaviour leading to an unprecedented extremely large apparent thermal hysteresis of 270 K: application for displays. *J. Mater. Chem.*, 7:857–858, 1997.

20. Jean-François Létard, Philippe Guionneau, and Laurence Goux-Capes. *Towards Spin Crossover Applications*, pp. 221–249. Springer Berlin Heidelberg, Berlin, Heidelberg, 2004.

21. Yousuf Raza, Florence Volatron, Simona Moldovan, Ovidiu Ersen, Vincent Huc, Cyril Martini, François Brisset, Alexandre Gloter, Odile Stéphan, Azzedine Bousseksou, et al. Matrix-dependent cooperativity in spin crossover Fe (pyrazine) Pt $(CN)_4$ nanoparticles. *Chem. Commun.*, 47(41):11501–11503, 2011.

22. Dan Qiu, Ling Gu, Xiao-Li Sun, Dong-Hong Ren, Zhi-Guo Gu, and Zaijun Li. Sco@SiO_2@Au core-shell nanomaterials: enhanced photo-thermal plasmonic effect and spin-crossover properties. *RSC Adv.*, 4(106):61313–61319, 2014.

23. Carlos Bartual-Murgui, Eva Natividad, and Olivier Roubeau. Critical assessment of the nature and properties of Fe (II) triazole-based spin-crossover nanoparticles. *J. Mater. Chem. C*, 3(30):7916–7924, 2015.

24. Juan Manuel Herrera, Silvia Titos-Padilla, Simon J. A. Pope, Isadora Berlanga, Félix Zamora, Juan José Delgado, Konstantin V. Kamenev, Xiao Wang, Alessandro Prescimone, Euan K. Brechin, and Enrique Colacio. Studies on bifunctional Fe(II)-triazole spin crossover nanoparticles: time-dependent luminescence, surface grafting and the effect of a silica shell and hydrostatic pressure on the magnetic properties. *J. Mater. Chem. C*, 3:7819–7829, 2015.

25. Ashley C. Felts, Ahmed Slimani, John M. Cain, Matthew J. Andrus, Akhil R. Ahir, Khalil A. Abboud, Mark W. Meisel, Kamel Boukheddaden, and Daniel R. Talham. Control of the speed of a light-induced spin transition through mesoscale core–shell architecture. *J. Am. Chem. Soc.*, 140(17):5814–5824, 2018.

26. Christina D Polyzou, Patroula Gkolfi, Christos T Chasapis, Vlasoula Bekiari, Ariadni Zianna, George Psomas, Malina Ondrej, and Vassilis Tangoulis. Stimuli-responsive spin crossover nanoparticles for drug delivery and DNA-binding studies. *Dalton Trans.*, 51(33):12427–12431, 2022.

27. Jean-François Létard, Nathalie Daro, and Sandie Auffret. Spin transition material, Patent WO 2009/007534, 2009.

28. Alexey Tokarev, Lionel Salmon, Yannick Guari, William Nicolazzi, Gábor Molnár, and Azzedine Bousseksou. Cooperative spin crossover phenomena in [Fe (NH_2 trz)$_3$](tosylate)$_2$ nanoparticles. *Chemical Communications*, 46(42):8011–8013, 2010.

29. Antoine Tissot, Jean-François Bardeau, Eric Rivière, François Brisset, and Marie-Laure Boillot. Thermo-and photoswitchable spin-crossover nanoparticles of an iron (ii) complex trapped in transparent silica thin films. *Dalton Trans.*, 39(33):7806–7812, 2010.

30. Lionel Salmon and Laure Catala. Spin-crossover nanoparticles and nanocomposite materials. *Comptes Rendus Chimie*, 21(12):1230–1269, 2018. Spin crossover phenomenon / Phénomène de transition de spin.

31. Laure Catala, Daniela Brinzei, Yoann Prado, Alexandre Gloter, Odile Stéphan, Guillaume Rogez, and Talal Mallah. Core-multishell mag-

netic coordination nanoparticles: Toward multifunctionality on the nanoscale. *Angew. Chem. Int. Ed.*, 48(1):183–187, 2009.

32. Olivia N. Risset, Pedro A. Quintero, Tatiana V. Brinzari, Matthew J. Andrus, Michael W. Lufaso, Mark W. Meisel, and Daniel R. Talham. Light-induced changes in magnetism in a coordination polymer heterostructure, $Rb_{0.24}Co[Fe(CN)_6]_{0.74}@K_{0.10}Co[Cr(CN)_6]_{0.70}$ nH_2O and the role of the shell thickness on the properties of both core and shell. *J. Am. Chem. Soc.*, 136(44):15660–15669, 2014.
33. Yu-Xia Wang, Dan Qiu, Sai-Fei Xi, Zheng-Dong Ding, Zaijun Li, Yunxing Li, Xuehong Ren, and Zhi-Guo Gu. Iron (ii)-triazole core-shell nanocomposites: toward multistep spin crossover materials. *Chem. Commun.*, 52(51):8034–8037, 2016.
34. Alexey Tokarev, Jérôme Long, Yannick Guari, Joulia Larionova, Françoise Quignard, Pierre Agulhon, Mike Robitzer, Gábor Molnár, Lionel Salmon, and Azzedine Bousseksou. Spin crossover polysaccharide nanocomposites. *N. J. Chem.*, 37(11):3420–3432, 2013.
35. Alejandro Enriquez-Cabrera, Amalia Rapakousiou, Mario Piedrahita Bello, Gábor Molnár, Lionel Salmon, and Azzedine Bousseksou. Spin crossover polymer composites, polymers and related soft materials. *Coordination Chem. Rev.*, 419:213396, 2020.
36. Yue Zan, Lionel Salmon, and Azzedine Bousseksou. Morphological studies of composite spin crossover@SiO_2 nanoparticles. *Nanomaterials*, 11(12), 2021.
37. Yaowei Hu, Matthieu Picher, Ngoc Minh Tran, Marlène Palluel, Laurentiu Stoleriu, Nathalie Daro, Stephane Mornet, Cristian Enachescu, Eric Freysz, Florian Banhart, and Guillaume Chastanet. Photothermal switching of individual plasmonically activated spin crossover nanoparticle imaged by ultrafast transmission electron microscopy. *Adv. Mater.*, 33(52):2105586, 2021.
38. Sébastien Pillet. Spin-crossover materials: Getting the most from x-ray crystallography. *J. Appl. Phys.*, 129(18):181101, 05 2021.
39. Azusa Muraoka, Kamel Boukheddaden, Jorge Linarès, and F. Varret. Two-dimensional Ising-like model with specific edge effects for spin-crossover nanoparticles: A Monte Carlo study. *Phys. Rev. B*, 84:054119, Aug 2011.
40. Alexandru Atitoaie, Radu Tanasa, Alexandru Stancu, and Cristian Enachescu. Study of spin crossover nanoparticles thermal hysteresis using FORC diagrams on an Ising-like model. *J. Magn. Magn. Mater.*, 368:12–18, 2014.

41. H. Oubouchou, A. Slimani, and K. Boukheddaden. Interplay between elastic interactions in a core-shell model for spin-crossover nanoparticles. *Phys. Rev. B*, 87:104104, Mar 2013.

42. Ahmed Slimani and Kamel Boukheddaden. An electro-elastic theory for the mechanically-assisted photo-induced spin transition in core–shell spin-crossover nanoparticles. *Phys. Chem. Chem. Phys.*, 20(45):28583–28591, 2018.

43. Hassane Oubouchou, Yogendra Singh, and Kamel Boukheddaden. Magnetoelastic modeling of core-shell spin-crossover nanocomposites. *Phys. Rev. B*, 98:014106, Jul 2018.

44. Yogendra Singh, Hassane Oubouchou, Masamichi Nishino, Seiji Miyashita, and Kamel Boukheddaden. Elastic-frustration-driven unusual magnetoelastic properties in a switchable core-shell spin-crossover nanostructure. *Phys. Rev. B*, 101:054105, Feb 2020.

45. Ahmed Slimani, Hamadi Khemakhem, and Kamel Boukheddaden. Structural synergy in a core-shell spin crossover nanoparticle investigated by an electroelastic model. *Phys. Rev. B*, 95:174104, May 2017.

46. Massimiliano Cavallini. Status and perspectives in thin films and patterning of spin crossover compounds. *Phys. Chem. Chem. Phys.*, 14(34):11867–11876, 2012.

47. A Il'ya, Gábor Molnár, Igor O Fritsky, Lionel Salmon, and Azzedine Bousseksou. Synthesis of [Fe $(hptrz)_3$]$(OTs)_2$ spin crossover nanoparticles in microemulsion. *Polyhedron*, 38(1):245–250, 2012.

48. Yoann Prado, Laurent Lisnard, Daniela Heurtaux, Guillaume Rogez, Alexandre Gloter, Odile Stéphan, Nada Dia, Eric Rivière, Laure Catala, and Talal Mallah. Tailored coordination nanoparticles: assessing the magnetic single-domain critical size. *Chem. Commun.*, 47(3):1051–1053, 2011.

49. Sophie Brossard, Florence Volatron, Laurent Lisnard, Marie-Anne Arrio, Laure Catala, Corine Mathonière, Talal Mallah, Christophe Cartier dit Moulin, Andrei Rogalev, Fabrice Wilhelm, et al. Investigation of the photoinduced magnetization of copper octacyanomolybdates nanoparticles by x-ray magnetic circular dichroism. *J. Am. Chem. Soc.*, 134(1):222–228, 2012.

50. Radu Tanasa, Jérôme Laisney, Alexandru Stancu, Marie-Laure Boillot, and Cristian Enachescu. Hysteretic behavior of Fe $(phen)_2$ $(NCS)_2$ spin-transition microparticles vs. the environment: A huge reversible component resolved by first order reversal curves. *Appl. Phys. Lett.*, 104(3), 2014.

51. Laure Catala, Florence Volatron, Daniela Brinzei, and Talal Mallah. Functional coordination nanoparticles. *Inorg. Chem.*, 48(8):3360–3370, 2009.

52. Helena J Shepherd, Gábor Molnár, William Nicolazzi, Lionel Salmon, and Azzedine Bousseksou. Spin crossover at the nanometre scale. *Eur. J. Inorg. Chem.*, 2013(5-6):653–661, 2013.

53. Florence Volatron, Laure Catala, Eric Rivière, Alexandre Gloter, Odile Stephan, and Talal Mallah. Spin-crossover coordination nanoparticles. *Inorg. Chem.*, 47(15):6584–6586, 2008.

54. Gautier Félix, Khaldoun Abdul-Kader, Tarik Mahfoud, Il'ya A Gural'skiy, William Nicolazzi, Lionel Salmon, Gábor Molnár, and Azzedine Bousseksou. Surface plasmons reveal spin crossover in nanometric layers. *J. Am. Chem. Soc.*, 133(39):15342–15345, 2011.

55. Mirko Mikolasek, Gautier Félix, William Nicolazzi, Gábor Molnár, Lionel Salmon, and Azzedine Bousseksou. Finite size effects in molecular spin crossover materials. *N. J. Chem.*, 38(5):1834–1839, 2014.

56. Gautier Félix, William Nicolazzi, Mirko Mikolasek, Gábor Molnár, and Azzedine Bousseksou. Non-extensivity of thermodynamics at the nanoscale in molecular spin crossover materials: a balance between surface and volume. *Phys. Chem. Chem. Phys.*, 16(16):7358–7367, 2014.

57. Ishtvan Boldog, Ana B Gaspar, Víctor Martínez, Pablo Pardo-Ibañez, Vadim Ksenofontov, Ashis Bhattacharjee, Philipp Gütlich, and José A Real. Spin-crossover nanocrystals with magnetic, optical, and structural bistability near room temperature. *Angew. Chem. Int. Ed.*, 47(34):6433–6437, 2008.

58. Carlos Bartual-Murgui, Amal Akou, Lionel Salmon, Gábor Molnár, Christophe Thibault, Jose Antonio Real, and Azzedine Bousseksou. Guest effect on nanopatterned spin-crossover thin films. *Small*, 7(23):3385–3391, 2011.

59. Ihor Tokarev and Sergiy Minko. Tunable plasmonic nanostructures from noble metal nanoparticles and stimuli-responsive polymers. *Soft Matter*, 8(22):5980–5987, 2012.

60. H Oubouchou, A Slimani, G Zribi, N Haine, M Zergoug, and K Boukheddaden. Thermal hysteresis behavior of spin crossover nanoparticles from a deformable lattice model: Monte Carlo investigations. *Sens. Lett.*, 11(8):1432–1439, 2013.

61. Orhan Yalçın, Rıza Erdem, and Songül Özüm. Origin of the martensitic and austenitic phase transition in core-surface smart nanoparticles with size effects and hysteretic splitting. *J. Appl. Phys.*, 115(5), 2014.

62. Gautier Félix, William Nicolazzi, Lionel Salmon, Gábor Molnár, Marine Perrier, Guillaume Maurin, Joulia Larionova, Jérôme Long, Yannick Guari, and Azzedine Bousseksou. Enhanced cooperative interactions at the nanoscale in spin-crossover materials with a first-order phase transition. *Phys. Rev. Lett.*, 110(23):235701, 2013.
63. Arnau Carne, Carlos Carbonell, Inhar Imaz, and Daniel Maspoch. Nanoscale metal-organic materials. *Chem. Soc. Rev.*, 40(1):291–305, 2011.
64. Renske M Van Der Veen, Oh-Hoon Kwon, Antoine Tissot, Andreas Hauser, and Ahmed H Zewail. Single-nanoparticle phase transitions visualized by four-dimensional electron microscopy. *Nat. Chem.*, 5(5):395–402, 2013.
65. Saioa Cobo, Gábor Molnár, José Antonio Real, and Azzedine Bousseksou. Multilayer sequential assembly of thin films that display room-temperature spin crossover with hysteresis. *Angew. Chem.*, 118(35):5918–5921, 2006.
66. Philipp Gütlich, Andreas Hauser, and Hartmut Spiering. Thermal and optical switching of iron(II) complexes. *Angew. Chem. Int. Ed. Eng.*, 33(20):2024–2054, 1994.
67. Martin B Duriska, Suzanne M Neville, Boujemaa Moubaraki, John D Cashion, Gregory J Halder, Karena W Chapman, Chérif Balde, Jean-François Létard, Keith S Murray, Cameron J Kepert, et al. A nanoscale molecular switch triggered by thermal, light, and guest perturbation. *Angew. Chem.*, 121(14):2587–2590, 2009.
68. Azzedine Bousseksou, Gábor Molnár, Lionel Salmon, and William Nicolazzi. Molecular spin crossover phenomenon: recent achievements and prospects. *Chem. Soc. Rev.*, 40(6):3313–3335, 2011.
69. Massimiliano Cavallini, Ilaria Bergenti, Silvia Milita, Jean Crispin Kengne, Denis Gentili, Giampiero Ruani, Ivan Salitros, Velimir Meded, and Mario Ruben. Thin deposits and patterning of room-temperature-switchable one-dimensional spin-crossover compounds. *Langmuir*, 27(7):4076–4081, 2011.
70. Massimiliano Cavallini, Ilaria Bergenti, Silvia Milita, Giampiero Ruani, Ivan Salitros, Zhi-Rong Qu, Rajadurai Chandrasekar, and Mario Ruben. Micro-and nanopatterning of spin-transition compounds into logical structures. *Angew. Chem. Int. Ed*, 47:8596–8600, 2008.
71. Paulo N Martinho, Tibebe Lemma, Brendan Gildea, Gennaro Picardi, Helge Müller-Bunz, Robert J Forster, Tia E Keyes, Gareth Redmond, and Grace G Morgan. Template assembly of spin crossover one-dimensional nanowires. *Angew. Chem.*, 124(48):12161–12165, 2012.

72. Anil Naik, Linda Stappers, Johan Snauwaert, Jan Fransaer, and Yann Garcia. A biomembrane stencil for crystal growth and soft lithography of a thermochromic molecular sensor. *Small*, 6(24):2842–2846, 2010.

73. Hong Li and Haonan Peng. Recent advances in self-assembly of spin crossover materials and their applications. *Curr. Op. Colloid Interface Sci.*, 35:9–16, 2018.

74. Ahmed Slimani, Fran çois Varret, Kamel Boukheddaden, Damien Garrot, Hassane Oubouchou, and Sumio Kaizaki. Velocity of the high-spin low-spin interface inside the thermal hysteresis loop of a spin-crossover crystal, via photothermal control of the interface motion. *Phys. Rev. Lett.*, 110:087208, Feb 2013.

75. Mouhamadou Sy, Damien Garrot, Ahmed Slimani, Miguel Páez-Espejo, François Varret, and Kamel Boukheddaden. Reversible control by light of the high-spin low-spin elastic interface inside the bistable region of a robust spin-transition single crystal. *Angew. Chem.*, 128(5):1787–1791, 2016.

76. Kamel Boukheddaden, Mouhamadou Sy, and François Varret. Emergence of dynamical dissipative structures in a bistable spin crossover solid under light: Experience and theory. *Adv. Theor. Simulations*, 1(9):1800080, 2018.

77. Ivan Šalitroš, NT Madhu, Roman Boča, Ján Pavlik, and Mario Ruben. Room-temperature spin-transition iron compounds. *Monatshefte für Chemie-Chemical Monthly*, 140:695–733, 2009.

78. Maria D Manrique-Juarez, Sylvain Rat, Lionel Salmon, Gábor Molnár, Carlos M Quintero, Liviu Nicu, Helena J Shepherd, and Azzedine Bousseksou. Switchable molecule-based materials for micro-and nanoscale actuating applications: Achievements and prospects. *Coordination Chem. Rev.*, 308:395–408, 2016.

79. María D Manrique-Juárez, Iurii Suleimanov, Edna M Hernández, Lionel Salmon, Gábor Molnár, and Azzedine Bousseksou. In situ afm imaging of microstructural changes associated with the spin transition in [Fe (htrz)$_2$ (trz)](BF$_4$) nanoparticles. *Materials*, 9(7):537, 2016.

80. Constantin Lefter, Sylvain Rat, José Sánchez Costa, María Dolores Manrique-Juárez, Carlos M Quintero, Lionel Salmon, Isabelle Séguy, Thierry Leichle, Liviu Nicu, Philippe Demont, et al. Current switching coupled to molecular spin-states in large-area junctions. *Adv. Mater.*, 28(34):pp–7508, 2016.

81. Victoria Shalabaeva, Sylvain Rat, Maria Dolores Manrique-Juarez, Alin-Ciprian Bas, Laure Vendier, Lionel Salmon, Gábor Molnár, and Azzedine Bousseksou. Vacuum deposition of high-quality thin films displaying spin transition near room temperature. *J. Mater. Chem. C*, 5(18):4419–4425, 2017.

82. Maria D Manrique-Juarez, Fabrice Mathieu, Victoria Shalabaeva, Jean Cacheux, Sylvain Rat, Liviu Nicu, Thierry Leïchlé, Lionel Salmon, Gábor Molnár, and Azzedine Bousseksou. A bistable microelectromechanical system actuated by spin-crossover molecules. *Angew. Chem.*, 129(28):8186–8190, 2017.

83. Maria D Manrique-Juarez, Sylvain Rat, Laurent Mazenq, Fabrice Mathieu, Isabelle Séguy, Thierry Leichle, Liviu Nicu, Lionel Salmon, Gábor Molnár, and Azzedine Bousseksou. Spin crossover materials for mems actuation: Film integration and characterization. In *2017 19th International Conference on Solid-State Sensors, Actuators and Microsystems*, pp. 1300–1303, 2017.

84. Jean-François Létard, Guillaume Chastanet, Philippe Guionneau, and Cedric Desplanches. Optimizing the stability of trapped metastable spin states. *Spin-Crossover Materials: Properties and Applications*, pp. 475–506, 2013.

85. Gregory J Halder, Cameron J Kepert, Boujemaa Moubaraki, Keith S Murray, and John D Cashion. Guest-dependent spin crossover in a nanoporous molecular framework material. *Science*, 298(5599):1762–1765, 2002.

86. E. König and B. Kanellakopulos. Mössbauer effect and magnetism down to 1.2 K in the triplet ground state of an iron(II)–phenanthroline complex. *Chem. Phys. Lett.*, 12(3):485–488, 1972.

87. Bernard Gallois, Jose Antonio Real, Christian Hauw, and Jacqueline Zarembowitch. Structural changes associated with the spin transition in bis(isothiocyanato)bis(1,10-phenanthroline)iron: a single-crystal x-ray investigation. *Inorg. Chem.*, 29(6):1152–1158, 1990.

88. Kamel Boukheddaden, Houcem Fourati, Yogendra Singh, and Guillaume Chastanet. Evidence of photo-thermal effects on the first-order thermo-induced spin transition of $[\{Fe(NCSe)(py)_2\}_2(m\text{-}bpypz)]$ spin-crossover material. *Magnetochemistry*, 5(2), 2019.

89. M. Carmen Muñoz and José A. Real. Thermo-, piezo-, photo- and chemo-switchable spin crossover iron(II)-metallocyanate based coordination polymers. *Coordination Chem. Rev.*, 255(17):2068–2093, 2011. Special Issue: 39th International Conference on Coordination Chemistry.

90. Killian Babilotte and Kamel Boukheddaden. Theoretical investigations on the pressure effects in spin-crossover materials: Reentrant phase transitions and other behavior. *Phys. Rev. B*, 101:174113, May 2020.
91. Yanling Li, Amina Benchohra, Buqin Xu, Benoît Baptiste, Keevin Béneut, Paraskevas Parisiades, Ludovic Delbes, Alain Soyer, Kamel Boukheddaden, and Rodrigue Lescouëzec. Pressure-induced conversion of a paramagnetic FeCo complex into a molecular magnetic switch with tuneable hysteresis. *Angew. Chem.*, 132(39):17425–17429, 2020.
92. Yann Garcia, Olivier Kahn, Jean-Pierre Ader, Alexandre Buzdin, Yann Meurdesoif, and Maurice Guillot. The effect of a magnetic field on the inversion temperature of a spin crossover compound revisited. *Phys. Lett. A*, 271(1):145–154, 2000.
93. A. Bousseksou, F. Varret, M. Goiran, K. Boukheddaden, and J. P. Tuchagues. *The Spin Crossover Phenomenon Under High Magnetic Field*, pp. 65–84. Springer Berlin Heidelberg, Berlin, Heidelberg, 2004.
94. Jim Long Her, Yasuhiro H. Matsuda, Motohiro Nakano, Yasuhiro Niwa, and Yasuhiro Inada. Magnetic field-induced spin-crossover transition in [Mn^{III}(taa)] studied by x-ray absorption spectroscopy. *J. Appl. Phys.*, 111(5):053921, 03 2012.
95. Andreas Hauser. Intersystem crossing in Fe(II) coordination compounds. *Coordination Chem. Rev.*, 111:275–290, 1991.
96. Jonas Krober, Epiphane Codjovi, Olivier Kahn, Francoise Groliere, and Charlotte Jay. A spin transition system with a thermal hysteresis at room temperature. *J. Am. Chem. Soc.*, 115(21):9810–9811, 1993.
97. Philipp Gütlich, Ana B. Gaspar, and Yann Garcia. Spin state switching in iron coordination compounds. *Beilstein J. Org. Chem.*, 9:342–391, 2013.
98. Houcem Fourati, Mamadou Ndiaye, Mouhamadou Sy, Smail Triki, Guillaume Chastanet, Sébastien Pillet, and Kamel Boukheddaden. Light-induced thermal hysteresis and high-spin low-spin domain formation evidenced by optical microscopy in a spin-crossover single crystal. *Phys. Rev. B*, 105:174436, May 2022.
99. Helena J. Shepherd, Patrick Rosa, Ian A. Fallis, Philippe Guionneau, Judith A. K. Howard, and Andrés E. Goeta. Structural origin of the gradual spin transition in a mononuclear iron(II) complex. *J. Phys. Chem. Solids*, 2(73):193, 2012.
100. Kazuyuki Takahashi, Hatsumi Mori, Hayao Kobayashi, and Osamu Sato. Mechanism of reversible spin transition with a thermal hysteresis loop in [$Fe^{III}(qsal)_2$][$Ni(dmise)_2$] $2CH_3CN$: Selenium analogue of the

precursor of an Fe(III) spin-crossover molecular conducting system. *Polyhedron*, 28(9-10):1776, 2009.

101. Cristian Enachescu, Jorge Linarès, and François Varret. Comparison of static and light-induced thermal hystereses of a spin-crossover solid, in a mean-field approach. *J. Phys.: Condensed Matter*, 13(11):2481, 2001.
102. Hélène Bolvin and Olivier Kahn. Ising model for low-spin high-spin transitions in molecular compounds; within and beyond the mean-field approximation. *Chemical Physics*, 192(3):295–305, 1995.
103. Kamel Boukheddaden, Isidor Shteto, Benoit Hôo, and Fran çois Varret. Dynamical model for spin-crossover solids. i. relaxation effects in the mean-field approach. *Phys. Rev. B*, 62:14796–14805, Dec 2000.
104. Masamichi Nishino, Kamel Boukheddaden, Yusuké Konishi, and Seiji Miyashita. Simple two-dimensional model for the elastic origin of cooperativity among spin states of spin-crossover complexes. *Phys. Rev. Lett.*, 98:247203, Jun 2007.
105. C Enachescu, M Nishino, S Miyashita, A Hauser, A Stancu, and L Stoleriu. Cluster evolution in spin crossover systems observed in the frame of a mechano-elastic model. *Europhys. Lett.*, 91(2):27003, 2010.
106. W Nicolazzi, S Pillet, and C Lecomte. Two-variable anharmonic model for spin-crossover solids: A like-spin domains interpretation. *Phys. Rev. B*, 78:174401, Nov 2008.
107. William Nicolazzi, J Pavlik, S Bedoui, Gábor Molnár, and Azzedine Bousseksou. Elastic Ising-like model for the nucleation and domain formation in spin crossover molecular solids. *Eur. Phys. J. Spec. Top.*, 222:1137–1159, 2013.
108. Nicolas di Scala, Nour El Islam Belmouri, Miguel Angel Paez Espejo, and Kamel Boukheddaden. Three-dimensional electroelastic modeling of the nucleation and propagation of the spin domains in spin-crossover materials. *Phys. Rev. B*, 106:014422, Jul 2022.
109. Nicolas di Scala, Nour El Islam Belmouri, Miguel Angel Paez Espejo, and Kamel Boukheddaden. Explaining the origin of the orientation of the front transformation in spin-transition crystals. *Phys. Rev. B*, 106:144107, Oct 2022.
110. Rong-Jia Wei, Qian Huo, Jun Tao, Rong-Bin Huang, and Lan-Sun Zheng. Spin-crossover Fe_4^{II} squares: Two-step complete spin transition and reversible single-crystal-to-single-crystal transformation. *Angew. Chem.*, 38(123):9102–9105, 2011.

111. Daniel Chiruta, Catalin Jureschi, Jorge Linares, Pierre Richard Dahoo, Yann Garcia, and Aurelian Rotaru. On the origin of multi-step spin transition behaviour in 1D nanoparticles. *Eur. Phys. J. B*, 88:1–5, 2015.

112. Catalin Jureschi, Jorge Linares, Aurelian Rotaru, and Yann Garcia. Multistep in 3D spin crossover nanoparticles simulated by an Ising model using entropic sampling Monte Carlo technique. *Magnetochemistry*, 2(1):13, 2016.

113. Eric Milin, Sabrina Belaïd, Véronique Patinec, Smail Triki, Guillaume Chastanet, and Mathieu Marchivie. Dinuclear spin-crossover complexes based on tetradentate and bridging cyanocarbanion ligands. *Inorg. Chem.*, 55(17):9038–9046, 2016.

114. Takashi Kosone, Itaru Tomori, Chikahide Kanadani, Toshiaki Saito, Tomoyuki Mochida, and Takafumi Kitazawa. Unprecedented three-step spin-crossover transition in new 2-dimensional coordination polymer Fe^{II} (4-methylpyridine)$_2$ [Au^I $(CN)_2]_2$. *Dalton Trans.*, 39(7):1719–1721, 2010.

115. Masayuki Nihei, Hirotaka Tahira, Nobukazu Takahashi, Yusuke Otake, Yasuhisa Yamamura, Kazuya Saito, and Hiroki Oshio. Multiple bistability and tristability with dual spin-state conversions in [Fe $(dpp)_2$][Ni $(mnt)_2]_2$ $MeNO_2$. *J. Am. Chem. Soc.*, 132(10):3553–3560, 2010.

116. Abdellah Kaiba, HJ Shepherd, Dalila Fedaoui, Patrick Rosa, AE Goeta, N Rebbani, Jean-François Létard, and Philippe Guionneau. Crystallographic elucidation of purely structural, thermal and light-induced spin transitions in an iron (ii) binuclear complex. *Dalton Trans.*, 39(11):2910–2918, 2010.

117. Maksym Seredyuk, Kateryna O Znovjyak, Joachim Kusz, Maria Nowak, M Carmen Munoz, and Jose Antonio Real. Control of the spin state by charge and ligand substitution: two-step spin crossover behaviour in a novel neutral iron (ii) complex. *Dalton Trans.*, 43(43):16387–16394, 2014.

118. HP Zhang, LC Yang, LJ Fu, Q Cao, DL Sun, YP Wu, and R Holze. Core-shell structured electrode materials for lithium ion batteries. *J. Solid State Electrochem.*, 13:1521–1527, 2009.

119. Aurelian Rotaru, Francois Varret, A Gindulescu, Jorge Linares, Alexandru Stancu, Jean-François Létard, Thibaut Forestier, and Céline Etrillard. Size effect in spin-crossover systems investigated by FORC measurements, for surfacted [Fe $(NH_2\text{-trz})_3](Br)_2$ $3H_2O$ nanoparticles: reversible contributions and critical size. *Eur. Phys. J. B*, 84:439–449, 2011.

120. Daniel M Pajerowski, Matthew J Andrus, Justin E Gardner, Elisabeth S Knowles, Mark W Meisel, and Daniel R Talham. Persistent photoinduced magnetism in heterostructures of Prussian Blue analogues. *J. Am. Chem. Soc.*, 132(12):4058–4059, 2010.

121. Daniel M Pajerowski, Bruce Ravel, Carissa H Li, Matthieu F Dumont, and Daniel R Talham. X-ray absorption study of structural coupling in photomagnetic Prussian Blue analogue core@shell particles. *Chemistry of Materials*, 26(8):2586–2594, 2014.

122. Ashley C Felts, Matthew J Andrus, Elizabeth S Knowles, Pedro A Quintero, Akhil R Ahir, Olivia N Risset, Carissa H Li, Isabelle Maurin, Gregory J Halder, Khalil A Abboud, et al. Evidence for interface-induced strain and its influence on photomagnetism in Prussian Blue analogue core-shell heterostructures, Rb_a Co_b $[Fe\ (CN)_6]_c$ m H_2O @ K_j Ni_k $[Cr\ (CN)_6]_l$ n H_2O. *J. Phys. Chem. C*, 120(10):5420–5429, 2016.

123. Yao Chen, Jian-Gong Ma, Jing-Jing Zhang, Wei Shi, Peng Cheng, Dai-Zheng Liao, and Shi-Ping Yan. Spin crossover-macromolecule composite nano film material. *Chem. Commun.*, 46(28):5073–5075, 2010.

124. Victor Martinez, Ishtvan Boldog, Ana Belen Gaspar, Vadim Ksenofontov, Ashis Bhattacharjee, Philipp Gütlich, and Jose Antonio Real. Spin crossover phenomenon in nanocrystals and nanoparticles of $[Fe\ (3\text{-}Fpy)_2\ M\ (CN)_4]$ (M^{II} = Ni, Pd, Pt) two-dimensional coordination polymers. *Chemistry of Materials*, 22(14):4271–4281, 2010.

125. Miguel Rubio, Rebeca Hernández, Aurora Nogales, Anna Roig, and Daniel López. Structure of a spin-crossover Fe(II)-1,2,4-triazole polymer complex dispersed in an isotactic polystyrene matrix. *Eur. Polym. J.*, 47(1):52–60, 2011.

126. Antoine Tissot, Cristian Enachescu, and Marie-Laure Boillot. Control of the thermal hysteresis of the prototypal spin-transition Fe^{II} $(phen)_2$ $(NCS)_2$ compound via the microcrystallites environment: Experiments and mechanoelastic model. *J. Mater. Chem.*, 22(38):20451–20457, 2012.

127. Joulia Larionova, Lionel Salmon, Yannick Guari, Alexeï Tokarev, Karine Molvinger, Gábor Molnár, and Azzedine Bousseksou. Towards the ultimate size limit of the memory effect in spin-crossover solids. *Angew. Chem., Int. Ed*, 47(43):8236–8240, 2008.

128. Silvia Titos-Padilla, Juan Manuel Herrera, Xiao-Wei Chen, Juan José Delgado, and Enrique Colacio. Bifunctional hybrid SiO_2 nanoparticles showing synergy between core spin crossover and shell luminescence properties. *Angew. Chem.*, 123(14):3348–3351, 2011.

129. Christophe Faulmann, Joe Chahine, Isabelle Malfant, Dominique de Caro, Benoît Cormary, and Lydie Valade. A facile route for the preparation of nanoparticles of the spin-crossover complex [Fe $(Htrz)_2$ (trz)](BF_4) in xerogel transparent composite films. *Dalton Trans.*, 40(11):2480–2485, 2011.

130. Pierrick Durand, Sébastien Pillet, El-Eulmi Bendeif, Cédric Carteret, Mohamed Bouazaoui, Hicham El Hamzaoui, Bruno Capoen, Lionel Salmon, Sylvie Hébert, Jaafar Ghanbaja, et al. Room temperature bistability with wide thermal hysteresis in a spin crossover silica nanocomposite. *J. Mater. Chem. C*, 1(10):1933–1942, 2013.

131. Haonan Peng, Simon Tricard, Gautier Félix, Gábor Molnár, William Nicolazzi, Lionel Salmon, and Azzedine Bousseksou. Re-appearance of cooperativity in ultra-small spin-crossover [Fe (pz){Ni $(CN)_4$}] nanoparticles. *Angew. Chem.*, 126(41):11074–11078, 2014.

132. Gábor Molnár, Sylvain Rat, Lionel Salmon, William Nicolazzi, and Azzedine Bousseksou. Spin crossover nanomaterials: from fundamental concepts to devices. *Adv. Mater.*, 30(5):1703862, 2018.

133. Ahmed Slimani, Kamel Boukheddaden, Francois Varret, Hassane Oubouchou, Masamichi Nishino, and Seiji Miyashita. Microscopic spin-distortion model for switchable molecular solids: Spatiotemporal study of the deformation field and local stress at the thermal spin transition. *Phys. Rev. B*, 87:014111, Jan 2013.

134. Laure Catala and Talal Mallah. Nanoparticles of Prussian Blue analogs and related coordination polymers: From information storage to biomedical applications. *Coordination Chem. Rev.*, 346:32–61, 2017.

135. Shin-ichi Ohkoshi, Hiroko Tokoro, and Kazuhito Hashimoto. Temperature-and photo-induced phase transition in rubidium manganese hexacyanoferrate. *Coordination Chem. Rev.*, 249(17-18):1830–1840, 2005.

136. Michel Verdaguer, A Bleuzen, V Marvaud, J Vaissermann, M Seuleiman, C Desplanches, A Scuiller, C Train, R Garde, G Gelly, et al. Molecules to build solids: high T_c molecule-based magnets by design and recent revival of cyano complexes chemistry. *Coordination Chem. Rev.*, 190:1023–1047, 1999.

137. M Presle, J Lemainque, J-M Guigner, E Larquet, I Maurin, J-P Boilot, and T Gacoin. Controlled growth of core@shell heterostructures based on Prussian Blue analogues. *N. J. Chem.*, 35(6):1296–1301, 2011.

138. Olivia N Risset and Daniel R Talham. Effects of lattice misfit on the growth of coordination polymer heterostructures. *Chemistry of Materials*, 27(11):3838–3843, 2015.

139. Matthieu F Dumont, Elisabeth S Knowles, Amandine Guiet, Daniel M Pajerowski, Ariel Gomez, Stefan W Kycia, Mark W Meisel, and Daniel R Talham. Photoinduced magnetism in core/shell Prussian Blue analogue heterostructures of K_j Ni_k [Cr $(CN)_6]_l$ n H_2O with Rb_a Co_b [Fe $(CN)_6]_c$ m H_2O. *Inorg. Chem.*, 50(10):4295–4300, 2011.

140. Nada Dia, Laurent Lisnard, Yoann Prado, Alexandre Gloter, Odile Stéphan, François Brisset, Hala Hafez, Zeinab Saad, Corine Mathonière, Laure Catala, et al. Synergy in photomagnetic/ferromagnetic sub-50 nm core-multishell nanoparticles. *Inorg. Chem.*, 52(18):10264–10274, 2013.

141. Morgane Presle, Isabelle Maurin, Fouad Maroun, Robert Cortès, Lili Lu, Rodaina Sayed Hassan, Eric Larquet, Jean-Michel Guigner, Eric Riviere, Jonathan P Wright, et al. Photostrictive/piezomagnetic core–shell particles based on Prussian Blue analogues: Evidence for confinement effects? *J. Phys. Chem. C*, 118(24):13186–13195, 2014.

142. Olivia N Risset, Tatiana V Brinzari, Mark W Meisel, and Daniel R Talham. Light switchable magnetism in a coordination polymer heterostructure combining the magnetic potassium chromiumhexacyanochromate with the light-responsive rubidium cobalthexacyanoferrate. *Chemistry of Materials*, 27(18):6185–6188, 2015.

143. Corey R Gros, Marcus K Peprah, Ashley C Felts, Tatiana V Brinzari, Olivia N Risset, John M Cain, Cauê F Ferreira, Mark W Meisel, and Daniel R Talham. Synergistic photomagnetic effects in coordination polymer heterostructure particles of Hofmann-like Fe (4-phenylpyridine)$_2$ [Ni $(CN)_4$] 0.5 H_2O and $K_{0.4}$ Ni [Cr $(CN)_6]_{0.8}$ n H_2O. *Dalton Trans.*, 45(42):16624–16634, 2016.

144. Tohru Kawamoto and Shuji Abe. Thermal hysteresis loop of the spin-state in nanoparticles of transition metal complexes: Monte Carlo simulations on an Ising-like model. *Chem. Commun.*, (31):3933–3935, 2005.

145. Alexandru Atitoaie, Radu Tanasa, and Cristian Enachescu. Size dependent thermal hysteresis in spin crossover nanoparticles reflected within a Monte Carlo based Ising-like model. *J. Magn. Magn. Mater.*, 324(8):1596–1600, 2012.

146. Gautier Félix, Mirko Mikolasek, Gábor Molnár, William Nicolazzi, and Azzedine Bousseksou. Tuning the spin crossover in nano-objects: From hollow to core–shell particles. *Chem. Phys. Lett.*, 607:10–14, 2014.

147. Victoria Shalabaeva, Mirko Mikolasek, Maria D Manrique-Juarez, Alin-Ciprian Bas, Sylvain Rat, Lionel Salmon, William Nicolazzi, Gábor

Molnár, and Azzedine Bousseksou. Unprecedented size effect on the phase stability of molecular thin films displaying a spin transition. *J. Phys. Chem. C*, 121(45):25617–25621, 2017.

148. JA Nasser. First order high-spin/low-spin phase transition induced by acoustic-phonons. *Eur. Phys. J. B*, 21:3–10, 2001.
149. JA Nasser, K Boukheddaden, and J Linares. Two-step spin conversion and other effects in the atom-phonon coupling model. *Eur. Phys. J. B*, 39:219–227, 2004.
150. K Boukheddaden, S Miyashita, and M Nishino. Elastic interaction among transition metals in one-dimensional spin-crossover solids. *Physical Review B*, 75(9):094112, 2007.
151. Kamel Boukheddaden, Masamichi Nishino, and Seiji Miyashita. Molecular dynamics and transfer integral investigations of an elastic anharmonic model for phonon-induced spin crossover. *Phys. Rev. Lett.*, 100(17):177206, 2008.
152. Cristian Enachescu, Laurentiu Stoleriu, Alexandru Stancu, and Andreas Hauser. Model for elastic relaxation phenomena in finite 2D hexagonal molecular lattices. *Phys. Rev. Lett.*, 102(25):257204, 2009.
153. Mirko Mikolasek, William Nicolazzi, Férial Terki, Gábor Molnár, and Azzedine Bousseksou. Surface transition in spin crossover nanoparticles. *Chem. Phys. Lett.*, 678:107–111, 2017.
154. Mirko Mikolasek, William Nicolazzi, Férial Terki, Gábor Molnár, and Azzedine Bousseksou. Investigation of surface energies in spin crossover nanomaterials: the role of surface relaxations. *Phys. Chem. Chem. Phys.*, 19(19):12276–12281, 2017.
155. Laurentiu Stoleriu, Alexandru Stancu, Pradip Chakraborty, Andreas Hauser, and Cristian Enachescu. Analysis of first order reversal curves in the thermal hysteresis of spin-crossover nanoparticles within the mechanoelastic model. *J. Appl. Phys.*, 117(17), 2015.
156. Cristian Enachescu, Radu Tanasa, Alexandru Stancu, Antoine Tissot, Jérôme Laisney, and Marie-Laure Boillot. Matrix-assisted relaxation in Fe $(phen)_2$ $(NCS)_2$ spin-crossover microparticles, experimental and theoretical investigations. *Appl. Phys. Lett.*, 109(3), 2016.
157. A Slimani, K Boukheddaden, F Varret, M Nishino, and S Miyashita. Properties of the low-spin high-spin interface during the relaxation of spin-crossover materials, investigated through an electro-elastic model. *J. Chem. Phys.*, 139(19), 2013.
158. A Slimani, K Boukheddaden, and K Yamashita. Thermal spin transition of circularly shaped nanoparticles in a core-shell structure investi-

gated with an electroelastic model. *Physical Review B*, 89(21):214109, 2014.

159. A Slimani, K Boukheddaden, and K Yamashita. Effect of intermolecular interactions on the nucleation, growth, and propagation of like-spin domains in spin-crossover materials. *Physical Review B*, 92(1):014111, 2015.

160. Gautier Félix, Mirko Mikolasek, Gábor Molnár, William Nicolazzi, and Azzedine Bousseksou. Control of the phase stability in spin-crossover core–shell nanoparticles through the elastic interface energy. *Eur. J. Inorg. Chem.*, 2018(3-4):435–442, 2018.

161. CP Slichter and HG Drickamer. Pressure-induced electronic changes in compounds of iron. *J. Chem. Phys.*, 56(5):2142–2160, 1972.

162. Lise-Marie Lacroix, Natalie Frey Huls, Don Ho, Xiaolian Sun, Kai Cheng, and Shouheng Sun. Stable single-crystalline body centered cubic Fe nanoparticles. *Nano Lett.*, 11(4):1641–1645, 2011.

163. Dan Qiu, Dong-Hong Ren, Ling Gu, Xiao-Li Sun, Ting-Ting Qu, Zhi-Guo Gu, and Zaijun Li. Spin crossover-graphene nanocomposites: Facile syntheses, characterization, and magnetic properties. *RSC Adv.*, 4(59):31323–31327, 2014.

164. Christophe Gatel, Francisco Javier Bonilla, Anca Meffre, Etienne Snoeck, Bénédicte Warot-Fonrose, Bruno Chaudret, Lise-Marie Lacroix, and Thomas Blon. Size-specific spin configurations in single iron nanomagnet: from flower to exotic vortices. *Nano Lett.*, 15(10):6952–6957, 2015.

165. Miroslaw Sawczak, Rafal Jendrzejewski, Dominik Maskowicz, Yann Garcia, Astha C. Ghosh, Maria Gazda, Jakub Czechowski, and Gerard Sliwinski. Nanocrystalline polymer impregnated [Fe(pz)Pt(CN)4] thin films prepared by matrix-assisted pulsed laser evaporation. *Eur. J. Inorg. Chem.*, 2019(27):3249–3255, 2019.

166. Dominik Maskowicz, Miroslaw Sawczak, Ashta C. Ghosh, Katarzyna Grochowska, Rafal Jendrzejewski, Aurelian Rotaru, Yann Garcia, and Gerard Sliwinski. Spin crossover and cooperativity in nanocrystalline [Fe(pyrazine)Pt$(CN)_4$] thin films deposited by matrix-assisted laser evaporation. *Applied Surface Science*, 541:148419, 2021.

167. Dominik Maskowicz, Rafal Jendrzejewski, Wioletta Kopec, Maria Gazda, Jakub Karczewski, Pawel Niedzialkowski, Armin Kleibert, Carlos A. F. Vaz, Yann Garcia, and Miroslaw Sawczak. Thin films of nanocrystalline fe(pz)[Pt$(CN)_4$] deposited by resonant matrix-assisted pulsed laser evaporation. *Materials*, 14(23), 2021.

168. Miguel Paez Espejo, Mouhamadou Sy, and Kamel Boukheddaden. Unprecedented bistability in spin-crossover solids based on the retroaction of the high spin low-spin interface with the crystal bending. *J. Am. Chem. Soc.*, 140(38):11954–11964, 2018.

169. Mouhamadou Sy, François Varret, Kamel Boukheddaden, Guillaume Bouchez, Jérôme Marrot, Santoshi Kawata, and Sumio Kaizaki. Structure-driven orientation of the high-spin–low-spin interface in a spin-crossover single crystal. *Angew. Chem. Int. Ed.*, 53(29):7539–7542, 2014.

170. Mirko Mikolasek, Maria D. Manrique-Juarez, Helena J. Shepherd, Karl Ridier, Sylvain Rat, Victoria Shalabaeva, Alin-Ciprian Bas, Ines E. Collings, Fabrice Mathieu, Jean Cacheux, Thierry Leichle, Liviu Nicu, William Nicolazzi, Lionel Salmon, Gábor Molnár, and Azzedine Bousseksou. Complete set of elastic moduli of a spin-crossover solid: Spin-state dependence and mechanical actuation. *J. Am. Chem. Soc.*, 140(28):8970–8979, 2018.

171. Elodie Tailleur, Mathieu Marchivie, Jean-Paul Itié, Patrick Rosa, Nathalie Daro, and Phillippe Guionneau. Pressure-induced spin-crossover features at variable temperature revealed by in situ synchrotron powder x-ray diffraction. *Chem. Eur. J.*, 24(54):14495–14499, 2018.

172. J. Jung, F. Bruchhäuser, R. Feile, H. Spiering, and P. Gütlich. The cooperative spin transition in $[Fe_x Zn_{1-x}(ptz)_6](BF_4)_2$: I. elastic properties: An oriented sample rotation study by Brillouin spectroscopy. *Zeitschrift für Physik B Condensed Matter*, 100:517–522, 1996.

173. H. Spiering, K. Boukheddaden, J. Linares, and F. Varret. Total free energy of a spin-crossover molecular system. *Phys. Rev. B*, 70:184106, 2004.

174. A. Bousseksou, C. Place, J. Linares, and F. Varret. Dynamic spin crossover in $[Fe(2\text{-}BIK)_3](ClO_4)_2$ and $[Fe(Me_2\ 2\text{-}BIK)_3](BF_4)_2$ investigated by Mössbauer spectroscopy. *J. Magn. Magn. Mater.*, 104-107:225–226, 1992.

175. Peter Adler, Andreas Hauser, Andreas Vef, Hartmut Spiering, and Philipp Gütlich. Dynamics of spin state conversion processes in the solid state. *Hyperfine Interactions*, 47:343–356, 1989.

176. Peter Adler, Hartmut Spiering, and Philipp Gütlich. Thermodynamics and kinetics of spin state conversion processes studied by pressure dependent Mössbauer spectroscopy. *J. Phys. Chem. Solids*, 50(6):587–597, 1989.

177. Laurentiu Stoleriu, Masamichi Nishino, Seiji Miyashita, Alexandru Stancu, Andreas Hauser, and Cristian Enachescu. Cluster evolution in molecular three-dimensional spin-crossover systems. *Phys. Rev. B*, 96:064115, Aug 2017.

178. Masamichi Nishino, Cristian Enachescu, Seiji Miyashita, Per Arne Rikvold, Kamel Boukheddaden, and François Varret. Macroscopic nucleation phenomena in continuum media with long-range interactions. *Sci. Rep.*, 1(1):162, 2011.

179. Seiji Miyashita, Yusuké Konishi, Masamichi Nishino, Hiroko Tokoro, and Per Arne Rikvold. Realization of the mean-field universality class in spin-crossover materials. *Phys. Rev. B*, 77:014105, 2008.

180. Melvin Avrami. Kinetics of phase change. I General theory. *J. Chem. Phys.*, 7(12):1103–1112, 1939.

181. Melvin Avrami. Kinetics of Phase Change. II Transformation-Time Relations for Random Distribution of Nuclei. *J. Chem. Phys.*, 8(2):212–224, 1940.

182. Melvin Avrami. Granulation, Phase Change, and Microstructure Kinetics of Phase Change. III. *J. Chem. Phys.*, 9(2):177–184, 1941.

183. Vinod K.S. Shante and Scott Kirkpatrick. An introduction to percolation theory. *Adv. Phys.*, 20(85):325–357, 1971.

184. M F Sykes, M Glen, and D S Gaunt. The percolation probability for the site problem on the triangular lattice. *J. Phys. A: Math. Nucl. Gen.*, 7(9):L105, 1974.

185. D. Stauffer and A. Aharony. Percolation. In Robert A. Meyers, editor, *Encyclopedia of Physical Science and Technology (Third Edition)*, pp. 655–669. Academic Press, New York, third edition, 2003.

Chapter 8

Helicenes as Springs: A DFT Estimate of Their Deformation Constants

Timur I. Lukmanov,[a] Arslan F. Akhmetov,[b] and Denis Sh. Sabirov[a]

[a] *Laboratory of Mathematical Chemistry, Institute of Petrochemistry and Catalysis UFRC RAS, 141 Prosp. Oktyabrya, Ufa, 450075, Russian Federation*
[b] *Department of Oil and Gas Technology, Ufa State Petroleum University, 1 Kosmonavtov st., Ufa, 450064, Russian Federation*
burf@destructio.ru, tngrusoil@mail.ru, diozno@mail.ru

This chapter is devoted to assessing the unsubstituted helicenes, approximated as classical springs, and the corresponding spring properties (spring constant and elongation). For this purpose, we have chosen the appropriate density-functional theory (DFT) method for describing the spiral structure of [n]helicenes with the known experimental structural parameters of the molecules (when n = 6, 7, 10, and 11). The calculation methods have been tested in two modes, with or without dispersion correction. We have used the pitch of the helicene spiral as a key parameter for the method selection because it is very sensitive to changes in computational methodology. The structures we got with the r^{2}SCAN-3c and PBE0-D/6-31G** methods have the helical pitches closest to the experimental ones. Then we calculated the spring

2D and 3D Nanostructures: Structure, Properties, and Applications
Edited by Jean-Claude Serge Levy

ISBN 978-981-5129-97-7 (Hardcover), 978-1-003-74871-7 (eBook)
www.jennystanford.com

DOI: 10.1201/97810037487-8

constants of [n]helicenes with n = 6, 12, 18, 24, and 30. The spring constants decrease with the spring length, which shows that helicenes behave similar to a classical spring in this regard. The similarity is incomplete because of the asymmetry of the deformation response and the fact that the value of the spring constant per coil turn grows with the helicene size.

8.1 Introduction

Helicenes are polycyclic aromatic hydrocarbons consisting of fused benzene rings with the closest connection pattern that results in a spiral structure [1]. This structure is unusual for sp^2-hybridized and aromatic hydrocarbons (which molecules are typically flat). The helicity of the molecules leads to a set of specific optical, mechanical, and electronic properties [2]. Helicenes are used in several areas, among them enantioselective synthesis [3], semiconductors [4], a budding field of organic electronics [5]. There are ideas to use them in molecular machinery [1]. Note that there is an increasing fundamental interest in helicenes, because they were predicted as possible compounds of interstellar and circumstellar media [6] and then detected in the laboratory simulations of circumstellar envelopes [7]. In this regard, the properties required for their detection in the astrophysical environments are computed [8].

Helicenes are commonly designated as [n]helicene, where n is the number of benzene rings in the molecule. Some theorists start counting helicenes from phenanthrene as [3]helicene [9], but the helical structure starts manifesting from $n = 4$. The first and the second turns of the spring are finished respectively at $n = 6$ and 12. The benzene rings do not overlap completely, as the π-stacking results in the additional twist for each helical turn in the helicene molecules [10]. Herewith, the interactions between the p-orbitals of the different coils of the helix are poorly described with the common quantum-chemical methods. Indeed, post-Hartree–Fock techniques overestimate the electron density between the layers, which results in a compacted helix. In contrast, DFT methods underestimate the electron density and, hence, predict a more expanded structure as compared with the X-ray data [8]. In previous works, the best results

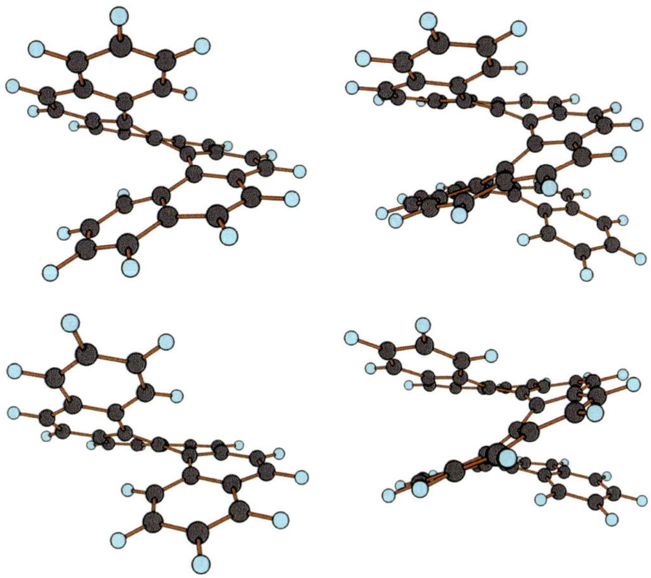

Figure 8.1 The helicenes, used for the method selection (the geometries are optimized with the PBE0-D3/6-31G(d,p) method; the color code hereinafter: black and cyan correspond to carbon and hydrogen).

were obtained using the DFT methods with dispersion corrections (DFT-D) [11].

Note that for the method selection procedure, we chose the helicene molecules with $n = 6$, 7, 10, and 11 (Fig. 8.1) due to the availability of the corresponding experimental data on their structures (hereinafter we use the "clockwise" (P) helicene enantiomers).

The spring properties of helicenes attract chemists and physicists. The carbon skeletons of helicenes look like springs and behave like them. Flexibility of various helicenes was observed experimentally [12–14]. Even if we consider only the class of unsubstituted helicenes, there are several works, whereby the spring constants of helicenes were calculated theoretically [11, 15, 16]. Axial and twist deformations of helicenes with assessing the Young's moduli were simulated in periodical quantum-chemical computational techniques [10, 17, 18].

designated with a minus sign. The distance change happened over 13 steps.

When describing [n]helicenes, it is sometimes convenient to use the triple helical model. In this model, all carbon atoms are considered belonging to one of the three helices: (i) inner helix made up with ($n + 1$) atoms, (ii) median helix constructed with ($n + 1$) atoms, or (iii) outer that is built with $2n$ atoms [31].

8.3 Results and Discussion

Helical pitch is the main structural parameter we focus on. Since the presentation of the experimental values follows the triple helical model, so do our calculated pitches. We define the helical pitch of [7]-, [10]-, and [11]helicenes as the average distance between the overlapping atoms belonging to the same virtual spiral on the neighboring turns of the helix. The pitch of [6]helicene is defined almost the same, except the "non-overlapping" atoms, but those which would have been overlapping if it were [7]helicene.

The experimental and calculated helical pitches are presented in Table 8.1. The value of 6.04 Å for the outer spiral from [32] looks like an outlier and we disregarded it. As follows from Table 8.1, the DFT methods systematically overestimate the helical pitch of helicenes, whereas RI-MP2 underestimates it. The inclusion of the dispersion corrections in the DFT calculations noticeably decreases the helical pitch and the calculated values become close to the experimental ones for [7]helicene and [10]helicene. These results confirm the findings of Rulichek et al. [11]. For [6]helicene, we see a mismatch between the calculated and measured rotation degree of the terminal rings of the helicene molecule, if we define this value as the difference between the pitches of the inner and outer spirals. A similar underestimation is seen for the outer spiral of [11]helicenes, for which PBE0-D3 predicts the pitch equal to 3.751 Å, which is smaller than the X-ray value 3.94 Å. Here the uncorrected PBE/Λ03 method provides the closer value (4.089 Å).

The increase in the number of basic functions in the quantum-chemical calculations results in a small contraction of the helicene spring, which, however, becomes more pronounced with the increase in the molecular size of helicene.

Table 8.1 The experimental and calculated helical pitches for selected [*n*]helicenes

Method	Helical pitch (Å)		
	Inner spiral	Medium spiral	Outer spiral
	[6]helicene		
Experimental [33]	3.26	4.59	5.68
Experimental [32]	3.26	4.57	6.04
Experimental [34]	n/a	4.622	n/a
PBE/3ζ	3.067	4.625	5.891
PBE/Λ22	3.066	4.614	5.872
PBE/Λ03	3.066	4.616	5.875
PBE0-D3/6-31G(d,p)	3.003	4.486	5.682
RI-MP2/3ζ	2.863	4.196	5.222
r^2SCAN-3c	2.986	4.489	5.616
	[7]helicene		
Experimental [33]	3.17	3.88	4.41
Experimental [11]	n/a	n/a	4.50
PBE/3ζ	3.314	4.241	5.038
PBE/Λ22	3.312	4.234	5.024
PBE/Λ03	3.308	4.230	5.020
PBE0-D3/6-31G(d,p)	3.157	3.861	4.431
RI-MP2/3ζ	2.984	3.421	3.711
r^2SCAN-3c	3.140	3.821	4.358
	[10]helicene		
Experimental [33]	3.22	3.67	3.94
Experimental [32]	3.22	3.62	3.88
PBE/3ζ	3.457	3.928	4.210
PBE/Λ22	3.421	3.888	4.163
PBE/Λ03	3.419	3.883	4.157
PBE0-D3/6-31G(d,p)	3.273	3.622	3.808
RI-MP2/3ζ	3.080	3.356	3.511
r^2SCAN-3c	3.269	3.626	3.780
	[11]helicene		
Experimental [33]	3.22	3.64	3.94
PBE/3ζ	3.474	3.885	4.141
PBE/Λ22	3.440	3.845	4.091
PBE/Λ03	3.439	3.843	4.089
PBE0-D3/6-31G(d,p)	3.286	3.590	3.751
RI-MP2/3ζ	3.078	3.339	3.492
r^2SCAN-3c	3.284	3.600	3.748

Another thing to note is the importance of setting tight convergence criteria during the helicene geometry optimization. This is especially important when calculating values, sensitive to the geometry change, because noticeable geometry changes in helicenes may occur without the accompanying change in total energy. With the normal convergence thresholds, even a minor input geometry difference or a different optimization algorithm may result in different geometries having the same energy. For example, the optimization of [7] helicene with the DRIVER algorithm results in the total energy of the molecules equal to −1152.8000241 Hartree and helical pitch of 4.433 Å. At the same time, the TROPT algorithm leads to other results for the same method (−1152.7999875 Hartree and 4.553 Å, respectively). A simple optimization algorithm change with the same convergence thresholds has more effect on the geometry than a fivefold increase of the basis set (from 3ζ to L03).

We used r^{2}SCAN-3c for the spring calculations. It is expected to be a more accurate method than PBE0-D/6-31G(d,p) [28], while having better performance, despite it predicting structures marginally further removed from the experimental data.

Helicenes do not follow Hooke's law exactly, i.e., the relationship between the energy and deformation is asymmetrical (Fig. 8.3), but it is still possible to accurately describe it with the second-order polynomial (R^2 is always larger than 0.955). Getting the spring constant value from such a polynomial is trivial. The results are shown in Table 8.2.

Table 8.2 Calculated values of spring constant of helicenes

		Spring constant (kJ/mol × Å^2)		
Molecule	Polynomial	Compression	Stretching	Average for Compression and Stretching
[6]helicene	148.65	217.18	79.15	148.17
[12]helicene	97.59	124.59	70.68	97.64
[18]helicene	72.62	83.86	61.46	72.66
[24]helicene	59.49	65.11	53.88	59.49
[30]helicene	50.59	53.89	47.18	50.53

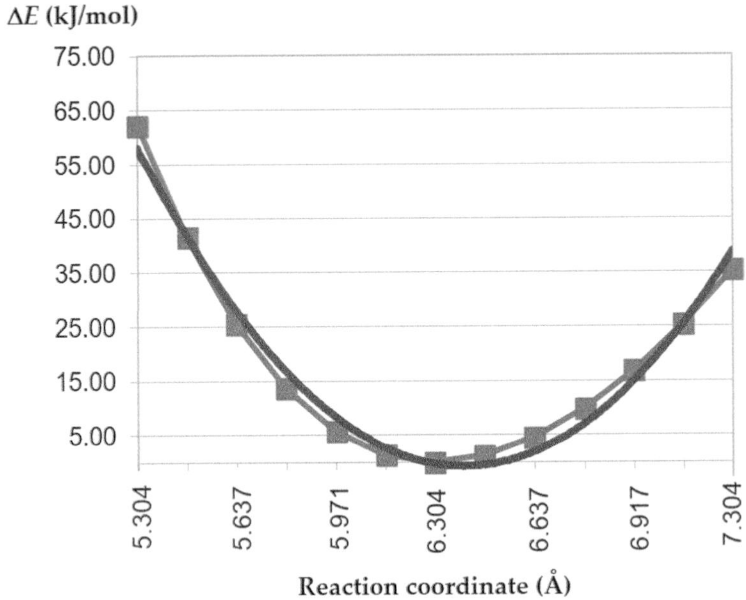

Figure 8.3 A demonstration of the deformation energy asymmetry on the example of mechanical manipulations with [12]helicene.

Before discussing the obtained constants, we want to note that the absolute value of the spring constant is very sensitive to the relaxed scanning parameters and choice of the fixed atoms, i.e., reaction coordinate for the section of the potential energy surface of the mechanical process. The most extreme example is the spring constant for [6]helicenes: it equals 18.89 kJ/mol $\times$ Å^2 when obtained with fixing the atoms of the outer helix but 201.57 kJ/mol $\times$ Å^2 if the scanning goes along the atoms of the inner helix (the displacement in both cases is $+1 \ldots -1$ Å, as previously). Therefore, it is hard to make a reasonable comparison between differently measured spring constants of helicenes if the difference in the molecular size does not equal six benzene rings.

We calculated the spring constants in different modes. Since the deformation energy of the helicene spring is asymmetric, we calculated separately the constants for spring compression and stretching. Comparing these numbers, we can see that the value of the average of the compression and stretching spring constants

Table 8.3 Quantities associated with the spring constant of helicenes

Molecule	R^2		$K = k \times N$	
		Polynomial	Compression	Stretching
[6]helicene	0.955	148.65	201.57	82.90
[12]helicene	0.982	195.17	238.88	146.28
[18]helicene	0.994	217.86	246.55	189.39
[24]helicene	0.997	237.98	255.82	219.25
[30]helicene	0.999	252.94	269.44	235.88

closely follows the polynomial. It means that calculating the spring constant for any helicenes, one can skip the relaxed scanning and directly compare the energies of two helicene structures, relaxed and deformed. Anyway, we deduce from these numbers that the spring constant indeed gets smaller with the helicene size. In this regard, helicenes behave similarly to Hookean springs.

To measure this similarity, we calculated two additional values: the coefficient of determination (R^2) of the polynomial approximation and the value of $K = k \times N$, where k is the spring constant and N is the number of coils in the helix. These values are presented in Table 8.3.

Since, per our calculations, the force/deformation ratio is constant for a single deformation direction, the value of R^2 only shows how symmetrical the helicene's response to deformation is. The increase in the value with the helicene size demonstrates that the asymmetry of the deformation energy gets smaller and, in this respect, helicenes behave close to the ideal spring since [18] helicene.

The K value is the other measure of "Hookeanness" – it must remain constant for the ideal spring. In preceding work [15], Jalaie et al. found that the K values converged toward [24] helicenes. However, our present calculations do not reproduce this behavior, especially in the case of the K values, associated with the spring stretching constants.

All these derivative values get closer but do not converge completely with the helicene growth. The ratio between the compression and stretching constants varies significantly from 2.74

for [6]helicene to 1.14 for [30]helicene. However, within the current spring constant calculation setup, we cannot say with certainty whether this failure to converge is the natural behavior or it is a consequence of the helicene twisting that affects the direction of the spring deformation and, in turn, the very sensitive spring constant value.

8.4 Conclusions

The geometry parameters of helicenes are sensitive to the theoretical approximation, whereas the basis set has a minor impact. For a better reproduction of the experimental results, the dispersion interactions must be accounted. Herewith, none of the tested methods was able to describe all helical pitch values exactly. The remaining pitch differences between the gas phase calculations and X-ray analysis may relate to the crystal packing energies, because the pitch change of 0.2 Å results in the energy change of ~1.5 kJ/mol.

The behavior of helicene springs can be described with the quadratic polynomial (i.e., Hooke's law) with a high determination coefficient for all investigated molecules ($R^2 > 0.95$). We found that the spring constant decreases with the helicene growth, but does not converge at [30]helicene, contrary to previous works. The spring constants of helicenes are noticeably asymmetric, unlike those of ideal springs. However, the compression and stretching constants approach each other with increasing the molecular size. The value of the spring constant per coil must be constant for the system of ideal springs in series, but it increases in helicenes. We consider that larger helicenes are very similar to ideal springs but cannot be described as a system of ideal springs in series.

On a methodological note, we would like to add that despite the noticeably asymmetry in the energy/deformation relationship of helicenes, it can still be closely approximated by a quadratic polynomial. Therefore, it is not necessary to perform a full relaxed scan to calculate the spring constants. A single calculation for each deformation direction is sufficient.

References

1. Shen, Y., and Chen, C. (2012). Helicenes: Synthesis and applications, *Chem. Rev.*, **112**, 1463–1535.
2. Gingras, M. (2013). One hundred years of helicene chemistry. Part 3: Applications and properties of carbohelicenes, *Chem. Soc. Rev.*, **42**, 1051–1095.
3. Narcis, M. J., and Takenaka, N. (2014). Helical-chiral small molecules in asymmetric catalysis, *Eur. J. Org. Chem.*, 21–34.
4. Hatakeyama, T., Hashimoto, S., Oba, T., and Nakamura, M. (2012). Azaboradibenzo[6]helicene: Carrier inversion induced by helical homochirality, *J. Am. Chem. Soc.*, **134**, 19600–19603.
5. Sahasithiwat, S., Mophuang, T., Menbangpung, L., Kamtonwong S., and Sooksimuang, T. (2010). 3,12-Dimethoxy-7,8-dicyano-[5]helicene as a novel emissive material for organic light-emitting diode, *Synth. Met.*, **160**, 1148–1152.
6. Sabirov, D. Sh., Garipova, R. R., and Cataldo F. (2018) Polarizability of isomeric and related interstellar compounds in the aspect of their abundance, *Mol. Astrophys.*, **12**, 10–19.
7. Zhao, L., Kaiser, R. I., Xu, B., Ablikim, U., Lu, W., Ahmed, M., Evseev, M. M., Bashkirov, E. K., Azyazov, V. N., Zagidullin, M. V., Morozov, A. N., Howlader, A. H., Wnuk, S. F., Mebel, A. M., Joshi, D., Veber, G., and Fischer, F. R. (2019) Gas phase synthesis of [4]-helicene, *Nat Commun.*, **10**, 1510.
8. Oña-Ruales, J. O., Ruiz-Morales, Y., and Alvarez-Ramírez, F. (2021). The helicenes: Potential carriers of diffuse interstellar bands, *ACS Earth Space Chem.*, **5**, 381–390.
9. https://www.sigmaaldrich.com/RU/en/product/aldrich/695114.
10. Domnin, A. V., Porsev, V. V., and Evarestov, R. A. (2023). Effect of the armchair and zigzag edge terminations on the properties of nanohelicenes: First-principles study, *Comput. Condens. Mat.*, **36**, e00826.
11. Rulíšek, L., Exner, O, Cwiklik, L., Jungwirth, P., Starý, I., Pospíšil, L., and Havlas, Z. (2007). On the convergence of the physicochemical properties of [*n*]helicenes, *J. Phys. Chem. C*, **111**, 14948–14955.
12. Tanaka, K., and Kitahara, Y. (1999). Synthesis, structure and properties of bridged and unbridged heterohelicenes, *Phosphorus Sulfur Silicon Relat. Elem.*, **153**, 399–400.

13. Liang, J., Ju, C.-W., Zheng, W., Wagner, M., Qiu, Z., Weil, T., and Müllen, K. (2021). Helical molecular springs under high pressure, *ChemRxiv.*, DOI: 10.26434/chemrxiv-2021-w3r3j.

14. Karaka, P., and Choudhury, Y. (2022). Conformationally flexible hetero-helicenes as stimuli-controlled soft molecular springs, *Chem. Sci.*, **13**, 11163–11173.

15. Jalaie, M., Weatherhead, S., Lipkowitz, K. B., and Robertson, D. (1997). Modulating force constants in molecular springs, *Electron J. Theor. Chem.*, **2**, 268–272.

16. Nakakuki, Y., Hirose, T., and Matsuda, K. (2018). Synthesis of a helical analogue of kekulene: A flexible π expanded helicene with large helical diameter acting as a soft molecular spring, *J. Am. Chem. Soc.*, **140**, 15461–15469.

17. Porsev, V. V., Bandura, A. V., and Evarestov, R. A. (2022). Ab initio modeling of helically periodic nanostructures using CRYSTAL17: A general algorithm first applied to nanohelicenes, *Comput. Mater. Sci.*, 203, 111063.

18. Porsev, V. V., and Evarestov, R. A. (2023) Current state of computational modeling of nanohelicenes. Nanomaterials, **13**, 2295.

19. Laikov, D. N., and Ustynyuk Y. A. (2005). PRIRODA-04: A quantum-chemical program suite. New possibilities in the study of molecular systems with the application of parallel computing, *Russ. Chem. Bull.*, **54**, 820–826.

20. Aprà, E., Bylaska, E. J., de Jong, W. A., Govind, N., Kowalski, K., Straatsma, T. P., et al. (2020). NWChem: Past, present, and future, *J. Chem. Phys.*, **152**, 184102.

21. Neese, F. (2012). The ORCA program system, *WIREs Comput. Mol. Sci.*, **2**, 73–78.

22. Neese, F., Wennmohs, F., Becker, U., and Riplinger, C. (2020). The ORCA quantum chemistry program package, *J. Chem. Phys.*, **152**, 224108.

23. Perdew, J. P., Burke, K., and Ernzerhof, M. (1996). Generalized gradient approximation made simple, *Phys. Rev. Lett.*, **77**, 3865–3868.

24. Adamo, C., and Barone, V. (1999). Toward reliable density functional methods without adjustable parameters: the PBE0 model, *J. Chem. Phys.*, **110**, 6158–6170.

25. Weigend, F., Häser, M., Patzelt, H., and Ahlrichs, R. (1998). RI-MP2: optimized auxiliary basis sets and demonstration of efficiency, *Chem. Phys. Lett.*, **294**, 143–152.

26. Grimme, S., Antony, J., Ehrlich, S., Krieg, H. (2010). *J. Chem. Phys.*, **132**, 154104.
27. Laikov, D. N. (2005). A new class of atomic basis functions for accurate electronic structure calculations of molecules, *Chem. Phys. Lett.*, **416**, 116–120.
28. Grimme, S., Hansen, A., Ehlert, S., and Mewes, J.-M. (2021). r^2SCAN-3c: A "Swiss army knife" composite electronic-structure method, *J. Chem. Phys.*, **154**, 064103.
29. Sabirov, D. Sh., Ori, O., Tukhbatullina, A. A., and Shepelevich, I. S. (2022). Structural descriptors of benzenoid hydrocarbons: A mismatch between the estimates and parity effects in helicenes, *C – J. Carbon Res.*, **8**, 42.
30. Lukmanov, T., Akhmetov, A. F., and Sabirov, D. Sh. (2022). Polarizability of kekulene, septulene, and nearest non-planar polycyclic aromatic hydrocarbons, *C – J. Carbon Res.*, **8**, 61.
31. Navaza, I., Tsoucaris, G., Le Bas, G., Navaza, A., de Rango, C. (1979). General models for helicenes, *Bull. Soc. Chim. Belg.*, **88**, 863–870.
32. Wittek, M. (1982). PhD. Thesis, Univ. Bonn 1982.
33. Laarhoven, W. H., and Prinsen, W. J. C. (1984). Carbohelicenes and heterohelicenes, *Top. Curr. Chem.*, **125**, 63–130.
34. Dračínský, M., Storch, J. Církva, V., Císařová, I., and Sýkora J. (2017). Internal dynamics in helical molecules studied by X-ray diffraction, NMR spectroscopy and DFT calculations, *Phys. Chem. Chem. Phys.*, **19**, 2900–2907.

Index

For Product Safety Concerns and Information please contact our EU representative GPSR@taylorandfrancis.com Taylor & Francis Verlag GmbH, Kaufingerstraße 24, 80331 München, Germany